617.95 C31B 2285444
CAUWELS, JANICE M.
THE BODY SHOP

THE BODY SHOP

BIONIC REVOLUTIONS IN MEDICINE

Janice M. Cauwels, Ph.D.

THE C.V. MOSBY COMPANY

ST. LOUIS 1986

Manuscript editor: CRACOM Corporation
Book design: CRACOM Corporation
Cover design: Diane Beasley
Illustrations: Christopher Burke

For information contact The C.V. Mosby Company, 11830 Westline Industrial Drive, St. Louis, Missouri 63146

Printed in the United States of America

Library of Congress Cataloging-in-Publication Data

Cauwels, Janice M.
The body shop.

Includes bibliographies and index.
1. Artificial organs. 2. Implants, Artificial.
3. Medical innovations. I. Title. [DNLM: 1. Artificial Organs—popular works. 2. Implants, Artificial—popular works. 3. Transplantation—popular works.
WO 660 C375b]
RD130.C38 617′.95 86-2344
ISBN 0-8016-0944-5

AC/D/D 9 8 7 6 5 4 3 2 1 11/B/139

CONTENTS

ACKNOWLEDGMENTS

Like the various devices that eventually became the Jarvik-7 artificial heart, this book is really the work of over 200 people: physicians, engineers, patients, research subjects, manufacturers' representatives, government officials, businesspeople, attorneys, public relations staffs, librarians, and others who supplied information. I wish particularly to thank the researchers and body part recipients who took the time to be interviewed and to review pertinent manuscript pages. I received valuable suggestions from several researchers who read entire chapters and from Pierre M. Galletti, M.D., Ph.D., who read the entire manuscript promptly and thoroughly.

For providing, checking, or advertising my need for information; answering questions; scheduling interviews; and offering encouragement, I wish to thank the staffs of the American Society for Artificial Internal Organs and the International Center for Artificial Organs and Transplantation (especially Cecilia Kasavich), Arthur Anderson, Brian Richard Boylan, Linda Broadus, John Dwan, Pamela Fogle, Rosemary Halun, Mary Johnson, Phil Oxley, Dorothy Pirovano, Mel Schroeder, Thomas Sommers, John Stigi, Sarah Stratton, Greg Taylor, and Betty Yarmon.

Vincent A. DeLeo, M.D., has repeatedly gone out of his way to expedite my local research. During the months that I traveled, Maureen Frey took over an important responsibility, and Colleen Kennealy relieved the tedium of suitcase life with kind hospitality. When a computer devoured half the manuscript and its operator gave up in disgust, Mirtille and Tom Lyons transcribed the final draft on their respective typewriters.

Finally I wish to thank Marie-Denise Kratsios, my literary matchmaker; Dorothy Harris, my editor; and Carole Abel, my agent.

PREFACE

The Body Shop is an optimistic and much-needed book. The concept of spare parts for the human body has surreptitiously entered our consciousness through the practice of specialty medicine and—with more fanfare—the folklore of television, yet until now no serious introduction to this field has been available for the general public. In libraries we can find elementary textbooks of human anatomy and physiology to learn about the structure and functions of the various organs in our bodies. Similar information is not readily found when these organs are replaced by an implant or an electromechanical device. We know intuitively that the replacement parts described in *The Body Shop* are simplistic substitutes for their sophisticated natural counterparts. We instinctively appreciate that our bodies will not necessarily accept the intrusion of man-made spare parts. Perhaps some of us feel guilty about interfering with the order of nature, whether it proceeds from intelligent creation or blind evolution. Above all we have not known where to turn for level-headed information.

Janice Cauwels tells us that even though spare parts medicine may not be as advanced or uniformly successful as popular media suggest, we should not be ill at ease in discussing what it has to offer to medical care. Millions of people have received implants, and their opinions should be heard. Sensational reports about feats of technology should not obscure the fact that less publicized implants such as cardiac pacemakers or middle ear prostheses have entered the standard practice of medicine. Janice Cauwels gives voices to the patients, who are often the true pioneers in the development of medical technology. She tells anecdotes of everyday life with prostheses and writes in terms a candidate for an unorthodox form of treatment can understand and relate to, yet the up-to-date accuracy of her account recommends it to medical professionals as well.

Artificial organ science is still too young to have developed a coherent set of general principles applicable to all medical conditions. Similarities among the various types of implants are often more apparent than real. The prospect of some surgical procedures is loaded with enormous emotion, whether we deal with life-threatening situations such as in the case of the artificial heart, private functional handicaps such as those addressed by penile prostheses or artificial sphincters, or purely aesthetic yearnings such as may justify mammary augmentation prostheses. Janice Cauwels accepted these limitations. She listened carefully to confront the problems of investigators and implant candidates. She recognized the uneasy position of artificial organ research in our medical culture. She identified nonscientific and nonmedical considerations and made a place for them in her account.

I encouraged her to write this book as an outsider looking in. She gives us an upbeat yet balanced report of what she saw and heard. She may not have covered all topics evenly: the field of artificial organs is too wide and disorganized to allow it. More important, she conveys the spirit of research and the mood of pioneer designers and adventurous patients. She has made a timely contribution to a debate that will be with us for a generation.

Pierre M. Galletti, M.D., Ph.D.
Professor of Medical Science
Vice-President (Biology
and Medicine)
Brown University
July 9, 1985

THE BODY SHOP

BIONIC REVOLUTIONS IN MEDICINE

SECTION

I

A LANDMARK SCIENCE

1

Behind the Bionic Hype

Picture a father seated in a darkened movie theater with his two children. The man, who seems to be in his early forties or thereabouts, is a scientist. Perhaps he has both an M.D. and a Ph.D., for he is intrigued by the engineering and clinical processes involved in making mechanical gadgets perform the functions of the human body. Because of his background, he is amused by the sophisticated technology splashed across the screen during a matinee of *The Empire Strikes Back* (but Darth Vader's mechanical face gives him the subtle creeps).

The climactic fight between Vader and Luke Skywalker finally occurs. Just as the children grow restless, a swish of Vader's laser gun neatly bisects Luke's forearm. A few scenes later, this minor disability has been cured: Luke is seated in a spaceship infirmary, where robot surgeons have attached a perfectly natural-looking artificial hand.

The scientist, who puts in long hours and tends to think about his work in the shower, suddenly feels old. He lowers his chin to one hand and groans softly. First the Six Million Dollar Man. Then the Bionic Woman. Now this.

His son has meanwhile leaned up to his father's ear. "Dad," he whispers, "Isn't that what you do?"

What Dad does is research in *bionics,* a term defined differently among reference sources but generally referring to the application of biological principles to engineering problems. *Biomedical engineering* (or just *bioengineering*), a synonym used in academic circles, likewise varies in definition but generally means the application of engineering principles (or equipment like artificial organs) to biology or medicine. *Bionic* has taken on this meaning in popular usage, confusing the terminology further.

For simplicity, all these terms are used here to mean a science in which knowledge of engineering and physiology is used to produce artificial devices or systems that benefit humans. Some of these devices are implants or prostheses—artificial substitutes for missing body parts. (*Prosthesis* is also used loosely: the Inflatable Penile Prosthesis substitutes not for a penis but rather for a natural erection.) Bionics research has also produced external devices that temporarily attach to the body, implantable pumps that deliver medications, mechanical creatures that substitute for humans in training sessions, and fluids that keep organs alive.

Most of us first encountered artificial human parts when we were children listening to tales of Long John Silver or Captain Hook. Years ago, peg legs or hooks were sure signs that their fictional owners were disreputable or downright villainous. More recently, popular entertainment has shown a reversal in this attitude. The Six Million Dollar Man and the Bionic Woman became superheroes when the creators of their respective television series endowed them with bionic arms, legs, and senses. With our love of superheroes, our fascination with gadgetry, and our faith in technology (recently affirmed by the microcomputer revolution), we welcomed them into our imaginations.

Our scientist is just one of many researchers upset by such goings-on. Most people do not recognize that real bionic persons have existed for over 40 years. Hundreds of thousands of people are alive thanks to artificial kidneys or heart-lung machines; many others are more comfortable and productive because they have artificial joints, heart valves, or pacemakers. Several people who appear in these pages have investigational devices that make disease or disability much easier to bear. Their adventuresome courage and that of their predecessors has helped establish a revolutionary era in medicine. Recipients of the Jarvik-7 artificial heart are corroborating what the Tin Woodman finally learned: one does not need a natural heart to love and be loved. (Given the popularity of film creatures like R2D2, C3PO, E.T., and Gizmo, we might wonder whether we are not rebounding from the natural to a mechanical extreme.)

Today artificial parts—whether external, percutaneous (placed through the skin), or implantable—are a billion dollar worldwide industry. In 1980 one estimate stated that some 2 to 3 million "artificial or prosthetic parts" manufactured by hundreds of companies were implanted in Americans each year. Research on these devices is underway at universities, Veterans' Administration (VA) hospitals, the National Institutes of Health (NIH), and other government organizations. Private companies involved in biomedical engineering range from tiny ones hastily assembled to provide desperately needed funds, to giants of their respective industries like Johnson and Johnson, Pfizer, 3M, Dow Corning, and Revlon. Nobody—not even the Food and Drug Administration (FDA)—has a completely up-to-date list of *all* the manufacturers who are testing artificial body parts.

These days we are bombarded by media hype describing the latest breakthroughs in artificial limbs, joints, organs, nerves, features, and senses. A reader who follows bionic "breakthroughs" in the popular press might naively fantasize, "If I must be crippled or disfigured in an auto accident, let it happen in front of a bionics research center." In the past few years literally hundreds of articles on bionics have appeared in the major newspapers as well as in popular and science magazines. Unfortunately, much of this information is highly sensationalized. "I don't even read press accounts of our work because all of them, even the good ones, are partly incorrect," says William C. DeVries, M.D., a heart surgeon in private practice in Louisville, Kentucky, and the Director of the Artificial Heart Program at Humana Heart Institute. "If I read all that stuff, I get defensive and upset and worry about what the patients' reactions will be if they see it. It's not even worth commenting on."

Bionics researchers write more accurately and even more prolifically than contributors to the popular press. A quick look at EMBASE or other biomedical and technological data bases will reveal that thousands of articles on implants and prostheses are published each year. The *List of Journals Indexed in Index Medicus 1984* has nine entries under "Artificial Organs" and 31 under "Biomedical Engineering." One young bionics researcher has confessed that he had started to write a book on his *specialty* but gave up because he found too much material.

Despite all this enthusiasm, many people—including physicians—do not know what types of implants and prostheses are available or being clinically tested. (Several of the patients we will meet in this book expressed eagerness to inform others about new devices that saved or greatly increased the quality of their lives.) Although the first artificial organ, the kidney dialysis machine, is 40 years old, in the minds of most people the Jarvik-7 artificial heart sprang from a void, so it attracted much attention, analysis, and criticism from a startled public. Artificial hearts, including the Jarvik-7, are far better evaluated within the context of *all* bionic parts.

People who learn about bionics from the media often burden researchers like our scientist with eager and unrealistic expectations of success. Such publicity can be cruelly disappointing to disabled people who investigate further only to find that cures for blindness or paralysis do not yet exist. "What most troubles me about media reports is that they create false hopes," says Woodie C. Flowers, Ph.D., Associate Professor of Mechanical Engineering at the Massachusetts Institute of Technology (MIT). "Many people who are justifiably anxious to have improved devices think that technology is going to make it happen tomorrow. It's a disservice to them, and they deserve more truth."

By encouraging us to await miracles, hype blinds us to the fact that the real science of biomedical engineering is already established as one of the most revolutionary and important advances in modern medicine. Were its status more widely known, people would better understand the significance of the human implantations of the Jarvik-7 artificial heart, for example. "The clinical tests of the Jarvik-7 heart have been public from the beginning, and people have been expecting miracles without recognizing that it will take time for things to go well," says Dr. DeVries. "The press and the public want to know how long a patient would have lived without the artificial heart and whether it will increase his quality of life. If he has complications, they say that the implantation wasn't worth doing. They emphasize the theoretical, therapeutic aspects of the operation rather than its experimental aspects. Many times people submit to experiments without gaining improved quality of life or the therapeutic benefits they hoped for. Meanwhile physicians have learned a tremendous amount that will benefit the patients who come after them. Regardless of outcome, these artificial heart patients have all contributed greatly to the future of humankind—a really important cause."

Bionic parts make such contributions not only in and of themselves, but also because they can spin off other exciting, practical technologies. Even more important, artificial body parts are unique research tools for studying physiology and disease states. Physicians did not know, for instance, that kidney failure victims eventually became anemic until they were able to keep patients alive for long periods on dialysis machines (the kidneys may produce a hormone necessary for red cell renewal). Nor were they aware that left ventricular dysfunction leading to shock could be reversed by prompt, vigorous perfusion until they began testing assist devices in the artificial heart program. The Jarvik-7 artificial heart has enabled tests to be run on its recipients that would not be possible with other patients. Artificial organs have also given researchers new information about atherosclerosis, carcinogenesis, and complex immunological processes.

Traditionally, when surgeons removed damaged organs, scientists learned from the rest of the body what their function had been. From a purely scientific viewpoint, artificial organ technology provides information in just the opposite way: researchers replace a body part with a device they think does the same tasks. By keeping patients alive, they find out that the organ did more than they had originally supposed (and the substitute can be repeatedly improved, the better to approximate its natural counterpart). "This substitutive approach of replacing body parts in a global fashion is philosophically a very important landmark in the thinking about biological science that very few people appreciate," says Pierre M. Galletti, M.D., Ph.D., Professor of Medical Science and Vice-President for Biology and Medicine at Brown University. In this sense artificial organs make a circular contribution to medicine: they teach investigators more about the natural organs and disease conditions, enabling them to return to direct treatment with a better perspective.

Progress in bionics is not nearly as smoothly progressing or as rapidly accelerating as this discussion would suggest. Any bioengineer will point out that we are several million years behind nature in terms of design. Perhaps bionics will eventually change this situation. Obviously our bodies do not evolve to keep pace with modern technology, or we would not be disturbed by the suffocation of air pollution or the scream of jet engines. We may be destined to evolve into closer, more effective interaction with our own technology, with a little help from our friends the scientists. Researchers are now speeding up our movement through the chain of humankind—replacing the lengthy, wasteful, trial-and-error process of natural evolution with more efficient scientific inquiry. They are influencing, even directing, the processes of change and perhaps altering the bases of evolution itself. "There's no *a priori* reason that life should be based on water and carbon just because we evolved that way," says Dr. Galletti. "In the past 50 years, scientists have developed an organic chemistry of silicone that is as refined and unique as organic compounds made of carbon. There are other ways to do things." Rather than implying that we can build bionic superpeople, the possibility of alternative solutions to the same biological problems should remind us how intricate and miraculous our own bodies really are.

The body is superbly equipped to combat invasion; unfortunately, however, it is unable to distinguish friend from foe. As we will see later, the biggest scientific problem confronting bionics researchers is that of finding materials that will not be rejected when they are implanted in the body. Once a material is determined to be biocompatible, the next problem is that of establishing its interface with its living surroundings. "We are entering an era in which the limiting factor of a prosthesis is its interrelation with the environment," says Dr. Galletti. "We don't know how to interface for biological reasons, particularly at the cellular level."

Not only the cells but entire organs and body systems still remain mysteries to medical researchers. The complexity of trying to imitate nature is implied in the example that throughout the world literally hundreds of different artificial knee joints are in use. Many researchers have found that they need not thoroughly understand a particular organ or system to mimic its contribution to human life. Others even design devices unlike their natural models when this makes more sense. "It's a weak argument of tradition to say that designing an organ a certain way is better because that's how nature does it," says Dr. Galletti. "God created the heart, for example, as a pulsatile blood pump—it's made of muscle, which must pulse, and He or She had no other choice." A key criterion for the design of any type of prosthesis is that it be as simple as possible.

Anyone disturbed by this apparent guesswork may be even more dismayed to learn that although some artificial organs like the artificial kidney have been used successfully for decades, *their* functions (like those of many standard medications) are still not completely understood. It is clear what the blood oxygenator does, but that machine performs only one of several functions of the natural lung that are just now being defined.

Some investigators believe that as these mysteries are solved, prostheses can be made superior to their natural counterparts. "In certain areas, we can supersede nature," says Willem J. Kolff, M.D., Ph.D., Director of the Institute for Biomedical Engineering and the Division of Artificial Organs at the University of Utah. "Someday a marathon runner will be disqualified because he has an artificial heart. It's quite feasible that in time we'll produce artificial hearts that are more effective than natural ones, although that doesn't mean that people would prefer them." Most researchers, however, believe that they will never be able to match, much less outdo, the human body. "The first thing people must recognize about bionics is that the devices we make are at best pathetic imitations of the real thing," says Robert Stephen, M.D., Research Associate Professor in the Division of Artificial Organs at the University of Utah. "The cells themselves make millions of judgments every day—that's far more information processing than we could duplicate."

The complexity of natural systems is one reason that implants face competition from transplants, of which kidney transplants are the most widespread. Now and then the media publicize the story of someone who faces death because he or she needs a donor organ that is not available. In June 1984 the U.S. House of Representatives overwhelmingly passed legislation to establish a computerized registry that would match organ donors with potential recipients. While banning the buying and selling of organs, the plan also authorized $30 million over the next 2 years to help transplant patients pay for immunosuppressive drugs and $40 million over the next 4 years for grants to regional organ procurement organizations. The U.S. Senate had already passed a measure to establish a group to study transplants.

This legislation is well meant, but society still has not answered all the legal, ethical, and moral questions surrounding the transplantation of organs removed from humans or animals. Most people do not die in accidents or seem willing to donate their healthy organs to others. Even if they did, physicians would still face the problems of storing the organs and getting them to potential recipients promptly. "With a 1-year survival rate of over 80% and a 5-year rate close to 60%, heart transplantations are absolutely fantastic," says Donald B. Olsen, D.V.M., Research Professor of Surgery and Pharmaceutics at the Institute for Biomedical Engineering at the University of Utah. "But we need ample lead time for evaluation, tissue typing, and location of donor organs." Transplant recipients face the possibility that their bodies will reject the new organs; they must take immunosuppressive drugs, which weaken their immune systems, for the rest of their lives. A transplanted organ can develop any disease that could affect its predecessor. "Transplantation of living organs is a stopgap measure right now," says Emil P. Paganini, M.D., Head of the Section of Dialysis and Extracorporeal Therapy in the Department of Hypertension and Nephrology at the Cleveland Clinic Foundation. "Since we'll never mimic the original organs anyway, it might be better if we could offer artificial implants to patients."

Implants have disadvantages also, however. They must work perfectly without imposing excessive weight, bulk, or toxic materials on the recipients. They must connect with the body's communication lines and energy supplies in a symbiotic interface. Prototypes are expensive and do not regenerate like natural organs. Like transplants, implants can develop the original organ's disease; they also come down with disorders that result uniquely from their artificiality. An artificial lung version of pulmonary edema can occur, along with other implant disorders like thrombosis, aneurysmal dilation or dissection, calcification, and infection. Although the implants themselves may be standardized, not everyone is a candidate for them, often because of complications that would seem to be irrelevant.

Despite these drawbacks, both transplants and implants hold considerable promise. With the use of cyclosporine and other medical advances, transplants are becoming increasingly successful. "We're learning so much about how to control rejection that transplants and implants will be in a race over the next 10 years or so," says Carl F.W. Wolf, M.D., Associate Clinical Professor of Pathology at the Cornell University Medical College. "It may be more pertinent to ask not if implants will be possible but whether they will be necessary."

Implants are, however, holding their own against transplants. Free of many of the legal, ethical, and moral dilemmas posed by the use of living organs, they can be manufactured in large quantities and in various sizes and easily stored until use. Implants are less likely than transplants to be rejected by the body, and they do not age and die. In the long run they may prove the less expensive treatment alternative. It costs $100,000 to $150,000 to maintain a total artificial heart patient for the first year as opposed to $60,000 to $100,000 for a transplant patient. The second year, however, the maintenance of the implant patient costs half that of the transplant patient. Implant recipients may also show a better survival rate. "We expect a 2-year survival rate on our total artificial heart of 80%, which is probably better than the transplant rate," says Yukihiko Nosé, M.D., Ph.D., Chairman of the Department of Artificial Organs at the Cleveland Clinic Foundation.

The comparison between implants and transplants may ultimately prove meaningless. To date they have been complementary as well as alternative treatments. As we will see, patients are being kept alive by left ventricular assist devices, dialysis machines, and plasmapheresis until they can have heart, kidney, or liver transplants. "Many people from the transplantation field do not realize that it was our version of Kolff's artificial kidney which made the first kidney transplantation possible," wrote the late John P. Merrill, M.D., Professor of Medicine at the Harvard Medical School and Director Emeritus of the Renal Division of Brigham and Women's Hospital, Boston, in a recent article.

Another argument against comparison is more basic to the field of bionics. When we compare transplants and implants, as Dr. Galletti points out, we refer to the replacement of organs by organs. This comparison obscures the fact that bioengineers make devices to replace vital *functions*, not individual organs or other body parts. Moreover, unlike the transplant, the implant can also treat or even eliminate the need for that function.

This is why Dr. Nosé believes that the ultimate mission of artificial organs will be that of preventive measures to make their present use obsolete. "It's terrible to think that thousands of people are living with crude, man-made spare parts necessitated by end-stage organ failure because we in the medical profession failed to cure their diseases," he says. "We should use our knowledge so that we don't need to use artificial organs. It's ridiculous to wait until a patient is in end-stage heart or renal failure to provide an implant, although this is when it is justified. Unless and until we make artificial organs as good as natural ones, our mission is to use artificial organ technology early on to prevent organ failure and assure longer life of better quality."

We will see that Dr. Nosé is referring to a process called cryofiltration that he developed with Paul S. Malchesky, D.Eng., Director of the Metabolic Assist Program of the Department of Artificial Organs at the Cleveland Clinic Foundation. Diseases like atherosclerosis, cancer, and autoimmune illnesses are associated with gradual accumulations of large toxic molecules in the bloodstream. Because we have no system designed to filter these molecules, they clog organs and cause them to fail. Cryofiltration, which like plasmapheresis is an offshoot of kidney dialysis, is a way to clean out and overhaul the human body so that it can last closer to 100 years. The procedure will prove a major battleground in the fight against medical conservatism. "It will take much education to convince people to use artificial organs as preventive measures," says Malchesky.

Organ transplants are beyond the scope of this book, as are external aids for the handicapped, computerized medicine, or ubiquitous bionic parts like dentures, contact lenses, or various types of shunts. The devices discussed here are attached to or implanted within the body either permanently or for a considerable or crucial period of time. They are being developed or used in America; research in foreign countries has been excluded because certain issues facing bionics researchers are unique to the United States. Specific products are discussed *not* to endorse them but simply to offer representative samples of the types of research underway in different branches of bionics. Similarly, illustrations of bionic parts are generic unless otherwise indicated.

This book is an introduction to bionics for the lay audience and a general reference work for health care professionals inside and outside the field. Patient histories are included to provide a human perspective on the science; engineers in particular may be interested to learn how people have reacted to and benefited from their bionic creations. Most of the patients we will meet here happened to be ill in the right research center at the right time to try investigational devices. Just about anyone may someday be able to choose a bionic destiny, the implications of which will be suggested.

Fictional man-made creatures usually fail, causing some interesting havoc along the way. But bionics researchers create parts, not people (with some notable exceptions whom we'll meet later), and are unlikely to further the tradition left them by Dr. Victor Frankenstein. The body has long been a model for electronic circuits, so bionic hardware is logically suited for use inside it. The parts that already exist to enhance people's lives are in this instance greater than the whole body, which has failed them. Traditionalists may be alarmed that bionics has further split the body-mind dichotomy into a body-spare part dichotomy. The various bionic people who appear in these pages, however, are mostly ordinary folks who prove that the mysteries of the human spirit will always elude the technicians.

There are bionics brewing that we will not hear about for several years. "I feel that major developments in this area have moved from the academic into the commercial arena," says Dr. Wolf. "The leading companies now have technology that will remain proprietary for the time being because that's where their competitive edge lies." The secrecy guarded by bionics manufacturers may be just as well, at least according to William F. Bernhard, M.D., Professor of Surgery at the Harvard Medical School. "We need a proper perspective on these devices and their limitations," he says. "Aside from a few prostheses like joints, valves, and pacemakers that have established their place in the management of disease, bionics is investigational. People interested in these devices must sort through a lot of inflammatory rhetoric intended to sell shares in companies and protect the careers of bureaucrats in funding agencies. In cases in which investigators are officers or shareholders in small companies producing these devices, I wouldn't believe everything they have to say."

The fact that companies must sell shares and bureaucrats must cover their bottoms ought not to diminish our optimism about the future of bionics. Joints, valves, and pacemakers themselves were once innovative devices and the subjects of enthusiastic rhetoric. Miraculous devices are not necessarily around the corner—like all medical innovations, they need time in which their developers can prove their safety, reliability, and value. "We all know that the Six Million Dollar Man will be here, that we'll be able to replace joints, organs, limbs, senses and so on," says Dr. DeVries. "We all know that bionic parts will work—it's just a matter of time. We're taking a small step in that direction right now."

2

The World of Bioengineering

The first step in the direction of bioengineering, of course, was taken by the man who started the science of bionics itself. His career is a fitting introduction to the field as a whole.

Willem Johan Kolff, M.D., Ph.D., known to his friends as "Pim" (the Dutch diminutive for "Willem"), calls himself "the oldest artificial organist." Although his father was a physician (as are his sons Jack, Cornelis, and Therus), he had reservations about studying medicine because he did not think that he could bear to watch people die. Rather than being a liability, Dr. Kolff's sensitivity inspired him to revolutionize the field. As we will see later, his sympathy for a young man dying of kidney failure in the Netherlands prompted him to build the first clinically useful artificial kidney in 1943. "My mind is occupied with people dying in misery," he says, "and when I see a way to do something about it, I follow through. I know very little about pharmacology and tend to see mechanical solutions instead. We don't have to understand all about the body in order to treat the patient—all we need do is get a result that works." Rather than just handling the patient's blood outside his or her body or replacing defective parts with bionic devices, however, Dr. Kolff seeks ultimately to rehabilitate the whole patient.

By the time he started the science of bionics, Dr. Kolff had already established the first European blood bank in the Hague in 1941 (he happened to be there because his wife's grandfather had died; he never made it to the funeral). In 1949 Dr. Kolff sustained an animal with a heart-lung machine. The following year he emigrated to the United States, where he established the world's first Department of Artificial Organs at the Cleveland Clinic Foundation. In 1957 his group handled the first animal in the western world to receive an artificial heart; 4 years later Dr. Kolff was a co-inventor of the intra-aortic balloon pump.

In 1967 Dr. Kolff moved to the University of Utah, where he established the Institute for Biomedical Engineering with a staff of eight. The Institute is now staffed by over 100 people (although many of them are part-time employees who hold other academic positions). Six years later he had a major setback—a fire destroyed most of the Institute. Undeterred, Dr. Kolff solicited funds and rebuilt it. "Catastrophes always work out to my advantage," he says. "That includes the heart laboratory fire—now we have the superb new facility in Old St. Mark's Hospital. I'm not easily discouraged, and I think that is my greatest value. I analyze my failures and learn from them." This perspective is useful in everyday life as well as during its most dramatic moments. "I go to the laboratory even late at night, so the young researchers will know that I'm interested in them," says Dr. Kolff. "When they grow discouraged over consecutive equipment failures or the deaths of several calves, I remind them that they're experiencing correctable problems and that the basic principle works." Dr. Kolff was again proven right, of course, in 1982, when his group performed the first implantation of an artificial heart as treatment for a living human being.

In a *Festschrift* in Dr. Kolff's honor published last year, Benjamin T. Burton of the NIH refers to him as "the compleat Mosaic phenotype of pioneer-physician-surgeon-inventor-engineer-artist-economist-organizer-writer and raconteur par excellence." Dr. Stephen presents a similar list, then dismisses it as incomplete. "If one had to use a single word in preference to all others to describe W.J. Kolff," he writes, "that word would be 'catalyst.' " Dr. Kolff is gifted with the ability to recognize the potential of various new ideas, then collect an incongruous group of investigators to develop them. "Kolff has done more than anyone else in this country to assemble engineers and physicians and get them to work together," says Stephen C. Jacobsen, Ph.D., Director of the Center for Biomedical Design at the University of Utah.

Although some bionics researchers believe that one should have a wide background to work in the field, Dr. Kolff disagrees. "Specialists should not be removed from their own environments," he says. "It's a great mistake to encourage someone to follow a Ph.D. in physiology with one in mechanical or electrical engineering. By the time he's finished his training, he's 36 and ready for his first heart attack. Instead I take enthusiastic young people from different fields and bring them together to work on imaginative projects. That's probably what I'm best at."

Another of Dr. Kolff's talents is that of recognizing enormous creative potential that may not be expressed in academic credentials. Jacobsen, who has worked on a portable artificial kidney, a myoelectric arm, a subcutaneous peritoneal access device, and an electric drug injection system, among other projects, failed second grade because he could not learn multiplication and later was nearly kicked out of engineering school. Robert K. Jarvik, M.D., after whom the Jarvik hearts are named, attended an Italian medical school because he could not get into an American one. Several other members of the Utah group have been equally unconventional. Dr. Kolff structured his Institute as a comprehensive program to help them tackle several different problems at once, taking a product from concept through development and clinical trials directly to market.

Dr. Kolff sees no contradiction in his having stepped back in the prime of life so that other researchers could take over his work. Having enormously enjoyed the father-son relationships he had had with his professors in the Netherlands, he tries to continue it with his young co-workers. In his wallet he keeps a list of their names (and their wives' names, so he can address them politely) and calls them day or night. Dr. Kolff has also recognized that administrative duties, especially solicitation of funds, would keep him from the device fabrication he loves to do himself. "But I have no regrets that I've trained so many young people," he says. "Doing so has much more of an impact on artificial organs than my trying to run a one-man show, as many laboratories do. But now I'm trying to get rid of all the administration and return to making devices."

His inventions and the people he has trained are not Dr. Kolff's only contributions to medicine. He is the author of over 600 publications, not counting his three books and several book chapters. He has worked with researchers from Europe, Finland, Rumania, Greece, Turkey, England, Australia, Asia, India, Thailand, the Philippines, Korea, China, Japan, and Central and South America. In recognition of his accomplishments, Dr. Kolff has received 11 honorary degrees and 79 other honors and awards and holds orders from the Netherlands and Argentina. He is unable, however, to identify his greatest contribution to bionics. "I'm still doing active research and too close to my work to tell," he says.

As might be expected, other leading bionics researchers praise Dr. Kolff generously. "Nobody can challenge Kolff, although we try hard," says Dr. Nosé, who inherited the Cleveland Clinic Division of Artificial Organs from him. "He should receive a Nobel prize. He was a tough teacher, but a very paternal, charming individual. I admire and respect him. In 17 years he never gave me a compliment, but he recognized me and considered me his successor. One time, when he had brought my wife flowers, he said to her, 'Bonnie, I'm so happy—Yuki really outgrew me.'" Dr. Nosé reports that he and Dr. Kolff exchange advice; the difference, he notes cheerfully, is that he takes Dr. Kolff's advice, and Dr. Kolff usually does not take his.

"I saw the first Kolff kidney in Geneva in 1948 or 1949 when I was a young student," says Dr. Galletti. "I was fascinated by his having thought about living structures in a way that was totally revolutionary. In my mind Kolff ranks as highly as Watson and Crick* in the history of medicine and biology. He is truly a giant by virtue of his new views and very different mind."

At 74 Dr. Kolff is too busy making heart valves and other devices to consider retirement.

Like Dr. Kolff, other renowned researchers hesitate to define their contributions to the field. "We create devices and prove that they're possible, then someone else can produce them. Whether we've contributed anything to the field is for someone else to say, not me," says Dr. Nosé, who like Dr. Kolff seems to prefer striking out in new directions, then leaving a field when it gets crowded. "I do know that it's important for me to stimulate people to think about issues in bionics and see things globally rather than piecemeal," says Dr. Galletti. "And I must try to help them keep a positive, optimistic approach

*The geneticist and biophysicist, respectively, who discovered the double helix.

rather than a negative, self-centered, frightened approach to these issues."

Some researchers like Dr. Kolff or Dr. Jarvik (whose father underwent heart surgery) began work on artificial body parts because a specific experience made them determined to improve medical technology. Others have a more generalized altruistic bent along with a passion for engineering. Everybody stays in the field for the same reason—it is fascinating and fun. "In science, if you really mean it, you literally bet your life that something will come of your work," says Samuel P. Bessman, M.D., Chairman of the Department of Pharmacology and Nutrition and Professor of Pediatrics at the University of Southern California (USC) School of Medicine. "I'm not doing my research because I want to be invited to the next meeting or to get a medal, although that would be nice. I'm doing it because it's so interesting. The work our laboratory is doing is unbelievably exciting."

"We work in bionics because we want solutions," says J. Donald Hill, M.D., Chairman of the Department of Cardiovascular Surgery at the Pacific Presbyterian Medical Center in San Francisco. "The technology is much too expensive to develop without the potential for practical applications. But no one would work in this area unless he or she enjoyed it. In the beginning, we have to get our excitement out of little successes, like a calf with an artificial blood pump *standing up*."

This excitement and commitment help them over the rough spots in their careers. A young bioengineer may actually have trouble finding work. The National Institute of Handicapped Research finances rehabilitation engineering centers across the country, but only a dozen or so of these exist. Bioengineering graduates must look instead to universities, private industry, foundations, or VA medical centers for jobs.

Once employed, researchers must learn to ignore shortsighted peers and funding agencies who claim that their ideas will never work. They must remain searchers who are not discouraged when answers elude them. "Working in our field means struggling around, making many mistakes and being willing to spend a lot of time running into dead ends," says Jacobsen. "It's a tough, inefficient, expensive, frustrating process that takes hard work and many ideas. Innovations in this field don't evolve efficiently or represent flashes of genius—they require constant proximity to the action. Most of the researchers I've met are driven to continue."

At some point researchers bump up against the need to convince others that their work has value. Although crucially important, many bionic devices have small markets and so do not attract the attention of semiconductor and other technology companies. Prototypes are prohibitively expensive, and when a product is ready it will be last in line for the most advanced electronic systems, like microprocessor technology, and the best marketing. Medical personnel are eager for breakthroughs rather than mere improvements, and they must be persuaded to use new products once these become available. "I know someone who spent a lifetime working on a vision device that he developed commercially, but then the blind wouldn't buy it," says Carter C. Collins, Ph.D., Senior Scientist at the Smith-Kettlewell Institute of Visual Sciences in San Francisco. "I myself feel that my talents are best applied by doing basic research on several devices." Other researchers try to balance both approaches.

Practical concerns like these have to be kept in perspective or they will stifle creativity and activity. "We need to remain open-minded about the possibility that a competitive approach may be as good as or better than ours," says Paul Bach-y-Rita, M.D., Professor and Chairman of the Department of Rehabilitation Medicine at the University of Wisconsin, Madison. "The worst part of dealing with instrumentation as opposed to pure science is that we can become proprietary. This happens when devices rather than just theoretical considerations are involved." "We're scientists—once we get interested in commercialism, we're in trouble," says Dr. Nosé. "Anybody can produce our creations. We must live somewhat hungry, because without that, we can't create anything."

Successful researchers in the field face problems of their own. Some physicians and researchers just do not like their colleagues to get the upper hand in publicity or reputation because that means fame and money. Professional jealousy is just as common in medical experimentation as it is anywhere else, and those working in a high-technology, potentially lucrative field like bionics are likely to guard their positions carefully. "My situation puts a lot of pressure on me, particularly when I'm criticized by my colleagues," says Dr. DeVries. "It upset me a great deal, for instance, when the president of the American Medical Association [AMA] came out against the artificial heart. My colleagues should have been better informed, but they weren't, and they should have known better, but they didn't. That's very different from the man in the street being critical because he's ignorant."

"Basically, though," he adds, "I was well enough convinced that the artificial heart would work to put my life and reputation on the line and thoroughly trust the many people who helped me. I wasn't going to listen to any criticism. My critics have to know a little bit more than I do—and I know more about these experiments than anyone else in the world." Many bionics researchers are in similar positions of having to demonstrate the firmness of their convictions to funding agencies, manufacturers, FDA officials, and skeptical peers.

Women bioengineers face particular difficulties because they are tokens in a field dominated by men. A few women bioengineering students can be found at various universities (in the Merrill Engineering Building at the University of Utah, a student pounced on the author in a women's room and eagerly inquired whether she too was a student). This discouraging situation seems, however, to be changing. "This year half our students are women, and three of them did the year's best projects," says Dr. Galletti. The Utah 100 heart (not to be confused with the Jarvik heart) is built at the University of Utah in a fabrication department supervised by 31-year-old Pamela A. Dew. Interesting husband and wife teams like Larry and June Wilson Hench of the University of Florida, Joseph and Celia Bonaventura of Duke University, and William and Gladys Dobelle of Avery Laboratories share work in their respective disciplines.

We have already noted that together these disciplines constitute a philosophical landmark. The importance of bionics is evident in terms of the four treatment concepts that have evolved during the history of medicine: palliative (relief of symptoms), curative (interruption or reversal of illness), preventive (avoidance of illness), and now substitutive (replacement of damaged body structures by spare parts). "The substitutive concept developed in the 1950s like a painting school," says Dr. Galletti. "An intellectual ferment led people to think in a certain way and to start taking parts off shelves in order to treat illnesses. Though not rationalized out in advance, the movement was part of a logical process of discovery." In fact, the substitutive approach progressed in a kind of anarchy in which groups that did not communicate well were doing the same work. It has been made clear that this situation will not do; instead, as Dr. Galletti points out, we need a new social contract between patients, physicians, manufacturers, and government agencies.

Substitutive medicine can be further divided into stages. The first round of artificial organs tried to replace defective nature with purely artificial devices. As we will see, we have begun the next round of hybrid or bioresorbable organs, which are designed to cooperate with nature.

"The bionic approach gives physicians the best of both worlds," says Dr. Paganini. "They know what the natural body does and help it as much as possible, but they're not limited by it. As a last resort they can use an artificial device to replace one of the failing functions of a natural organ. The most knowledgeable people in this field treat whole patients, not just dysfunctioning organs, and are making intelligent decisions about when to use or not use artificial devices."

Like other scientists, bioengineers must clearly define specific goals, then work toward them within the context of an entire problem. In bionics an entire problem encompasses the biological, the physiological, and the mechanical, from basic research through FDA approval, cost-effective production and marketing, and legal and ethical problems raised by success. "This is one field in which nobody gets bored because we work with so many different disciplines," says Leonard Trudell, D.Sc., Assistant Director of the Artificial Organ Laboratory at Brown University. "It takes many people, specialties, and compromises to make a project succeed." The best researchers seem to be at least familiar with a large store of engineering and medical knowledge.

University departments of biomedical engineering may include biophysicists, biochemists, biologists, chemists, physiologists, veterinarians, mathematicians, microelectronics specialists, electrical engineers, chemical and mechanical engineers, metallurgists, computer scientists, various surgeons, anesthesiologists, hematologists, cardiologists, neurologists, endocrinologists, nephrologists, immunologists, or immunopathologists. Those in the engineering camp may have had an undergraduate major in that discipline along with graduate courses in anatomy, physiology, and physical and organic chemistry. Some universities have programs enabling students with doctorates in bioengineering to earn medical degrees in 18 months, although the reverse does not occur. "The M.D. degree provides more medical background than the basic courses a bioengineer would have taken while earning the Ph.D.," says Collins. "It's better to get the Ph.D. first in order to develop the habit of rigorous scientific thinking, which can be more difficult to learn later."

As might be imagined, this assembling of disparate specialists can create many problems. "C.P. Snow spoke of two cultures: science and nonscience," say Peter Richardson, Ph.D., D.Sc., Professor of Engineering and Physiology at Brown University. "But there are many subcultures in science that communicate so ineffectively that the same concepts are formulated differently and named separately by each discipline." Similar distancing occurs between the medical specialities. Of course, combining engineering and medicine can make communication break down entirely, as many researchers emphasize, and cause goal conflicts. While a scientist may be patiently and meticulously designing devices that represent logical thought, a physician may be taking a less disciplined, more intuitive approach in his or her anxiety to help the patient. Many researchers in bionics are, however, able to bridge this gap. "Research *per se* is absolutely fine, but it's just an outcropping of more important things, at least in medical research, whose goal is to help people. I want to see that goal fulfilled. On the other hand, we don't jump on bandwagons—we believe in doing important work, doing it well, and doing it slowly," says Dr. Paganini. "People can't be held critically to a schedule when they're trying to develop answers to extremely difficult questions," says Dr. Bernhard. "We were way off schedule in the artificial heart program—but so what, if we were making progress and learning new things?"

The success of a project depends on the ability of researchers to overcome these difficulties, communicate their ideas and desires, and work well together. The Department of Artificial Organs at the Cleveland Clinic Foundation, for instance, is structured to facilitate this process among researchers of various nationalities as well as different disciplines. Each of the Department's projects has a physician and an engineer as co-leaders. A researcher might be a co-leader on two projects and a worker on three. Gordon Jacobs, MSEE, the Director of Blood Pump Assist, spends about half his time as technical manager and half providing technical engineering input in mathematics, electronics. design, or physiological analysis as well as writing grant proposals and selling ideas. "To emphasize job titles, status, or levels in a matrix-style organization like this would be disastrous," he says. "We can't have mutual cooperation if one researcher claims that he or she is *the boss* and another researcher is not. I believe one of the secrets of leadership is to make each person feel that he or she is the one most important to the project's success."

One thing bionics researchers across the country do agree on is their goal: to improve the quality of life. "We're not trying to add to life span *per se*. We're trying to improve the quality of life for the appointed span—not to make a person immortal, but to make his or her body work better," says Trudell. "Any artificial organ we create should allow its recipient to live a normal life, or we're not doing a good job. We're getting there," says Dr. Nosé.

Dr. Kolff is not satisfied with mere normality. "The crucial question that motivates me is, 'Does an implant or prosthesis make the recipient happy?' " he says. "If it doesn't, then we should not use it. I have enormous respect for a happy life, but I see no sense at all in prolonging misery. This is why I was so incensed when people who had not seen Barney Clark said that we were prolonging his death. Why didn't they come here and talk to Barney Clark before passing judgment?"

On the face of it there seem to be important distinctions between prolonging life and increasing its quality. The former raises ethical questions; the latter is a nonethical issue of increasing a person's capabilities rather than intervening in a disease process. Quantity can be measured; quality cannot. These and other distinctions, however, are actually blurred. Bionic parts that now prolong life have much room for improvement. Increasing life's quality is likely to make a person live longer. Nature itself is rather stingy when it comes to determining life span. "To be biologically successful, all a creature need do is live until its young can protect themselves," says Harvey Sasken, M.D., M.P.H., Associate Pathologist at Rhode Island Hospital and Associate Professor of Pathology at Brown University. "But most people don't become productive until they are in their mid-twenties. Civilization needs people who are not necessary biologically, so some of our work tries to keep people viable and productive until senescence."

The world of bionics is a contradictory one in which to work. Its products are hailed by the media but often ignored by organizations with the money to fund them. Its practitioners are self-absorbed in their enjoyment of their work, but their ultimate goal is to benefit humankind. Like other sciences it combines logical discovery processes with plain good luck. "Our work is a matter of learning how to laugh at and be critical of ourselves while accepting a hell of a lot of serendipity," says Donald J. Lyman, Ph.D., Director of the Biomedical Engineering Center for Polymer Implants at the University of Utah. "Many times we back into something we weren't really looking for. We pose hypotheses just to get ourselves over the next little hill, then reform them." Bionics evolves through the logic of discovery, which really is not very logical at all. "One must be very brave to say that a concept is possible or even that a particular progression will occur," says Richardson. "Ultimately it depends on bright ideas, which are unpredictable—that's part of the charm and the challenge."

3

Biomaterials

Another part of the challenge is finding materials with which to implement new ideas. Unpleasant as it may sound, our bodies are essentially chemical swamps. Each of us is full of acids, saline solutions, and constantly shifting tissues that make an extremely hazardous environment for any type of implant. To make matters worse, the body mobilizes itself for attack whenever it is violated by a foreign object. It responds by checking its arsenal to find the deadliest solvent, by driving the invader out through the skin or into some nearby cavity, or by capturing it inside a wall of tissue and grudgingly ignoring it. "Not only is the body an extremely hostile, corrosive environment, it can direct its worst forces against a particular implant," says Dr. Sasken. "In our devices we're constantly trying to keep one step ahead of the attack." The biggest roadblock to creating successful implants has been that of finding biocompatible (or more accurately, *biotolerable*) materials.

A biomaterial can be defined quite simply as any material found in any implantable or percutaneous bionic part. Biomaterials scientists study the body's tissues, the properties of certain materials, and the interaction of the two to develop safe biomaterials with appropriate characteristics. Although hundreds of biomaterials are being discovered, created, and tested, applications for them greatly exceed the number of compatible materials available. "Purposeful implantation of foreign material with expectation of long-term success violates centuries of evolving surgical principles and indeed violates what we know of healing and body defense mechanisms," writes Thomas J. Krizek, M.D., Professor of Plastic Surgery at USC, in a recent essay. "What is surprising is not that it works so well but that it works at all."

For centuries physicians have tried using everything from horsehair to gold wire to repair the human body. The modern era of biomaterials really began in the late 1940s, inspired by military needs during World War II and postwar industry. It was fueled by the space program and really took off in the 1950s. During the next few decades biomaterials were passive—once implanted, they remained inert. "The number one tenet in biomaterials is that they should do no harm," says James M. Anderson, M.D., Ph.D., Professor of Pathology, Macromolecular Science and Biomedical Engineering at Case Western Reserve University. "Researchers did not understand the basic biological interactions of implants, but they were satisfied that the materials were safe." The next step, of course, is to try to do some good. Researchers are moving into an exciting area of functional biomaterials, which actively interact with the biological system in understandable, beneficial ways. "This has been the biggest change in the whole field of biomaterials," says June Wilson Hench, Ph.D., Program Manager of the Bioglass Research Center at the University of Florida. "While there will always be room for inert materials with specific properties and applications, the field of new surface-active materials will probably continue to expand."

Like the implants they form, biomaterials have until quite recently represented compromises. The body's tissues have remained state of the art over millions of years of evolution. To understand, accommodate, and approximate them, biomaterials scientists must consider all kinds of biological and technological interactions. "Biomaterials are so imperfect that it's depressing to listen to people describe them," says Dr. Bernhard, "but at least they're not hopeless." In fact, the materials that result from various studies may well perform new and unique functions of their own.

Although not everyone would agree with Dr. Bernhard that descriptions of biomaterials are depressing, readers may find this chapter tedious compared to the rest of the book. Be warned, however: the chapter's contents are basic to understanding bionics; skipping it will result in later confusion.

Problems of Biomaterials

Biomaterials are imperfect because they pose many problems to scientists trying to use them. Those proven safe, for example, might not have the right combination of other qualities. Perhaps they do not adhere to anything, including each other, or perhaps they are tough, strong, and wear-resistant when the application calls for soft, elastic flexibility. Specific materials may be alienated from other family members: two different metals implanted together, for instance, corrode each other at the points at which they touch. Biomaterials can be unpredictable, reacting idiosyncratically to purification, sterilization, molding, combination with other materials, long-term implantation, and other necessary procedures.

Biomaterial surfaces, especially those of polymers, can be very sensitive. Sometimes the side next to the mold for a device has properties different from the side that faced outward. Body cells look, grow, and act differently depending on the side of a material to which they stick. Another problem is the implant's shape. "We've known for years that if we implant anything in the body, its edges will become carcinogenic, and the rest of it will not," says Dr. Galletti. "A triangle is more carcinogenic than a circle. Nature seems to abhor acute angles."

Depending on the type of material used, an implant can weaken and loosen because of corrosion, pitting, leaching, or dissolution. These processes can release toxins, inflammatory agents, or carcinogens. If materials shed fibers or cause blood clots, the implant recipient risks having these substances wander to the brain or other organs with catastrophic results. If they are difficult to sterilize, biomaterials can cause serious infections. One of Dr. Anderson's continuing studies has found that while fewer than 5% of patients develop infections from artificial heart valves, 80% of those complications are fatal.

An outcome of all biomaterial implantations is encapsulation: a certain amount of scar tissue forms around every implant. A thick capsule can decrease circulation, fill up with biochemical junk, or nourish an infection. It may calcify and harden, causing tissue damage, pain, or paralysis. As localized stress changes, the implant can loosen and fracture. Like implant materials, the capsule itself can start breaking off fragments, which sail through the bloodstream looking for mischief.

Any of these problems can require a reparative surgical procedure, which is more difficult than an original implantation, more complicated if tissue damage or infection has occurred, and more traumatic for the patient.

In this book we will see several examples of how researchers who cannot beat nature try to join it, and biomaterials scientists are no exception. One way to handle encapsulation, for instance, is to make a virtue of a vice. Scientists can deliberately construct implants from materials that encourage fibrous or bony growth around them. The woven or porous surfaces of these materials promote ingrowth that permanently anchors an implant better and more naturally than sutures or cements. Meanwhile the body thinks that it has conquered the invader—as it indeed does when scientists implant slowly dissolving materials that prompt tissues to regenerate themselves.

Types of Biomaterials

Biomaterials fall into three groups: metals, ceramics, and polymers (composites are sometimes classified in a group by themselves).

METALS

Because of their strength, metals are particularly valuable as either temporary or permanent orthopedic implants. Bone plates, nails, screws, rods, sutures, and more sophisticated hardware can be implanted anywhere from a few days to several months to stabilize broken bones or body tissues until they heal. Skull plates and total joint replacements are implanted as permanent prostheses. As will be discussed later, joints with porous or coated surfaces into which tissue can grow will probably make bone cement obsolete. Metals are also used to make electrodes; to seal implantable energy packs; and to form heart valve components, maxillofacial and dental implants, and meshes for tissue repair. The metals most suitable for these purposes include titanium, stainless steel, platinum, tantalum, and cobalt-chromium alloys.

CERAMICS

The increasing popularity of ceramics in high-technology industries is represented also in the field of bionics. Ceramics have fully established their credibility as functional biomaterials. The first successful bioactive ceramic, which gave rise to a whole group of related materials, was Bioglass, developed in 1969 by Larry L. Hench, Ph.D., Director of the Bioglass Research Center at the University of Florida.

Bioglass is essentially a glass material designed to have a chemistry so similar to that of bone that the body does not recognize it as foreign. Instead the body forms a bond with Bioglass similar to that between bone and muscle or new bone. When Bioglass is implanted in a bone defect, a layer forms on its surface in which the body lays down collagen fibers, then the mineral structure; the resulting bond is stronger than either the glass or the bone. If Bioglass is implanted in soft tissue, which has no bone cells, it will develop just the collagen bond to hold it fast.

This property underlies the development of a Bioglass device now on the market: a substitute for the three tiny bones of the middle ear, which can deteriorate from chronic infection. The Henches are collaborating with various centers here and abroad to test other applications of Bioglass, including its use as a filler for tooth extraction sites to prevent resorbtion (deterioration) of the jawbone. In the future they hope to develop Bioglass composites and coatings for artificial joints, flexible Bioglass for soft tissue replacement, and a version to be used in cell sorting.

Once molded, ceramics can be made very hard and both heat and wear resistant; some types have unusual electrical and electromagnetic properties. New materials, particularly calcium phosphate ceramics like hydroxyapatite (HA) and tricalcium phosphate (TCP), have been appearing, most noticeably in the field of dentistry. Other ceramics can be anchored with bony ingrowth rather than bonding, and still others, like some calcium phosphates and calcium aluminates, are resorbed and slowly replaced by bone.

One ceramic-like material, carbon, is used to reinforce composite materials. Its enormous range of properties makes it a good contact surface. Textured carbon attaches to bone or tissue, making it suitable for percutaneous leads and perhaps the attachment of artificial limbs; smooth carbon precludes blood clotting. Pyrolytic (heat-treated) glassy carbons are used for heart valves, blood access and heart assist devices, tooth roots, artificial joints, ear prostheses, and skull plates. They are also found on the fabrics of heart valve sewing rings and vascular grafts. Carbon fibers, also used in composites, have been fashioned into artificial tendons and ligaments.

POLYMERS

Simple molecules called monomers react with each other to form large, long chains called polymers. Although they include gels and liquids, polymers are usually classified as either elastomers (rubbers), which can be stressed out of and returned to their original shapes, or plastics, which are rigid. The two types of plastics are thermoplastic (linear rows of monomers that soften when heated and can be reshaped) and thermosetting (monomers that have been chemically treated to form cross-linked patterns so that heat does not soften the material). Whatever their configuration, polymer molecules are usually described in terms of their resemblance to a plate of spaghetti.

Our bodies are composed largely of polymers in the forms of protein mucopolysaccharides, found in mucus, and polymer ceramic composites, found in bones. Collagen, another natural polymer, is a generic term for the body's framework of skin, tendons, cartilage, and connective tissues made of fibrous proteins. Pork collagen heart valves are chemically treated and used to replace human heart valves, while soluble collagen can be injected under the skin to smooth hollows, flaws, and wrinkles. Other chemically treated natural polymers include fabrics like cotton, wool, and silk and materials like rubber, leather, and resins.

More than any other type of biomaterial, polymers can be custom-made to show an enormous variety of properties. Researchers agree that the future of biomaterials for soft tissue applications will be in polymers and composites of polymer fibers, ceramics, blends, and alloys.

Plastics. Most people are aware that some of the man-made materials widely used in our homes and our clothing can be placed in our bodies as well. Acrylics like Plexiglass and Lucite are used to make intraocular lenses, soft contact lenses, and artificial eyeballs; polymethyl methacrylate powder is the standard ingredient of bone cement. Polyester (Dacron) is made into several implantable fabrics, especially for vascular grafts, which are anchored by tissue ingrowth. Vascular grafts can also be fashioned from polytetrafluoroethylene (Teflon). This material is also often used in ear prostheses. Researchers at the University of Florida are developing a new Teflon so slick that calcium, clots, and even bacteria cannot stick to it. They are beginning animal tests of urinary catheters made from the new material. (Catheters become encrusted with calcium; infections caused by their contamination kill 50,000 people each year.) The Teflon may also be used to make permanently sterile implants and supplies.

Other plastics include polyethylene, a remarkably biocompatible material used for articulation surfaces in joints, for ear implants, and for tissue repair mesh. Polysulfones, which are extremely hardy and heat resistant, are fashioned into neurological implants and components of heart valves and pacemakers; these materials may someday serve as bone replacements. As we will see later, a fluorocarbon fluid called Fluosol-DA can substitute indefinitely for human blood.

Elastomers. Polyurethanes, which include hundreds of different compounds, have had bad press because some of them have varying degrees of bioresorption—that is, they decompose inside the body. As the field of biomaterials evolves, however, this *functional* property gives polyurethanes new prestige, and improved versions of them are starting to show up everywhere. "The urethane field is coming up, perhaps just because we can make a greater variety of structures right now in this generic family," says Lyman. "In polymers, polyurethanes are certainly going to be the materials of the future," says Dr. Anderson. Another elastomer, polyolefin, has meanwhile established itself as a flexible material suitable for various joints and components of artificial hearts.

Any solid or liquid biomaterial that is oily, gel-like, or rubbery is probably one of the organosilicon polymers (silicones). Before their dangerous side effects became clear, silicone injections were used to enlarge breasts; some physicians still inject this substance under facial skin for cosmetic improvement. Silicone rubber implants are standard prostheses for reconstructing the face, ears, breasts, larynx, esophagus, genitals, and urinary tract; replacing small joints and tendons; and making drug-release capsules and leads, power sources, and antennae for pacemakers and neurostimulators. In at least some circles, however, silicone is losing its popularity: "Silicone was a wonderful, exciting material in the 1950s, but it's simply passé at this point," says Donald Leake, D.M.D., M.D., F.A.C.S., Director of the Dental Research Institute at the University of California, Los Angeles (UCLA), and Professor and Chief of Oral and Maxillofacial Surgery at the University of California, Torrance.

COMPOSITES

Composites are tailor-made to show desirable properties like strength and flexibility that are otherwise mutually exclusive in biomaterials. Polymer composites, for example, are being tested for dental applications. The Henches hope to develop a composite of Bioglass that combines its bonding power with the flexibility of polymers. Carbon fibers are combined with carbon, polyethylene, or Teflon to make implants suitable for use in maxillofacial and plastic surgery, ear and bone reconstruction, and joint coatings. (For this reason, carbon is sometimes classified as a polymer.)

New Developments in Biomaterials

Researchers are continually seeking new techniques and opportunities to develop, evaluate, and improve biomaterials. Dr. Anderson's group, for instance, has a well-established implant retrieval program evidenced by a huge histological slide collection and drawers full of artificial joints. The researchers are sent all the implants removed at the university hospital, enabling them to study various interactions with human tissues. Sophisticated measurement techniques are providing new insights as to how such interactions occur. "I can stain all the layers that form around an implanted fiber and measure the capsule thickness," says Dr. Sasken. "After 2 weeks, I can predict what the capsule will look like in a year. Using light microscopy, I can study fibers at various stages of biodegradation so we'll need fewer experiments in the future to fill in the gaps."

ALTERED SURFACES

Traditional work in biomaterials has sought and concentrated on those that offer suitable interfaces to their biological surroundings. Physicist Wolfgang Pauli is supposed to have said that "God made solids—but surfaces were the work of the Devil," and most biomaterials scientists would agree. Fortunately, however, many researchers are up to some devilry of their own. When appropriate, they can use functional biomaterials that encourage tissue ingrowth to anchor an implant, for example. A more flexible—and tricky—method is to develop a surface that forms a chemical bond with surrounding tissue to preclude implant deterioration.

Lyman has had considerable success in custom designing the surfaces of polymers, using a principle that could apply to other materials as well. His field is that of structure-property relationships and their possible variations; his group is able to make polymers from scratch and test them right through to human implantation. Lyman's method is simply that of manipulating single molecules or atoms, rearranging the chemical groups on the polymer surfaces according to how he wants the materials to perform. Theoretically this technique could make available an infinite number of biomaterials. "We'd need a large selection because of the variety of body tissues," says Lyman. "An implant that could succeed in one specific section of the body could fail anywhere else. But there are also many reactions common to all implantations, and we're trying to understand them better." Lyman's approach, which he has applied to materials for vascular grafts, has broad implications for cell growth, blood clotting, and diagnostic techniques. These will be discussed later.

So far Lyman has successfully created surfaces on which blood will not clot. The opposite effect, that of making polymers adhere to soft tissue, is somewhat trickier to achieve. "I worked in England for 5 years trying to do with polymers what we've done with the Bioglass system here, that is, modify the surface for tissue adhesion," says June Wilson Hench. "Researchers are only just starting to get results, because it's much more difficult to do with soft tissue." To replace tissue successfully, a material must imitate its mechanical properties; to form a tight bond, both the material and its surroundings must temporarily stay put. It is much easier to immobilize a hard implant in bone long enough for them to bond together than it is to keep in place a flexible material and soft tissue.

Another method applicable to any surface is that of *biolization*, developed at the Cleveland Clinic Foundation. In a recent article the researchers defined this process as "either the chemical or thermal treatment of natural tissue or protein (natural tissue derivative) by either coating or blending protein with other polymers."

For smooth materials like soft rubber, epoxy, or metals, the biolization must be preceded by salt casting to anchor the coating mechanically. The implant is first molded into a blood pump or whatever else it is supposed to be. Then the researchers apply a polyurethane coating mixed with salt. When the coating has reached the proper thickness, the device is dipped in water to remove the salt, leaving a textured surface with tiny pores.

To biolize the implant, the researchers then apply 5% purified gelatin (denatured protein) solution to the new surface, using a vacuum to fill all the pores. They cool the surface, then cross-link the gelled coating with aldehyde. The last procedure makes the gel insoluble in water, increases the number of cross-links in the protein chain for greater mechanical strength, and assures biocompatibility. The result is a smooth surface that does not attract platelets, fibrin, or calcium; white cells may visit for a few days but then detach. "We can use this procedure, for instance, on any surface that must be in contact with blood," says Shun Murabayashi, Ph.D., Scientific Director of the Biomaterials Section in the Department of Artificial Organs at the Cleveland Clinic.

The surfaces of biomaterials might be used in an important interface with *our* surface—the skin. Implantations of the Jarvik-7 artificial heart have made the public aware that percutaneous tubes have the potential to cause infections. Biomaterials may become a means of preventing such complications. "Whenever the skin surface is breached, the epithelium grows down around the leads," says Dr. Sasken. "Either the leads are extruded, or bacteria gain entrance. We need an interface so compatible with both the material and the skin that scar tissue joins with it to prolong the life of the device."

BIORESORBABLE MATERIALS

One way of avoiding the problem of biocompatible surfaces while having materials function beneficially is to create bioresorbable ceramics and polymers. Most people are familiar with sutures and meshes that dissolve inside the body. As we will see later, a bioresorbable material can be made into an artificial scaffold that guides the ingrowth of natural tissue. As the tissue weaves through it, the implant slowly degrades, leaving a new natural organ or tissue in its place. Long-term biocompatibility is not an issue. First, the material disappears at the end of its carefully defined life span. Second, the worst biological reactions are often caused by biomaterials with the longest stability; with temporary biomaterials, these reactions might be much better controlled.

Conceivably a bioresorbable material could be treated to induce different types of cells to grow in their original configurations, although this goal is still quite a distance away. Meanwhile researchers like Lyman and Dr. Sasken are experimenting with relatively "simple" structures like blood vessels. "A static material like a synthetic graft in a dynamic living situation like the body is archaic," says Lyman, and most of his colleagues agree.

Besides growing new vessels and organs, bioresorbable materials could also be used in timed release drug treatment. Drugs could be contained in microscopic spheres that either have tiny holes in them or disappear entirely.

"In the last 5 to 7 years we've seen a resurgence in this area because of research in cloning and gene-splitting," says Dr. Anderson. "We have many companies generating hormones, proteins, and other bioactive agents that are very short-lived in normal therapies but can be delivered over far longer periods with sustained release." One injection might provide a year's supply of birth control medication, antibiotics,or even chemotherapy. "We have many chronic, debilitating diseases in which drug levels necessary to treat the symptoms border on toxic levels," says Dr. Anderson. "Using timed release techniques on an outpatient basis, we can deliver stronger localized therapy." Dr. Anderson has also experimented with implants made of biomaterials impregnated with timed release antibiotics that prevent localized infections.

CELL SORTING

A material's surface determines whether cells will adhere to it and do so normally. Removing a cell from the body without disturbing it enables scientists to study its behavior on a particular surface. Cells placed on a Bioglass surface, for instance, act differently than they would on that of an ordinary glass. If a mixture of bone cells and connective tissue cells is placed on a Bioglass surface, the bone cells will grow, while the others drift into a kind of suspended animation. "We think we know why this happens and why we get a mineral rather than a scar tissue bond when we implant Bioglass in bone," says June Wilson Hench. "We want to find out if we can use bioactive surfaces to build a sorting system for cells, separating those that grow from those that rest."

Cells that hibernate are less important, however, than those that become malignant. Conveniently enough, normal cells will stick to some surfaces, although cancer cells will not. "Using surfaces, we could study beginning pre-changes in cell chemistry that have not yet affected its morphology or its triggers for other events," says Lyman.

The Future of Biomaterials

The biomaterials field must maintain a wide range of materials like those discussed here. Specific applications require the prescription and design of biomaterials with very carefully defined properties. Over the past decade biomaterials scientists have moved toward a better understanding of structure-property relationships from both a material and a biological point of view. To make longer lived implants, they must continue in this direction. "In a sense we're at the college level in understanding polymer structure-property relationships and in kindergarten when it comes to understanding the biological interactions and what controls them," says Dr. Anderson. Researchers grasp the "materials" side down to molecular and anatomical levels, but they are far less familiar with the "bio" side. Dr. Anderson, for one, believes that fundamental understanding of biological interactions should precede study of how molecular or macromolecular structural variations affect them. Rounding out their investigations, he believes, will give researchers design criteria for new and better materials.

"We still need to define both blood and cellular biocompatibility," he says. "Are complex interactions positive, negative, or synergistic? Are cells near a biomaterial activated or inactivated? Do they secrete products that might control the proliferation or metabolism of other cell types surrounding the material? All these questions are fundamental to our work."

This research will require two changes in the field of biomaterials. Medicine has low priority when scientific research funds are being handed out; biomaterials, basic as they are, are likewise neglected when medical research funds are distributed. "We have many inquiries from undergraduates and graduate students who would like to work in biomaterials, but they get discouraged. There are no jobs and no funds for expensive research, and it's a tough area in which to advance," says John Douglas Mackenzie, Ph.D., Professor of Engineering at UCLA. "A number of problems just on the verge of being solved won't be, because the support isn't there," says Dr. Leake. "Researchers who write papers on biomaterials get six to ten times as many reprint requests from Scandinavian countries, especially Sweden, as they do from Americans. Those scientists have the research money." Scientists fortunate enough to be funded by private companies are delayed by the expectation that they will concentrate on products to sell rather than important basic research.

Besides better funding, the biomaterials field needs to become more strongly interdisciplinary. "People in disparate fields haven't communicated with each other, and try as we might, we still find it difficult. But we need people from different fields to evaluate biomaterials," says Dr. Sasken. "I feel like the grease between two wheels," says Dr. Anderson, "but this is necessary for translation and communication, since often the biomaterials scientists and the surgeons may not grasp each others' meanings." This problem is true of the whole field of bionics, and its foundation is probably the best place in which to initiate change.

SECTION

II

RESTORED MOVEMENT

4

The Body's Scaffolding

If we were cardboard Halloween skeletons, we could get by with joints that moved at the pull of a few strings. The grace and strength of human movement, however, depend on a hierarchy of precisely structured composite tissues, including those surrounding and stabilizing the joints. Ligaments connect bones or cartilages, while tendons link bones to muscles. These tissues continually add or remove materials in response to demands made on their strength and resilience. This dynamic balance is one attribute that makes them very difficult to duplicate with nonliving synthetic materials.

Ligaments and Tendons

Torn ligaments or tendons often look like shredded mop ends and can be difficult to repair surgically. If they cannot stitch the ends together, surgeons often use another tissue within the joint as a graft. To replace the anterior cruciate ligament (ACL) of the knee, for instance, they might use the tough iliotibial band tissue that runs alongside the leg. A strip of this tissue can be run retrograde through the knee joint; a little block of bone left on one end can be tapped into the tibia (shin bone) to lock it in place. Similar use might be made of a central strip of the patellar tendon, which runs from the kneecap to the lower leg.

Results of such operations are highly variable, especially with older or more severe injuries. The graft is weaker than and otherwise different from the original tissue; eventually it can stretch out so that the joint becomes unstable. Meanwhile, the graft site tends to weaken as well. The tissue in either location may not grow back. Some joints cannot be repaired with grafts at all.

Scientists have therefore tried to create artificial ligaments and tendons for repairing these joint structures. In the early 1970s the ill-conceived marketing of a polyethylene implant turned out to be a disaster that gave ligament replacement a bad name for years. More recently some researchers have had better luck with implants made of Dacron with silicone attachments, polyester velour and woven tape, Gore-tex, and expanded Teflon. No permanent material used *alone*, however, is really strong enough to support stresses for a lifetime, particularly that of a young patient. This is especially true for knee joints.

Knee ligaments are crucially important because they act like rubber bands around the joint to hold it in place. Without one of these bands, the knee can give out, scraping the joint and causing arthritic pain.

One of the most serious sports injuries is a torn ACL. This ligament connects the femur (upper leg bone) to the tibia. It runs through the center of the knee and prevents abnormal forward motion while providing much of the stability necessary for running, jumping, and pivoting. It is torn more often than the posterior cruciate ligament (which lies just behind it and prevents abnormal backward motion of the knee) or the medial collateral ligaments (those on either side).

Reconstruction of the ACL may not return the knee to its original strength and stability. When other knee ligaments are torn, they just stay in place, but the ACL must be repaired quickly or it disappears. "The ACL goes through space, so to speak—it extends from bone to bone," says Robert C. Meisterling, M.D., an orthopedic surgeon in private practice in Stillwater, Minnesota. "When the ligament tears, it falls down and lies on the tibia, where the synovial fluid in the joint literally dissolves it away."

Repair of an acute ACL tear often requires an arthrotomy—an opening of the joint capsule. If the ligament is past repair, tightening and adjusting secondary supportive tissues can work satisfactorily for most patients. A new and difficult surgical procedure being followed at the University of Wisconsin, Madison combines portions of the patellar tendon, patella (kneecap), and tibia to replace the ACL. Engineers at the University of Minnesota, Minneapolis are meanwhile studying the loads and compensations of athletes with torn ACLs to see if this information can help improve rehabilitation procedures.

Ligaments are slow to heal because of their poor blood supply, and casts cause muscle atrophy and joint stiffness that lengthen rehabilitation time. An athlete with an ACL tear may be sidelined for a year, which can preclude his or her resuming the sport. Victims of chronic ACL insufficiency have unstable, painful knees for the rest of their lives. For these people 3M Orthopedic Products Division has been investigating the Ligament Augmentation Device (LAD) created by a Canadian physician, the late J.C. Kennedy.

THE LIGAMENT AUGMENTATION DEVICE

The LAD is a strip of polypropylene yarn—an inert, porous material that is very strong when braided. This is sewn to a strip of patellar tendon. The composite graft is then stitched to the shin bone, run through the knee joint, and stapled to the thigh bone. The implant permanently shares load and stress with the connective tissue, which becomes vascularized as a new ligament. The LAD implant is easy to handle surgically and seems to provide greater joint stability and require less rehabilitation time than standard repair procedures while helping to prevent degeneration of surrounding joint tissues.

The LAD was originally intended for chronic rather than acute ACL tears. The healing processes for these two types of injuries are altogether different.

Acute tears cause much more scarring than would occur if they were reconstructed. The potential for tightening a knee with an acute tear is much greater than it is with a chronic tear. "Foreign substances in knees cause unique complications," says Dr. Meisterling. "Implantation of a device in a knee with much proliferative scarring might cause inflammation and joint stiffness. If alternatives are available, they should be used. This is why I've been saving the implant as a salvage procedure for torn ligaments that won't improve otherwise."

More recently, however, Dr. Meisterling has begun arthroscopic implantation of the LAD as an augmentation to semitendinous tissue. The connective tissue used is less important than the arthroscopic procedure and the new application of the LAD to selected acute injuries. "By performing arthroscopies, we cause much less soft tissue destruction, much less insult to the joint, and therefore considerably less scarring," says Dr. Meisterling.

A person with chronic ACL insufficiency is likely to collapse when turning corners or going down stairs because the weak knee gives out. Participating in an experimental protocol, Dr. Meisterling has used the LAD for the past year on a dozen or so of these patients. "We've had no failures yet," he says, "but we need a few years to really tell. I understand that Dr. Kennedy had 90% to 95% good results in Canada over an 8- or 9-year period." Patients are permitted to go about their daily activities but are discouraged from playing highly rigorous sports. "Some patients will play sports no matter what we tell them," says Dr. Meisterling. "Medical advances occur when people insist against our advice on trying activities that seem beyond their capabilities. If we follow them closely and they succeed, then we learn that we can let other patients do more."

One person who would like to play more sports but *is* following the rules is 29-year-old Sharie McTeague, an engineering administrator from St. Paul, Minnesota. Sharie has a congenital problem with her left knee that began bothering her 10 years ago. On the advice of an orthopedist she gave up raquetball, tennis, and softball (her favorite sport) for a few years. When she resumed playing softball, her knee gave out one day while she was running across the field.

By the time she had gone through a few casts, two operations, and three orthopedists, Sharie had developed a serious condition. "I'd fall down stairs because my knee gave out," she says. "I'd be grocery shopping and fall into the produce—but at least I never landed in the freezer case." She began going back to therapy daily, but her knee gave out that often unless she wore her prescription metal brace. Finally she was referred to Dr. Meisterling and had the LAD implanted in April 1984.

"Basically I'm still recovering," says Sharie. "All I'm allowed to do is go bowling, and Dr. Meisterling has been very noncommittal about whether I can try other sports." Sharie's knee is still misaligned, although this condition has nothing to do with her implant. "The knee doesn't go out on me any more, but it crunches constantly," she says. "We think it may be related to the surgery and hope that it's just temporary." She still has some pain, and in light of her previous experience she cannot really speculate on the future.

"But there's absolutely no doubt that with the LAD my knee is better than it was," Sharie says. "I've already advised another of Dr. Meisterling's patients to have the surgery, and I'd definitely recommend it to others, particularly if they had chronic problems like mine. I wouldn't want anyone else to go through what I did." For the first time since 1980, Sharie can walk without having to worry about ending up on the floor someplace. "If I had any inkling 4 years ago what I would be going through," she says, "and if this device had been available then, I would have sold my soul for it."

If modified, the LAD could be used for other knee ligaments but probably would not be necessary for those in other weight-bearing joints. "Most ligament problems in other joints are not as symptomatic as they are in the knee, and there are other treatments available for them," says Dr. Meisterling. "Ligaments in non-weight-bearing joints like shoulders can be easily repaired, but the LAD may be used for elbow ligaments in the future."

THE CARBON FIBER IMPLANT

An alternative method of repairing ligaments and tendons has been developed by J. Russell Parsons, Ph.D., Associate Professor of Surgery; Harold Alexander, Ph.D., Professor of Surgery and Director of the George L. Schultz Laboratories for Orthopaedic Research; and Alexander B. Weiss, M.D., Professor of Orthopaedic Surgery at the University of Medicine and Dentistry in Newark, New Jersey.

In 1977, when they began their work, the researchers could not think of a material they trusted for a permanent tissue replacement. Instead they decided to make an implant that could serve as scaffolding for soft tissue ingrowth. Their material of choice, which had been previously tested in Wales and South Africa, was carbon fiber.

Carbon is extremely strong and biocompatible, but raw fiber is like stretched out insulating fiberglass—brittle, tough, and difficult to handle. Borrowing some ideas from the aerospace industry and some polymers from friends at Johnson and Johnson, the researchers began experimenting with various resorbable coatings for the fibers. They found that coating carbon fibers with polylactic acid (PLA)—or even better, with poly-ϵ-caprolactone (a copolymer), which has the same biological response—makes them more flexible while retaining their strength. The investigators form the coated fibers into a pliable ribbonlike structure that looks like a long black shoelace with surgical needles at both ends. A surgeon can use this ribbon in a lock-weave stitch to suture torn tissues.

The coating offers several advantages. It improves the implant by making the carbon much easier for the surgeon to handle. It keeps the carbon from breaking during the operation, scattering fibers throughout the tissues, and also prevents the fibers from sloughing off after the ribbon is implanted. Although carbon fibers are among the most biocompatible materials available today and in this case are not of a size that might promote carcinogenicity, their tendency to work their way into the lymph system worries researchers. The Newark investigators tested the effects of carbon particles in the synovial joint fluids of animals and found that if the implant broke inside a joint, the carbon would have little effect either mechanically or on microscopic structures. "Anyone who lives in a major city has lymph nodes loaded with carbon particles," says Parsons. "Although carbon fiber debris is somewhat different, and we don't consider its spread to be hazardous, we've minimized it anyway."

Both the PLA and the copolymer coatings are bioresorbable. In about 2 weeks they disappear, leaving the carbon fibers exposed. The implant is then locked in loose space-filling scar tissue. Fibroblasts (connective tissue cells) from the natural ligament or tendon invade the fiber scaffold and lay down new collagen. The implant provides initial mechanical strength; as the tissue rapidly grows in and is subjected to stress, it solidifies, remodels, and thickens. The researchers conjecture that eventually the carbon itself will break down mechanically into smaller segments as the load transfers to host tissue.

"We were lucky that the growth and resorption rates matched," says Parsons. "If the polymer didn't expose the fibers fast enough, the soft tissue would encapsulate the implant rather than growing into it in its natural form. I was struck by the amount, strength, and rapidity of tissue ingrowth in animals, even in highly stressed joints."

Experiments in tissue cultures with tendon and ligament cells showed that different fiber diameters and surfaces would give different results in cell form, structure, and layout on the fibers. The tissue response and the nice alignment it forms occur primarily because of the carbon's size, which resembles that of a collagen fiber. "Each fiber is too big to interest a macrophage and too small for tissue to try to wall it off with a soft capsule," says Parsons. "It's in the never-never land in which the fibroblast happily attaches and makes collagen. The result thoroughly resembles normal tendon and ligament tissue in many instances."

The implant can be used either extra-articularly or intra-articularly (outside or inside the joint). Outside the joint are tremendous numbers of fibroblast-like cells that attach to the fiber and make collagen. Inside the joint the environment is more hostile and fibroblasts are fewer, so the implant is used as a reinforcing agent for a soft tissue graft. The soft tissue provides the cells for ingrowth while the implant holds it together and later prevents it from stretching out. The tissue heals well and appears to grow thicker so that eventually it can withstand loads. "About half our knee patients are either completely intra-articular or a combination of intra- and extra-articular—that combination makes up about three fourths of our patient total," says Parsons. "Most of our knee patients have carbon inside their joints without experiencing difficulty. The type of surgery used depends on what the patient needs." The implant works for either chronic or acute repairs.

In early 1981 the Newark group began FDA trials of the implant. Since then about 1200 patients have received it in about 25 centers nationwide. The implant material is now being sold outside the United States. "We hope for FDA approval in about 6 months, so the implant might be available here in about a year," says Parsons.

The patients have had operations on their extensor mechanisms, knee ligaments, Achilles tendons, shoulder rotator cuffs (tough cuffs of tissue that surround and hold together the shoulder joints), and other soft tissues. Before the operation they are examined very thoroughly; afterward they are tested in the joint function analysis laboratory. This facility has both commercial and specially designed machines to measure the joints' stability, muscular strength, and range of motion compared to those of the natural joints. The computerized machines are more sophisticated than those at many other centers. "In quantitative terms ours is one of the most carefully controlled and monitored studies of ligament repair and replacement that has been done thus far," says Parsons. "We're very proud of that."

The researchers have data beyond 4 years on some patients; over 100 patients have passed the 2-year mark. The early trials have gone very well, especially in patients with repaired knee ligaments, the most common operation. "These results are very encouraging to us because the knee is an extremely complicated joint and because the first patients we implanted were really badly off," says Parsons. Most joint patients come in functioning at about 55% of their capacity. A year after the procedure they are up to 80% and tend to level off after that. They seldom regain 100% of functioning on knees but usually do on Achilles tendons—even the bad chronic cases with large defects.

One Achilles tendon patient who is happy to have been restored to normal is 60-year-old John Pendlebury, a salesman and government contract handler from Kearny, New Jersey. Four years ago John rose from his desk one day and turned to leave his office. He felt a snap in his left heel, and the front of his foot rose from the floor. "I found that I could no longer bend my ankle to point my toe," he says. "I had to walk as if I had a club foot, pushing my foot forward, straightening it, and stepping on it carefully."

When the condition persisted for a few weeks, John went to the VA hospital in East Orange, New Jersey, one of several with which Dr. Weiss is affiliated. There he learned that he had completely ruptured his left Achilles tendon. John speculates that the tear resulted from his having spent several hours the previous weekend leaning forward at an angle while tarring the roof of his summer bungalow. He suspects that his excess weight and the phlebitis in his left leg may have been contributing factors.

Dr. Weiss and several colleagues examined John's foot and recommended the carbon fiber implant. "I was surprised to hear about the implant, and a little doubtful," John says, "but Dr. Weiss thought it would be more successful than standard surgery, and I have great confidence in him." After his operation John spent 8 weeks in a cast, using crutches and then relying on a cane until his foot was completely healed.

"I'm fully amazed at how well my foot is doing compared to its condition before surgery," John says. "It's functioning normally, and I'm confident that it will stay that way, although I avoid taking chances of hurting it again by favoring it a little when I go up stairs. Since I need to lose weight, I'm tempted to start jogging and resuming my other sports, but I don't know if the doctors will permit anything more strenuous than walking."

Like Sharie, John recommends his implant to others. "I think that many people my age have similar problems but figure that they're getting old and nothing can be done for them," he says. "If someone inspired their confidence to have this surgery, there might be more operations and fewer people limping."

Other patients show continuous improvement measured by better stability, increased function, less pain, and improvement in other measured parameters. Unlike a permanent prosthesis, which might weaken over time, the implanted scaffold encourages gradual strengthening rather than just a buildup of scar tissue. In a recent article the Newark researchers write, "We believe the real advantage in the use of carbon fiber absorbable polymer material may lie in our ability to pursue an earlier, more vigorous rehabilitation protocol."

Because it can be fashioned into many shapes, the coated carbon fiber can be made into mesh sheets for hernia repair and other soft tissue defects. In operations that required drill holes in bones, the Newark group found that bone also would grow into the implant, although it took much longer than soft tissue.

"We consider this a very simple, versatile surgical article that is extremely useful for repairing soft tissue," says Parsons. "We don't view it strictly as a ligament implant. It's really a very strong, biocompatible, biologically attaching suture with very interesting properties." Although it is difficult to estimate, Parsons thinks that if accepted by orthopedic surgeons, the device could help perhaps 100,000 Americans a year.

Cartilage

We will see later that artificial joints have liberated many people from pain, disability, and deformity. These symptoms are caused by severe damage and loss of cartilage. Artificial cartilage does not yet exist.

Cartilage is tough, translucent, glossy, elastic tissue that cushions the bones from stresses and shocks. It does so in part by absorbing water from surrounding joint fluid so that a viscous lubricant remains. Its pliancy enables the cartilage to support a film of this lubricant even under stress. As pressure points on the cartilage change with joint movements, it continually absorbs and releases synovial fluid so that the joint parts are constantly lubricated. Artificial cartilage should have all these characteristics plus biocompatibility.

In modified procedures in which half a joint is replaced, the metal prosthesis eventually wears down the cartilage lining the part of the natural joint that remains. The prosthesis also prevents the cartilage from moistening itself with the joint fluid that nourishes it. Artificial cartilage would solve these problems completely because surgeons could replace only diseased cartilage rather than parts of entire joints.

The Newark researchers tried to develop such a material but ran into difficulties. "We can create material with the same compliance or 'give' of natural cartilage; it self-lubricates marvelously because it exudes water with pressure," says Parsons. "But it self-destructs. It develops extreme internal stress from the water absorption and has very poor mechanical properties. I think that in man-made hydrogels—water-absorbing polymers—we'll always have some mechanical trade-off between stability and correct properties."

Muscles

Toyoichi Tanaka, Ph.D., Professor of Physics, and his colleagues at MIT have developed substances that might someday act as artificial muscles—not immediately in humans, but rather in robots and artificial limbs. The team is working with microscopic gels made of loosely associated polymers (like polyacrylamide and polystyrene strands) arranged in a three-dimensional network. The researchers can add ionizable groups like acrylic acid or its salt to the polymer networks. If they then apply electricity or change the temperature, solvent composition, or acidity of the fluid medium, the gels expand or contract in a manner similar to that of natural muscles.

Later we will see that natural muscles are like electromechanical transducers: they translate electrical impulses from the nerves into mechanical energy. The muscles expand or contract in response to changes in their calcium ion concentrations (which act along with several specific proteins). Muscle action is essentially the shrinking and swelling of the array of natural polymers that control the muscles.

When electricity is applied to an ionized gel, the positive and negative ions are attracted to the oppositely charged electrode. If the negative ions are pulled toward the positive electrode with enough force, the stress induced by gradients of pH or other parameters causes the gel structure to collapse. When the electrical field is turned off, the gel resumes its original size.

Temperature changes also cause the gel to shrink or swell, depending on the particular substance. The gel maintains a balance between the interactions among the polymers (which collapse the network) and the pressure of the ions from the ionizable groups (which expand it). Temperature changes gradually and continuously alter the balance between these two forces. "The important points are first, when we add ionizable groups to the polymer network, the volume transitions become discontinuous and sudden, and second, we can control the size of volume change by varying the amount of ionizable groups," says Tanaka. "At certain temperatures the gel can contract or expand many hundreds of times." The speed of the change depends on the size of the gel itself. "By testing thermal fluctuations, we have obtained evidence that gels the size of muscle fibers would contract or expand in milliseconds—the same speed as natural fibers," says Tanaka.

Robotics specialists are particularly intrigued by these gels as possible substitutes for actuators, which are limited when made very small because they require tremendous amounts of current. In this discussion we are more interested in them as possible muscles for artificial limbs.* "To use a gel in this way, we would need to learn how to keep the gel from breaking under strong forces, how to connect it to 'bones' in the limb, and how to maintain a fluid environment similar to that in which natural muscles operate," says Tanaka. "Human implantation would be difficult because of biocompatibility and many other problems that it would raise and because unlike the natural muscle, which replenishes itself, the gel would have to be replaced periodically."

This prospect would be disappointing were it not for the fact that a different type of artificial muscle was tested successfully on animals only to have manufacturers decide that it was not an economically feasible product. A group of researchers at Battelle Columbus Laboratories in Ohio produced a muscle consisting of a crimped, elastic Dacron inner tube within a silicone rubber outer tube. The inner tube was longer so that its ends could attach the device to the tendons. Muscles come in pairs that pull against each other; the implant was intended to replace one of two antagonistic muscles.

It is possible that the enormous scientific and governmental interest in robotics will increase support for these and other approaches to artificial muscles and connective tissues.

Bones

In addition to protecting vital organs, the skeleton structures the body and enables its muscles to create various movements. Like ligaments and tendons, bone is dynamic tissue, the composition of which continually fluctuates in response to stress. Like the liver, bone is one of the few body tissues that actually regenerate rather than simply patching themselves up with scar tissue.

Bone is composed largely of HA (a calcium mineral) particles bonded with elastic collagen fiber. A long bone is a tube with a hard outer layer of compact tissue, a spongy inner layer of cancellous tissue, and a cavity filled with marrow containing erythroblasts, which make red corpuscles. Blood vessels, nerves, and some lymphatic vessels run through the bones. This structure makes bone very tough, strong, porous, lightweight, and resilient—and hard to duplicate with synthetic materials.

---

*Contractile polymers could be used more obviously to replace the sphincter muscle, which contracts and relaxes to control urination.

Scientists have tried to replace bone with all kinds of substances: transplanted bone, collagen, metals, metal powders and alloys, ceramics, glasses, carbon-based materials, polymers, and composites. Like those forming tendon or ligament implants, the materials in bone implants may be strong but are otherwise seriously mismatched with natural tissue. The implants are inflexibly designed, may not be biocompatible over the long term, and can loosen.

Bone plates, intermedullary rods, and other orthopedic implants are usually made of metal or polymers. They work very well in general but have several problems. The devices are impermeable to x-rays and respond to temperature changes at rates different from that of natural bone. They corrode and continually release metallic ions like cobalt, chromium, molybdenum, and nickel, which may be hazardous even though they are not as yet known to cause tumors or systemic problems. All these devices are much stronger than bone. They work well as the bone heals, but in time they take on most of its stress. An extremely active biological tissue, bone becomes demineralized under these conditions. When the device is removed, the weak bone in that area commonly refractures. "Most man-made orthopedic implants are on the near edge of disaster, with a very small safety margin, because of their high stresses and size limitations," says Parsons. "If the implants can be made very strong, the patient's bones may become osteoporotic and break."

Despite these difficulties, researchers have had considerable success in fashioning bone replacements and repair devices. Bone plates, for example, can be inserted temporarily, as in leg lengthening or bone grafting, or they can be used to patch holes in the skull left by injury or surgical procedure.

Bone implants are available to replace diseased parts of the upper leg and shin bones and the knee, hip, and shoulder joints along with much of the surrounding bone. Fractured bones can be temporarily held in place by a huge assortment of rods, pins, nails, and screws used mostly in the hip and thigh bones. Pins or screws can hold together the ball joint at a fracture site in the hip. Some are designed to be inserted through the skin and broken off at their proper length. Others are flexible enough to curve through fractured long bones. Some rods and nails are long enough to hold together an entire upper or lower leg bone that has fractured in the middle. Certain types of rods are stiff at one end and ductile at the other for greater stability. Even tiny plates and screws to fix finger bones are available.

Someone who needs a leg lengthened can be fitted with a device consisting of two telescoping tubes with adjustable length. This provides support until the new bone grows or is grafted in.

Implants are also used to repair the spine. Researchers have used metal mesh filled with bone cement to bridge missing sections of the spine. They have also implanted coiled springs flanking the spinal column to stabilize fractures. Extreme cases of spinal curvature can be corrected by cable stitching the spine straight.

A few years ago, Charles C. Edwards, M.D., an orthopedist at the University of Maryland Hospital, dramatically reconstructed 6 inches of the spine of a 33-year-old woman after removing a huge tumor. The specially made device he used consisted of a metal alloy cylinder implanted in front of the spinal cord, with two slender metal rods protecting it from the back. The device was held in place by retractable rods fastened with horizontal screws and cement. Strips of metal mesh along the front of the spine and bone grafted from the woman's pelvis provided additional support. Unfortunately the patient died of a kidney infection before a second planned bone graft could be completed.

Researchers are still searching for materials that might make better bone substitutes. The Newark group has contemplated using their coated carbon fiber material for fracture fixation as well as tissue ingrowth. Using relatively new technology, they take several layers of the carbon ribbon and heat it in a mold to fuse it. The resulting material, which is rigid as well as strong, can be used to fix bones. A bone plate made from carbon fiber gradually loses its rigidity as the polymer disappears. This eliminates stress protection atrophy because the bone slowly takes on more stress and remodels much more normally. The bone also makes callus around the fracture, as it would normally do to strengthen that area. Another advantage of the carbon fiber bone plate is that it need not be removed. The group is now testing this application in animal trials.

Other devices are being developed for permanent reconstruction on other parts of the skeleton. We will note in a later chapter that Dr. Leake's group has used Osteomesh, a cloth that acts as a mold for a bone graft; when the bone has healed, the mold may be removed.

About 6 million people suffer broken or fractured bones each year. Roughly 100,000 of these injuries do not heal thoroughly and are subject to chronic repeated fractures. Although not yet clinically proven, electrical stimulation for bone growth seems to help heal even the most stubborn fractures within 6 months. Several companies now make totally implantable bone growth stimulators that can be easily removed once they have served their purpose.

On a much smaller scale, chronic ear infections can cause the small bones of the middle ear—the hammer, anvil, and stirrup—to disintegrate. For many years scientists have used devices made of inert materials like polymers, metals, and porous substances to replace these tiny bones. Although these prostheses are generally successful in some 20,000 such procedures a year, they are surrounded by compliant scar tissue that can prevent sound vibrations from crossing the bone chain. Because it does not bond to soft tissue, a device made of inert material constantly rubs against the eardrum. Extending from beneath the eardrum on one side to what is left of the stirrup on the other, the implant can wear its way through.

The FDA has recently authorized the sale of a new middle ear device made of Bioglass. We have seen that in the absence of bone, Bioglass causes a collagen bond to form. In the ear this bond is not stressed; no buildup of scar tissue occurs. The implant adheres to the eardrum, preventing extrusion. The device looks like a thumbtack, the peg of which must be trimmed and molded to fit over the remaining part of the stirrup, which connects with the nervous system. "Surgeons are absolutely delighted to find that they can see through the implant, because that makes its modeling and fitting much easier," says June Wilson Hench.

Jawbones and Teeth

On September 9, 1984, the *New York Times* reported on the case of 19-year-old Aaron Isom of Cuyahoga Falls, Ohio, who was charged with aggravated robbery of a false tooth. Brandishing a gun, Isom had allegedly threatened to shoot 19-year-old Todd Montgomery unless the latter removed the tooth with a pair of pliers and handed it over. The police speculated that Isom was punishing Montgomery for an outstanding debt.

Montgomery may owe a bigger debt to his dentist unless he had a new tooth implanted as soon as possible.

We have seen that bone has the unusual ability to regenerate and restore itself to normal function. When an orthodontist places carefully calculated stresses and strains on a patient's jawbone, it remodels vigorously, allowing the teeth to move with pleasing aesthetic results. The jawbone responds to pressure, however, with permanent resorption.

The presence of teeth conveys proper stresses on the jawbone so that new bone is formed in response. A missing tooth causes functional forces to be transmitted to the jawbone as pressure, which can break down the alveolar ridge (the part of the jaw from which the teeth arise). Both the alveolar and the basal ridges can resorb from the loss of a single tooth—although Montgomery may be relieved to learn that this happens much more often in women and so may be related to hormonal changes. Most resorption occurs down the lingual (inner) side of the bone; the ridge loses height, narrows, and can develop a knife edge along its buccal (outer) side. Missing teeth can cause a person to lose 40% to 60% of the alveolar bone in 2 to 3 years. The process accelerates aging and the appearance of wrinkles, may cause overclosure and complicate dental restoration, and has negative psychological effects. If this atrophy of the mandible (lower jaw) is not corrected, the person will not be able to wear dentures, thus increasing the rate of resorption.

To avoid these complications of tooth loss, dentists perform ridge maintenance. Failure to do so requires ridge reconstruction using either a bone graft or some type of artificial bone. Grafting bone onto the jaw poses the same kinds of problems as any other type of body graft: it is painful surgery with a long recuperation period; it is expensive; enough bone may not be available to harvest; infection can occur; donor bone can be rejected. Grafted bone also resorbs and may create a poor ridge that is not uniformly level, so the patient cannot wear dentures anyway. For these reasons researchers are experimenting with various types of artificial bone. Used initially in dentistry, these materials may someday be applied to other parts of the skeleton.

CALCIUM PHOSPHATE CERAMICS

Hydroxyapatite. HA is a common mixture of calcium and phosphate that makes up about 65% of bone and 98% of tooth enamel. In its natural form it is a source of phosphate ions used in fertilizers, detergents, and other products. We consume its refined powder form in baked goods and table salt (in which it acts against caking) and as a calcium food supplement.

Synthetic HA is a polycrystalline apatite ceramic that is dense, nonporous, nonresorbable, and inert. As a permanent artificial bone it can be used as scaffolding to guide new bone formation. HA strongly bonds to and becomes integrated with living bone by natural mechanisms.

HA is available in granular and solid form for maintaining or reconstructing ridges and repairing periodontal defects. It is by far the most frequently used material for alveolar ridge work. In this application HA requires no added operation to harvest bone; reduces patient risks, time, and cost; and is readily available and easy to use. In its granular form HA can be poured into periodontal lesions and tooth extraction sites. One problem is that granular HA can wander off the ridge down further into the tissues.

Solid HA is used to make tooth roots for extraction sites that retard resorption and preserve the ridge. It poses little risk of complications and can support standard dentures immediately and comfortably. HA tooth roots will not support crowns, however.

Solid HA may someday be used to anchor an artificial arm or leg as well as for maxillary (upper jaw) and mandibular defects, cysts, and clefts. Bulk macroporous HA has been used in Europe and elsewhere to fill bone and soft tissue defects, especially in the maxillofacial area, with good results. Bony ingrowth is rapid, and the implants stay in place.

Tricalcium phosphate. TCP is another polycrystalline apatite ceramic. Like HA, it is completely nontoxic and naturally bonds to bone. Unlike HA, it is porous and bioresorbable, so it cannot be made into a permanent implant. This characteristic does not preclude its being used as a scaffolding, although attempts to do so have not been terribly successful.

TCP could possibly be used instead of an acrylic as grout for total hip joints. HA combined with a little TCP is being used to replace ear bones. "If the TCP in this mixture is replaced by bone, it will have some very exciting applications," says Dr. Leake.

PLASTIC

Hard tissue replacement (HTR) is a relatively new material being used in dentistry. Its granular form consists of polyhema-coated acrylic (PMMA) beads;

it is also available molded. It is easy to handle, strong, and bactericidal. Like HA, HTR is porous, nonresorbable, nontoxic, and inexpensive. HTR both promotes new bone formation and becomes integrated with blood clots to promote stability. Also like HA, granular HTR can be poured into extraction sites and periodontal or bony defects.

HTR is plastic rather than ceramic, however, and so less like natural bone. When heated, it can be molded. This form can be custom-tailored to match a tooth root and popped into an extraction site. Probably the most important use of HTR is for ridge augmentation. The dentist can mold it to replace the ridge rather than injecting granules to increase the ridge height. "Like other materials, HTR works well on the atrophic mandible, where we're concerned less with strength than with the height and the denture support base," says Dr. Leake. "Solid HTR already forms the desired ridge, and we can place the denture base over it immediately."

HTR has already been used in orthopedics, neurosurgery, plastic and reconstructive surgery, and ophthalmology. Although he has made a chin implant from it, Dr. Leake does not recommend the material for applications requiring strength. "I squeezed a solid mandible of HTR slightly, and it fractured," he says. "It hasn't the right characteristics for this application. If it can be made to induce bone growth, however, it would be revolutionary."

HTR shows promise but should receive more test follow-ups, better investigational work, and improved marketing and packaging of the granular form.

CERAMIC COMPOSITES

Some new materials are composites of the types already mentioned. ALCAP, a combination of aluminum, calcium, and phosphorous oxide, is a strong, resorbable ceramic that causes a tissue response similar to that of TCP. It has not been used as extensively as HA or TCP, but in limited clinical trials it has been successful in spinal fusion procedures and repair of cleft palate deformities.

An Oral Reintegration System developed by Neodontics, Inc., consists of two components. The core is a titanium implant coated with TCP, which bonds directly to the bone without forming scar tissue that could surround and loosen the implant. Once the device is implanted in the jaw and the bone has healed, the dentist attaches a threaded cone to support a fixed or removable bridge. The cone acts like tooth supportive tissues so that the jawbone keeps replenishing itself and holds the implant tightly.

Later we will discuss other techniques used to repair defects of facial bones and jawbones; meanwhile we turn our attention to the bones that help us move.

5

Balls, Sockets, Hinges, and Roller Coasters

Natural joints range in complexity from the simple hinges inside the fingers to the roller coaster rotating inside the knee. The major enemy of joints is arthritis: both osteoarthritis and the rheumatoid varieties. "We don't know why some joints become arthritic more often than others do," says Kenneth E. Marks, M.D., Head of the Section of Musculoskeletal Oncology in the Department of Orthopaedic Surgery at the Cleveland Clinic Foundation. "Certainly ligament strains and other injuries predispose the joints to arthritis. The disease relates somehow to joint geometry—the amount of surface area and loading characteristics. But we don't know why hips often become arthritic and ankles rarely do, since the ankle bears the same amount of weight as the hip."

Other conditions that damage the joints include fractures, dislocations, and deteriorations due to poor blood supply. When standard treatments fail, artificial joints can relieve pain and restore most functions. Artificial joints are notable for outdoing their natural alternatives (autogenous or cadaver bone grafts) because they can be tailor-made and heal much faster: a patient can be up and about in 3 to 4 months. Bone resections followed by arthroplasties (joint replacements) are even substituting for amputation as the treatment of choice for some bone and joint cancers.

The oldest and most widely used type of artificial joint is for the hip. Approximately 100,000 hip, 50,000 knee, and 3300 ankle arthroplasties are performed each year, most of them on the elderly. Prostheses are used less frequently to replace toes, fingers, wrists, elbows, and shoulders. As of 1982 the FDA had issued generic classifications for 49 types of joints, about evenly divided between Class II and Class III categories (about which we will learn more later).

Lower Extremity Joints

HIP

The hip is essentially a ball and socket joint, the ball being the head of the femur, the socket the acetabulum (a section of the pelvic bone). This structure gives it a wide range of motion in every direction.

Artificial hips must be extremely strong and durable, especially for younger patients, because they bear five times the recipient's weight with each step. The femoral stem is usually made of chromium-cobalt alloy, titanium alloy, or stainless steel, while the acetabular cup is made of high-density polyethylene. More recently the heads and cups have been made of aluminum oxide and other ceramics. Plastics and ceramics are used for the stem and cup to help minimize friction because an artificial joint lacks the synovial fluid that lubricates the natural joint's every movement. Like other joints, hips are designed to remain in place despite activities and stresses.

Surgeons had experimented with artificial hips in the nineteenth century but had little success until the 1960s, when British orthopedic surgeon Sir John Charnley invented the polyethylene socket and pioneered the use of PMMA cement to hold such implants in place. Now hip arthroplasties are routine operations that have a 95% success rate for periods up to 10 years, the remaining lifespans for most of their recipients. (After 10 years 25% of hip joints fail.) Besides resulting from arthritis and other conditions, hip arthroplasties (like those on the shoulder) sometimes follow resection of tumors in long bones.

The femoral component of a hip joint is cemented into the shaft of the thigh bone after the natural diseased femur has been removed. The acetabular component is cemented into the space in the pelvis by reaming the acetabulum. A complete prosthesis of this type is called a stem prosthesis.

Some stems are tapered so that the bone itself holds them in place (as will be discussed later, joints are also treated to encourage bony ingrowth). The stem and cup can be linked together or implanted as separate components. In another type of hip the stem has a pin that fits into one of several interchangeable balls to provide more rotation.

In hip arthroplasties the surgeon exposes the joint socket, removes all diseased tissues and cleans out the cavities. The surgeon then mixes powder and liquid to make the acrylic cement, which is squirted into the pelvic bone and pressed into the porous tissue. An appropriately sized socket is pressed into the cement until fixed. Similarly, the surgeon then removes the ball joint, cleans out the hollow center of the thigh bone, and injects cement. The stem is pressed into the cavity until it sticks.

In modified hip replacement the joint is simply resurfaced. The ball joint is scraped and capped with a metal head, and the socket is cleaned and lined with plastic. The outcome of this procedure is less predictable than that of total hip replacement because of possible loosening or fracture of the femoral neck. Replacing just the femur often wears down the cartilage and eventually causes pain.

Patients usually spend 2 weeks in the hospital and use crutches or canes for 3 months. For several years most can count on being free of pain and able to perform daily activities as well as enjoy certain sports like swimming, bicycling, or golf. They cannot run, play tennis, or otherwise pound the joints.

Jim Beaver, a 57-year-old investment officer from San Francisco, was only 49 years old when he had his first hip replacement. He suffered from arthritis that he believes he inherited from his mother, who had both hip joints opened and cleaned out back in 1955. "My arthritis finally reached the point at which I could hardly walk," Jim says. "The fog and cold winter weather aggravated the condition, and the pain kept me awake at night." In 1976 Jim had his left hip joint replaced; 2 years later he had a reoperation because some of the cement had eroded and was causing him pain. In 1983 Jim's right hip was replaced, and since then his implants, manufactured by Intermedics, Inc., have been fine.

Eight years ago hip arthroplasties were not performed as often as they are today. The only other patients Jim knew of were much older than he was, and he worried about and feared the operation. After he went through the first operation and knew what to expect, however, the latter two were much easier. Now he knows 20 or 30 people who have had hip implants.

Jim now has no pain whatsoever. He pursues most normal activities, including his favorite sports, duck hunting and golf. "I avoid running or activities in which I might risk falling," he says, "I do think that walking is good for an artificial hip recipient, because it strengthens leg muscles and bones that haven't been used for a while." Jim is not concerned about having had the surgery at a relatively youthful age. "If I live long enough, the artificial joints will give out," he says, "but by then something better will have come along."

Along with other patients to whom he has talked, Jim believes that an arthritis victim must reach the point at which the pain becomes intolerable before he or she is really ready for a surgical solution. "I believe that the medical profession waits for patients to come to them saying that they've had it and want the implants," he says. "By that point, they have nothing to lose and everything to gain. The catch is that patients feel so much better after their operations that they wonder why they didn't have them done long before. There's no reason not to have artificial hip joints if you need them."

Persons with recent or remote hip infections, current bladder infections, metabolic diseases that weaken bone, heart disease, respiratory problems, or excess weight are not good candidates for artificial hip joints. For those who have the operation, complications can include dislocation, abnormal bone formation, bone fractures, blood clots, or long-term failure due to fracture or wear. About 60% of hip arthroplasties are revisions of previous operations.

The worst complication of any joint arthroplasty is infection, especially among rheumatoid arthritis patients who take steroids. Several theories about what causes infection have been proposed. "My animal tests have shown that infection is probably related to the amount of bone that is killed by the joint surgery, because dead bone has little defense against bacteria," says Dr. Marks. "This finding has been borne out by the hinge type of knee prosthesis, which kills more bone and causes infection more frequently than knee resurfacing. To some extent, the tissue pressure around the implant may also contribute." Other theories are that bacteria can coat the metal joint surfaces or that the immune reaction produced by the artificial joint predisposes the body to infections.

Like all musculoskeletal infections, which are difficult to treat, an infected joint is a horrible problem. Often surgeons must remove the implants and leave the patients with no joints; sometimes they must fuse knee joints. Treatment requires extensive hospitalization.

To avoid infection, surgeons perform joint arthroplasties in highly filtered clean rooms; they wear impermeable space suits to prevent skin bacteria from entering the environment. The patients receive antibiotics during their operations; these drugs are added to the cement in high-risk cases. Despite these precautions, infections occur in about 1% of patients.

KNEE

The knee joint is the largest in the body and the most often injured. It is both powerful and vulnerable.

Like hip surgery, knee arthroplasty relieves pain, disability, and deformity. Although 90% of knee joint recipients experience pain relief, hip joints are more likely to restore a wide range of movement. The knee is not a simple hinge joint, as it appears: with each step the joint meanders from side to side, rolls, glides, and circles around a moving axis. The latest knee joints try to simulate this movement so that the muscles will not twist them loose.

Sir John Charnley revolutionized hip joint replacement by minimizing artificial joint friction; in the late 1960s an associate of his created a total joint whose range of movement made a breakthrough in knee arthroplasty. This was a polycentric prosthesis consisting of two arched runners, cemented at the base of the thigh bone, that moved in two polyethylene tracks perched atop the shin bone. Its successor, the geometric prosthesis, had the same design with only one runner and track instead of two.

Hundreds of different knee joints have appeared since then, some based on the polycentric design and some not. They can be roughly classified into two categories depending on whether just the load-bearing surfaces of the joint are removed or whether the surrounding ligaments are removed as well. An example of the latter type, which restrains movement the most, is the hinge joint, which rotates on a pin. The hinge is used only when removal of tumors and all the ligaments have left the surgeon no alternative; the prosthesis can break and loosen, causing extensive bone destruction. A less radical method is to resurface the joint using a peg or other devices to replace the ligaments.

In a partial resection only half the knee is resurfaced, usually not very successfully. The most successful type of surfacing is nonarticulated surface replacement, in which a plastic cap is placed on the shin bone and a metal cap on the thigh bone after the diseased tissue has been trimmed. The most popular knee joint for this procedure is the total condylar knee.

"Different surgeons and researchers will fight for their favorite total knee joints with almost a religious fervor, undiluted by science," says Dr. Marks.

"Most of them, however, are tending to use the total condylar knee. It's very difficult to choose a particular model of this type, however, because it must be used on thousands of knees to prove any difference in performance." Most total condylar knees are easy to insert, cause few complications, and restore sufficient range of motion for everyday use. They are suitable for the elderly but often fail in young people, who tend to forget that they have artificial knees once the joints stop hurting. "I had one young patient with two artificial knees who sprained and loosened one doing a wheelie on his motorcycle," says Dr. Marks. "Since then I've done revisions on both his total knee replacements."

A patient who takes better care of her artificial knees is Marjean Wallover, a 60-year-old homemaker from Beaver, Pennsylvania. Marjean was suffering terribly from pain in both knees resulting from 16 years of arthritis, but rheumatologists in Cleveland and Pittsburgh refused to refer her for total knee replacement because they felt that at 56 she was too young. "I argued with them for the quality of my life," Marjean says. "Why should I spend 10 years being crippled, I wanted to know, when I could be free of pain?" Finally physicians referred her to an orthopedist, who told her that she had no cartilage left and performed total arthroplasties on both her knees, 2 years apart.

Marjean found the operations to be a huge help. "I'd hate to go through that again, but I'm glad I did it," she says. "I've had 4 good years of leg function. The pain is relieved, and I've had no problem with the implants. I think it's wonderful." More recently, Marjean has had both feet "repaired" and four joints implanted in her right hand.

The contraindications and complications for knee joints are similar to those for hip joints. Knees cannot be revised as easily as hips can when they fail, however; instead the natural joint may have to be fused stiff once the implant is removed.

ANKLE

The ankle is a relatively simple, stable joint that has less to do than knees and hips but is subject to the same excessive loads. We have noted that ankles seldom develop arthritis, but they can be painful if afflicted with degenerative conditions like the complications of fractures. The treatment of choice for relieving ankle pain is joint fusion; arthroplasty is performed only on arthritic elderly patients or those few with rheumatoid arthritis and low demand on the joint.

"At the Cleveland Clinic, we used to do a large number of ankle joints," says Dr. Marks. "Like other major centers, however, we have now greatly decreased the number of total ankles we do because of poor long-term results with the devices."

TOE

Unlike the ankle, the toes are a common target for arthritis. Prostheses similar to those used in the fingers can be used in the toes without having to be cemented in place. They are usually made of some type of silicone elastomer.

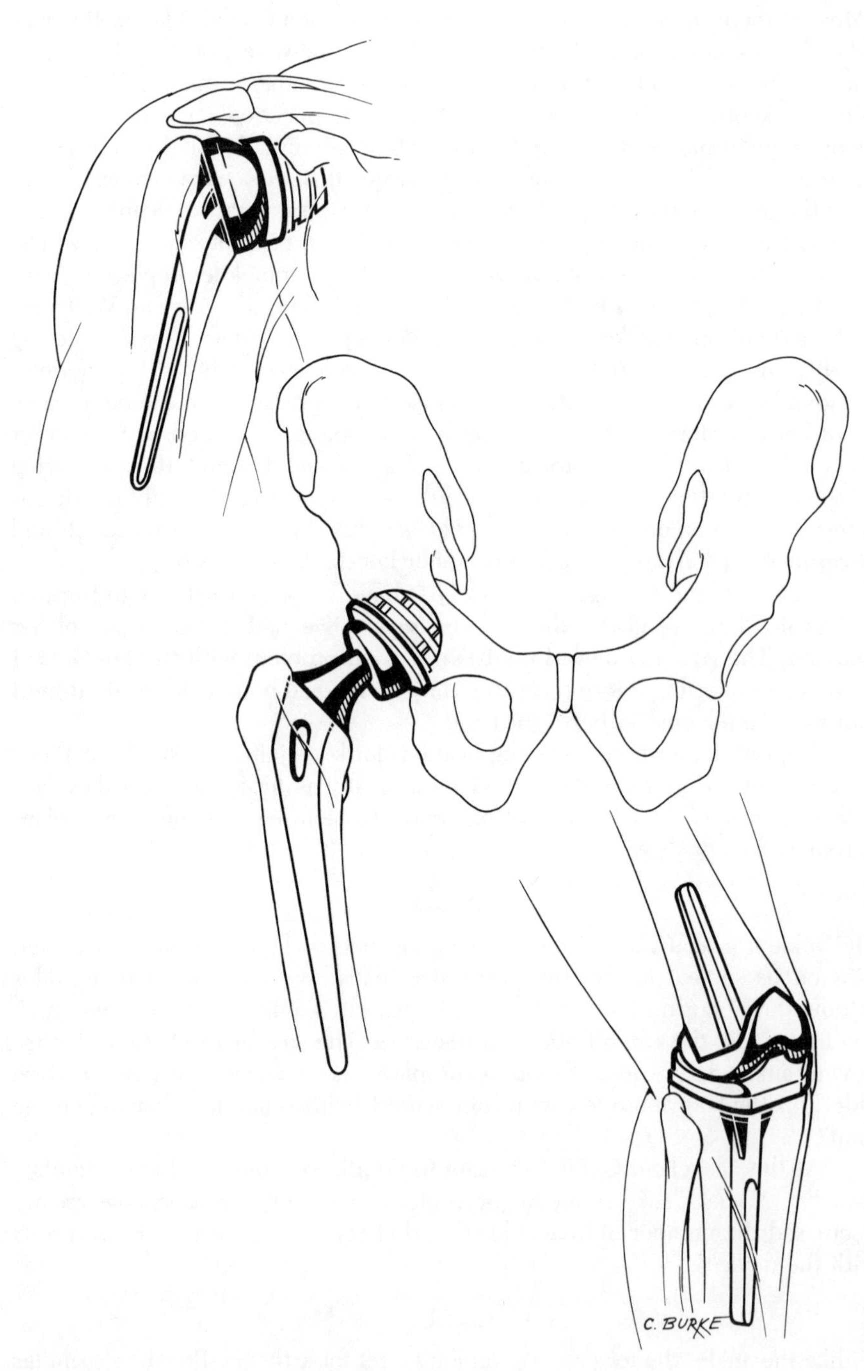

An artificial shoulder, hip, and knee.

Intermedics, Inc., has developed a newer type of toe (and finger) joint made of carbon, and Judy Cicero, a 40-year-old woman from Novato, California, who works in a school superintendent's office, was one of its early recipients. The big toe on Judy's right foot had been painfully afflicted with arthritis for several years and finally reached the point at which she could not walk normally or find shoes to fit. X-ray examination showed a diseased joint, a bone spur, and a tiny fracture. Because Judy was only 37 and needed a long-lasting prosthesis, her local orthopedist sent her to Stanley L. Kampner, M.D., an orthopedic surgeon at the University of California, San Francisco (UCSF), who was using the carbon implants. As discussed earlier, carbon is strong, biocompatible, and porous, thus it encourages bone ingrowth.

"The funny thing about Judy's surgery was that we all thought it would be an office visit type of thing and she'd be back to work in about 3 weeks," says her husband Michael. Instead Judy spent 5 days in the hospital, then was encased in a cast that reached halfway up her leg. "The pain wasn't bad until I got home and had to keep the foot elevated because when I lowered it the joint would throb and swell," she says. "I was out of work for 8 weeks. I'm so used to doing for myself that it was terribly depressing to be on crutches for a month. Michael and the kids helped out because they felt sorry for me."

The artificial joint has a shallow ball and socket serving as the hinge. "Dr. Kampner took out as much bone and tissue as he could, but what he left in still causes me a little pain sometimes," Judy says. Her toe is also shorter than its counterpart on the other foot. Judy says that this is a result of the surgery. Michael claims that *he* shortened the toe by tapping it while Judy was asleep.

"The pain I feel is nothing like it was," says Judy. "The toe is flexible, and Dr. Kampner says I can do any activity I want that isn't painful. I don't run anymore, but I do take aerobics classes. I'm careful not to hurt it."

Like Jim Beaver, Judy recognizes a discrepancy between having the operation as a final resort and wishing it had been done earlier. "I wouldn't want to go through it again, but if my other toe became arthritic, I'd have an implant put in," she says. "I see now that it was silly for me to put it off so long, because it really has made my life easier."

Joints of the smaller toes sometimes need replacement because of arthritis, dislocation, unsuccessful surgical procedures, stiffness, bone disease, or deformity. Implants are available to replace the first and second joints of the smaller toes and the articulating surface at the toe base.

Upper Extremity Joints

The upper joints are replaced far less often than the lower joints. They do not bear weight, and standard treatments for disease or fracture are often effective for them. Like upper extremity amputees, people with disabled shoulders, elbows, wrists, or fingers can often use their good natural joints to take over many functions. Moreover, their complexity makes these joints practically irreplaceable. The elbow joint is subject to torsional stress that makes it almost impossible to duplicate, and it is even more difficult to restore range of motion to the shoulder.

SHOULDER

The shoulder is the most mobile joint in the body: it can move forward, backward, up, down, and completely around. It is a ball and socket joint but much less stable than the hip. The glenoid cavity (socket) is shallow, allowing the large, curved head of the humerus (upper arm bone) to rotate in all directions. We have already noted that shoulders have rotator cuffs: bands of muscle and tendons that act as soft sockets for the joints. Compared to the deep, bony socket of the hip, the shallow, flexible shoulder socket presents a real design challenge.

The standard shoulder prosthesis is that developed by Charles S. Neer, M.D., of Columbia Presbyterian Medical Center in New York City. It consists of a slightly curved ball blended at the stem into four fins, which lock the prosthesis and keep it from rotating. One fin has two holes that allow it to be sutured in place. The stem itself is grooved and slightly tapered for snug fit with acrylic cement. The shallow socket is made of high-density polyethylene fixed into the shoulder blade by a triangular keel that sticks out from the back and is cemented in place. The keel has a hole through which a wire can be run to secure the device. The socket comes in one standard size, the stem in several. The manufacturer, 3M Orthopedic Products, is currently running clinical trials of an artificial rotator cuff.

Because shoulder function depends so much on soft tissue, arthroplasty must be followed by extensive rehabilitation to restore as much muscle function as possible. Such therapy may also help preclude dislocation of the prosthesis or fracture of the natural bones.

ELBOW

The elbow is a weight-bearing joint in the sense that lifting objects strains the joint much more than it does the hand. This joint is even more complex than the knee in terms of its abilities to raise the hand, extend the arm, and angle or rotate quickly or simultaneously. The elbow is actually two joints at the juncture of three arm bones: the humerus and the radius and ulna (forearm). Each joint bears separate stresses and rotates differently. A poorly functioning elbow joint is extremely disabling, but recipients of artificial elbow joints can regain motion without corresponding strength and stability.

The most commonly used elbow is a semiconstrained device called the Ewald elbow. This joint gives stability through the elbow without linking it, so the stresses that travel up and down the arm are not transmitted to the cement.

A sophisticated elbow joint has been developed by Robert G. Volz, M.D., of the College of Medicine at the University of Arizona. This elbow replaces both sections of the joint. The humeroulnar joint has three planes of motion, while the radial head component keeps the forearm properly aligned so that the joint bears much less stress. The tissues surrounding the joint also help cushion stresses. The joint allows the elbow to flex and extend the arm quite naturally with little risk of loosening.

WRIST

The wrist joint consists of eight carpal bones that are arranged roughly in two rows. Like the knee joint, the wrist particularly depends on the ligaments that hold it together. If the ligaments are damaged or the joint attacked by rheumatoid arthritis, the wrist can be fused, but a prosthesis may be a better choice if it successfully preserves some motion.

Wrist implants are made of either silicone elastomer or metal. Single surfaces like the arm bones or adjoining wrist bones can be capped with silicone elastomer. One type of artificial wrist can be fixed without cement; the component is linked across the joint.

Nonlinked implants replace the joints linking the radial bone with the wrist and thumb. One version is a "precentered" prosthesis that has metal stems for the radial bone and that of the middle finger. The radial stem ends in a polyethylene ball, the socket of which is in the finger stem. The ball's axes of motion are offset from those of the stems to make the joint move more like the natural wrist. The prosthesis comes in styles for the right and left hands.

FINGER

Finger joints are a favorite site for arthritis, whose symptoms can be relieved with artificial joints made of either silicone elastomer or a polypropylene-polyester compound. These joints are basically hinges that join two stems; since the finger bones are hollow, the stems can be slid into place without cement. Some researchers believe that any movement of the stems inside the bones simply allows more natural finger movement, but some types of implants have stems covered with Dacron mesh to encourage bone ingrowth. The implants can be used in either first or second finger joints.

Silicone hinges can fracture and inflame the joint, so researchers have been experimenting with other materials, including implants made of polyolefin elastomers combined with rigid metals. New designs are likewise resulting in implants that can flex thousands of times without breaking. One model of a thumb joint is a captured ball and socket that provides full range of motion without this risk.

Research in Artificial Joints

BONE CEMENT

Joints have traditionally been fixed with acrylic cement, which mechanically attaches to the bone but does not adhere directly to the tissue. When the cement is mixed and applied, the chemical reaction releases heat, which destroys some bone tissue; as the cement cools, it shrinks slightly. Some of the monomer remains on the surface of the cement mixture, where it slowly kills more tissue. Bone adjacent to the implant is not stressed, so it starts to atrophy. By causing the implant to loosen, these factors contribute to the main problem with artificial hip replacements.

A loose artificial joint can split the bone; in younger patients it requires revision of the operation. Removing the old cement during revision destroys yet more bone, so that allografts or local bone must be used to replace it. "After a few revisions, the joint area starts to look like a canyon," says Dr. Marks.

Surgeons are trying to apply bone cement more effectively by using pressurization to improve its penetration, centrifugation to increase its strength and reduce voids, cement spacers to even out its thickness along the bones, and metal-backed components to redistribute stress.

Bone cement is still standard, and depending on activity level and projected life span, it is the treatment of choice for patients over 65. "I use cement for patients over 65 because the bone grows more slowly and can weaken itself, and a cemented implant should last the rest of their lives," says Dr. Marks. "For those from 60 to 65, it depends on the patient. For patients under 60 who have good quality bone, I use porous implants."

POROUS JOINTS

Porous joints are the opposite of conventional ones in that instead of the cement being used to fill the interstices of the bone, the bone grows into the pores of the prosthesis. The joints have special coatings of cobalt-chromium alloy or similar metal formed into either crushed wire or tiny metal balls that are sintered onto the artificial joint. The joint is fastened either by a press fit, if it is a hip, or by pegs driven into the adjoining bones, if it is a knee. The bone grows into the coating pores to anchor the implant three-dimensionally.

Coating the joint completely would cause stress shielding and bone resorption; selective coating, on the other hand, distributes stresses more effectively and makes the prosthesis easier to remove if necessary. "We don't know yet just how the joints should be coated," says Dr. Marks. "This is new technology; it's very hard to simulate in animals or on computers, so we do what we think is logical, then follow patients very closely. That enables us to refine the details of how much surface should be coated, and where."

To date porous coatings are pretty much limited to artificial hips, knees, and a few shoulder joints. The ball-coated prostheses seem less likely to break loose and easier to shape than those coated with wire, although some researchers argue that the pore size of the wire coatings makes for better ingrowth. Since both types seem to be working, the prevalence of one or the other may end up being a manufacturing rather than a clinically based decision.

Researchers at the Cleveland Clinic Foundation are presently conducting clinical trials with two types of porous artificial hips, both of which have sintered ball coatings. One is a straight-stemmed prosthesis; the other is an anatomical hip.

A multicenter study of hip bones found that people of different sizes have bones of varying proportions and that growth does not occur proportionally in all dimensions with increasing size. The Clinic researchers tried to develop a prosthesis to accommodate these findings. The anatomical hip has a curved stem that is unique for its left and right components, which follow the curves of the thigh bones. The stem has a posterior bow and expands into a three-dimensional wedge at the top. Its larger size and better fit may reduce undesirable stresses while improving stress distribution, ease of insertion, stability, and range of motion.

The anatomical hip has pores larger than those in the straight stem joint. The cup, which is implanted in the pelvic bone, has a polyethylene liner that is replaceable after years of wear without removing the joint; the ball can likewise be changed.

The benefits of porous implants are less clearly defined in the knee because its anatomy makes loosening less of a problem than it is with the hip. While the hip is subject to shear stresses, the knee receives concussive stresses.

Patients must have healthy bones to receive porous joints. The joints cause more postoperative swelling than cement fixation, and the rehabilitation time may be doubled to 6 months or so. These and other factors make porous joints about twice as expensive as conventional ones. The porous joints offer significant advantages, however. Bones are loaded favorably, and the pain relief is just as good. The press fit, which is like driving a stake into the ground, requires less resection of bone. Should the prosthesis loosen, the body will react by making more bone rather than losing it. Porous joints are especially good for younger patients, whose implants would probably loosen with time and require reparative procedures. If necessary, they can be cemented in just like conventional joints.

"Here we've had patients wearing porous joints for 3 years, while elsewhere they have done so for 5 years or more, with good results," says Dr. Marks. "The problem is that we must wait for patients to live long enough to show us definitive results. I think it's the technology of the future, but I always tell our patients that we're not yet sure."

BIOGLASS COATINGS

Researchers are exploring coatings other than porous ones for fixing artificial joints. "We have patented processes for putting Bioglass coatings on both alumina ceramic and on metals used in orthopedic devices, for noncement fixation," says June Wilson Hench. The advantage of the old process is that bone cement works like a gasket, and the surgeon has some freedom to perfect the alignment before it sets. Bioglass fixation does not provide this opportunity. "Bioglass bonding over a whole joint would be a much less forgiving technique for the difficult biomechanics of orthopedic fixation," says Hench. "The surgery and alignment would have to be perfect the first time. We have more research to do in biomechanics before beginning clinical trials."

The Henches also hope to explore the possibility of resurfacing joints rather than removing them; this would change arthroplasty to a two-stage operation. By resurfacing part of the joint with an alumina shell coated with Bioglass, the surgeon could buy time for a younger patient before replacing the entire joint.

CAD/CAM

Used for years in the production of aircraft, automobiles, and skyscrapers, computer-assisted design/computer-assisted manufacture (CAD/CAM) is presently being applied to the customization of artificial joints. Most of the time a surgeon uses a commercially available prosthesis; the patient's bone is cut so that it will fit. To tailor an implant for a most appropriate fit, the surgeon makes an x-ray examination of the joint and sends the pictures to a manufacturer. Often the surgeon cannot read the drawings and templates of the proposed joint that the manufacturer sends back for evaluation, so this time-consuming and expensive procedure can result in a prosthesis that fits no better than a standard size. The trial and error process can take months.

A CAD/CAM system has a library of standard designs with software that allows the surgeon to modify them. Working with the surgeon, an engineer places a computer-assisted tomography (CAT) scan of the joint on a special screen where the computer can read it. In response to the picture and other data about the patient, the computer recommends a standard design for the joint. To further customize the implant, the engineer punches in instructions on the computer terminal. Within a few minutes the computer can draw new designs or three-dimensional blueprints.

When the final design has been perfected, the computer records it on paper tape that is fed into the manufacturing machinery. Sophisticated cutting and honing devices shape any of various orthopedic metals and polymers into the precisely designed joint.

Because they fit better, joints made by CAD/CAM last longer and perform more effectively than standard ones. Their manufacturing time is cut from weeks to days. Since subcontracting costs are eliminated, joints made by CAD/CAM are 25% cheaper than other customized joints. Computer-assisted prosthesis selection (CAPS) can enhance the system further by allowing the surgeon to choose or design a joint without the help of an engineer by answering simple questions posed by the computer.

Only the Hospital for Special Surgery in New York City presently uses CAD/CAM for making artificial joints. It is licensing the technology for the CAD/CAM/CAPS system to Johnson and Johnson, which plans to set up computer terminals in other hospitals and manufacture the prostheses they request. Linking these terminals would enable hospitals across the country to pool their memories for further precision in designing joints. Meanwhile, commercial manufacturers are using similar but less sophisticated systems to custom-design joints.

"They've been talking about using CAD/CAM for so long that I'll believe it when I see it," says Dr. Marks. "I would have great use for it. Custom joints for patients at the far ends of the anatomical spectrum would really help. We could better fit an odd anatomy while saving more of the patient's natural tissue and bone."

These and other innovations should further expand the use of artificial joints, already a routine and successful alternative to pain and limited mobility. Like other forms of substitutive medicine, however, joint prostheses are not a cure for the degenerative diseases that make them necessary. "Artificial joints don't restore people to normal," says Dr. Marks. "But recipients who accept their limitations and use them for everyday activities can do very well." In this respect the recipients' prognoses resemble those of many persons who have lost entire limbs.

6

"Camille" and other Electric Arms

The Six Million Dollar Man and the Bionic Woman made having artificial limbs look easy. Amputees know differently, however. Upper extremity amputees fitted with the "dynamic duo"—the Utah arm and the Otto Bock hand, both of which are electrically powered—may sometimes feel less like bionic superhumans than like the title character played by Peter Sellers in *Dr. Strangelove,* whose artificial hand generally misbehaved and even tried to throttle him.

Take Alice Olson's experiences, for instance. Alice's fellow employees sometimes forget to warn her when they turn on their walkie-talkies, which cause her artificial arm to go up and down by itself. She sets off buzzers at airport security checkpoints and gets strange looks from guards when she tells them that she has a prosthesis (at the San Francisco airport, she was pulled to the side and frisked). When carrying luggage, Alice switches off her arm so that the hand will not open and drop the bag. A cab driver once started a tug of war with her over a suitcase until Alice explained that she had to turn on her hand to let go.

Ken Huff, who has two Otto Bock hands, might be considered a threat to public safety. Last winter Ken was carrying a heavy sack when he slipped on some ice. His left socket and hand popped off and went sliding in different directions. "I wasn't hurt, but some poor guy standing nearby almost had a heart attack," he says. Then there was the time Ken was horsing around with his 16-year-old son and grabbed the boy's arm. The weak battery in Ken's hand picked that moment to die, so he could not let go. His wife and daughter had to help pry his son loose.

Amputees agree that incidents like these require a sense of humor. Anyone who complains of "needing a hand" within earshot of these folks is likely to have a disconnected Otto Bock waved in his or her face. When not using her artificial arm, named "Camille," Shauna Bingham usually carries it slung over one shoulder. She has been known to plant the arm or hand where unsuspecting people can happen across them.*

Artificial limbs do, however, offer some advantages. Once Alice Olson was stopped for speeding. In traffic court she pointed out to the judge that airlines

*Lower extremity amputees can be equally mischievous: most of Tom Andrew's neighbors did not know that he was an above-knee amputee the Halloween he took his children trick-or-treating dressed as a pirate with a peg leg. Andrew has since built himself a skateboard prosthesis based on a prototype by Flowers.

ask her not to use her arm while in flight because it interferes with the plane's radar. Since she was *not* going as fast as the officer claimed, Alice noted, perhaps his radar was affected by interference from her arm. The judge reduced her fine from $55 to $20.

These and other amputees are making the public more aware that amputation does not have to be a crippling disability. Most amputees are anxious to return to work and resume productive lives. In the farming industry, loss of an arm is considered an occupational hazard, as is leg loss in the logging industry; it is *not* considered an excuse to spend the rest of one's life sitting behind a desk.

The entire country is on a health kick, and amputees are no exception. They joined other disabled persons in the 1984 International Games for the Handicapped in New York. Studies show that about 20% of people who have lost a leg still play basketball and volleyball, 10% play tennis, and 5% jog. Recently 22-year-old Jeff Keith, who lost his right leg to bone cancer 10 years ago, ran from Boston to Marina del Rey, California (3300 miles) in 8½ months.

Sports favored by amputees include fishing, swimming, bowling, hunting, golf, baseball, and horseback riding. "There's a big trend, especially with younger amputees, toward giving them back what they lost," says Jeffrey J. Yakovich, C.O., Chief of Orthotics and Prosthetics at the Cleveland Clinic Foundation. "As prosthetists, we have a commitment to help people be active." Yakovich, who lost his left leg to bone cancer, still fishes and swims at the beach.

Most amputees with this attitude are using equipment that is less than ideal. As we will see here and in the next chapter, prosthetics is one of the best examples of how adequate research funds could help scientists make a big difference in the quality of people's lives. The technology is expensive because building a better artificial limb is an incredibly complicated task. "In my 30 years of research and development, in fields as disparate as missiles, computers, consumer products, and capital equipment, I have not experienced a field as complex and intractable as that aimed at ameliorating amputation," writes Robert W. Mann, Ph.D., Whitaker Professor of Biomedical Engineering at MIT, in a recently published lecture. Limb prostheses require effective interaction between humans and machines, which presents an especially difficult challenge to the researcher. The prostheses must provide tolerable, controllable imitations of human movement.

The arm well illustrates why this is a nearly impossible task. The natural arm can make a wide variety of movements and often must do so simultaneously. The shoulder, elbow, wrist, and fingers can bend, rotate, raise, and lower. The muscles controlling these movements are composed of millions of fibers, each in turn controlled by a nerve ending. An arm movement results when the brain pictures it and sends commands to the nerve fibers attached to the appropriate muscles. Within a split second all the necessary muscle fibers are fired simultaneously by nervous stimulation. This is quite a feat to try to duplicate in a relatively small, lightweight device.

Conventional mechanical arms are operated by cables and pulleys that require enormous concentration. The amputee must use both shoulders and various body contortions to activate the cables and locks. Perhaps 10% to 15% of amputees actually use cable arms, and only about 1% do so successfully. One is 48-year-old Larry Campbell, a warehouseman from Roy, Utah. In 1977 Larry lost his left arm about halfway between his shoulder and elbow in an accident on his job for the Union Pacific Railroad. Before getting a Utah arm he wore a cable arm for 3 years and still uses it for work in which it could be damaged or become wet.

Larry is skillful enough with his cable arm to pick up a mound of ash from an ashtray and set it back in without disturbing the other ashes. But he dislikes the mechanical prosthesis. "It always makes me feel awkward and clumsy," he says. "I feel like I have St. Vitus' dance because I have to be moving constantly to work it. I always have to think ahead where I want my hand—up, down, relaxed, close to my body, or operating." Larry knows a writer who can write with a cable arm but suspects that he feels the same way.

These discomforts have led researchers to experiment with electrically powered limbs. The nerve messages that move muscles are actually electrical impulses, the voltage of which increases with the amount of muscle tension required. Small portions of these impulses, called electromyographic (EMG) signals, reach the skin, where they can be read by specially designed instruments.

The nervous system of an amputee still produces myoelectrical signals; they just reach the end of the stump and have no limb to move. While researchers have succeeded in capturing these signals and using them to power artificial arms, few of these devices have withstood the rigors of daily life or reached commercial development. The most successful myoelectric arm is the Utah arm, developed by Jacobsen. Individually calibrated electrodes in the arm's socket read myoelectrical signals from the amputee's stump; all he or she must do to move the arm is think about it.

The Utah Arm

Introduced in October 1981, the Utah arm is the only totally electric elbow and hand in the world. It is made of injection-molded, flexible plastic composed mostly of glass-reinforced nylon and carbon fibers. The Otto Bock electric hand has a realistic cosmetic glove. The arm is modular for easy customization, improvement modifications, and convenient repair by mail.

Electrodes in the arm's socket are fit on the amputee with a myotester so that the circuits can be individualized. The electrodes can read either the biceps and triceps muscles or the deltoid muscles of the shoulder. The arm has smart electronics that distinguish muscle signals from other body impulses or from outside electrical interference. A high-performance differential amplifier reads only signals that are different among the electrodes (identical signals are coming from outside).

A microcomputer in the elbow commands the arm's battery to power the movements. The arm has been adapted for use with the Otto Bock hand or other electrical terminal devices; locking the arm transfers the power to the hand, which can then be opened and closed. Both hand and arm have gain adjustments—as muscles strengthen over time or tire with prolonged use, the amputee can compensate by turning a little dial.

"We can't copy all arm functions like feeling, healing, or regeneration, nor have we tried to," says Barry K. Hanover, a mechanical engineer at the University of Utah Center for Biomedical Design. "But we have copied functions that are most important to people, using the basic form and weight of the natural arm." The motions the prosthesis can perform include passive side to side movement of the upper arm, powered flexion of the elbow, and, as noted, opening and closing of the hand. "In laboratory experiments using a research version of the arm, we've successfully controlled three powered functions, two extremely well. We've done the fine control it takes to join shoulder and elbow movements simultaneously to make diagonals and circles and to change speeds. We hope to include more powered motions in later versions of the Utah arm," says Hanover. "Although the arm is very sophisticated, the electronics have been its most reliable part," says Andrew, Director of Prosthetics at Motion Control, Inc. "It has exceeded engineers' expectations for maintenance and reliability."

The group uses CAD/CAM to design the four-bar linkage that drives the arm's elbow and the new terminal devices they are working on. "We believe in CAD/CAM, but we also believe in POD/POM (plain old design/plain old manufacture), which requires thought," says Jacobsen. "Having these capabilities doesn't relieve us of the responsibility of enlightened thinking."

The Utah arm has limitations, of course. The amputee cannot lift it high up in the air or get it very close to the body when raised. He or she cannot drink with it holding the glass, because the elbow must be locked to use the hand.

Its advantages, however, make the Utah arm far superior to conventional ones. It is much easier to use, even by people with no limb remnants, because it is physiologically based. It has more functions and a wider range of motion than conventional arms and requires less restrictive suspension systems. The stump muscles become healthier and do not atrophy because they are actively used. The arm looks and feels natural and swings freely while its owner walks. It is quiet and lightweight: the arm weighs 2 pounds, 1 ounce, the hand 1 pound. It is rugged and easily maintained, and the battery pack is readily recharged or replaced.

Jacobsen's group is working on several improvements that can be added to the arm easily. They hope to develop a lighter hook and to move the hand's motor to the forearm, where it can be remotely actuated and will better distribute the weight. Eventually the prosthesis should also have powered upper arm and wrist rotation, a passive lockable two-axis shoulder module (with two degrees of freedom: abduction and flexion), and a microprocessor-based digital control module. While researchers hope eventually to build control systems for prostheses into single microchips on which they can be customized and reprogrammed, this may not be in the Utah arm's immediate future. "Presently the volume of arms we manufacture does not make it cost-effective to have the control system on one integrated circuit," says Andrew. "Changing the circuit would require expensive changes in the chip, and repairs would be complicated."

The group's long-range goals include attachment of the arm to the skeleton, proportional and simultaneous control of several joints, and grip feedback.

Getting the Utah arm from laboratory to patients requires a team of researchers, prosthetists, marketing personnel, and instructors for other prosthetists. The arm is manufactured by Motion Control, Inc., a company created by Jacobsen and Dr. Kolff as a receptacle for this and other technologies that will be discussed later. Over 200 Utah arms have been fitted on patients so far at a cost of $25,000 to $38,000 each; the arms are considered well worth the price by their owners. Besides the Salt Lake City firm, over 85 different fitting centers for the arm have been established in the United States, Canada, the United Kingdom, Norway, Sweden, and Italy.

The first person to have been fitted with the Utah arm–Otto Bock hand combination was Alice Olson, a 35-year-old head desk clerk at the Newport (Oregon) Hilton. In 1977 Alice's left arm was caught between two rollers on a glue spreader in a plywood mill. The arm had to be amputated 2 inches above her elbow.

Having tried a conventional arm, Alice decided against wearing one. She heard about the Otto Bock hand on a television program and eventually found the group at Utah, who were then fitting the first experimental arms. "When they fitted me, I got tears in my eyes," she says. "The arm's response was so natural, it was almost like having my own arm back. It was incredible—so easy and graceful—that I was overjoyed. I had excellent muscle control and was a real natural for the arm—and I was really jazzed for it."

Alice believes that every disabled person should have the opportunity to try such new technology. "This particular prosthesis is not for everyone, but they should decide that from experience," she says. "The cosmesis alone, not to mention the function, is well worth the money spent. The arm is so reflexive that it's a part of me—I can do all kinds of things I couldn't do with conventional prosthesis."

Alice uses the arm as she would her natural left arm. "I can do 99% of what most people do, and I literally could not do my job without this prosthesis," she says. Alice can hold eggs or Styrofoam cups of coffee, open doors, carry luggage, push carts, tear down folios, crack walnuts, slice tomatoes, peel vegetables, open jars, knit, ride a bicycle, and put on makeup.

Shauna Bingham uses makeup too: her hobby is manicuring her long nails. When she woke up in a hospital and learned that her left arm had been amputated, her first groggy response was, "Well, now I can get my nails done for half price."

An 18-year-old student from Roy, Utah, Shauna had a job a few summers ago running a roller coaster at a local amusement park. She was cleaning a trailer underneath the ride when the roller coaster came through and yanked her arm behind her back, tearing the muscles and ripping all the nerves from the spinal cord. The arm was amputated at the shoulder.

Shauna's Utah arm, "Camille," now sports a few fake gold fingernails just like the ones on her natural hand. Camille helps Shauna prepare meals or do needlepoint and is easier to use than a conventional arm because of Shauna's phantom limb pain.* "I don't wear the arm if I'm sitting around the house, if it's really hot or when I'm driving, because I drive a stick [steering with both knees], and it gets in the way," Shauna says. She often forgets that she has lost an arm. "The prosthesis is fun to have; it makes my clothes fit better, and it's really worth any aggravation in wearing it," she says. "Things could be a lot worse."

Larry Campbell has not practiced lifting ashes with his Utah arm but much prefers it to his cable arm. "It's much easier to manipulate," he says. "I can sit very still and just flex my muscles to operate it. That's worth a lot to me." Larry is not interested in adding powered wrist or elbow rotation to his arm because he considers it easier to twist the prosthesis with his other hand than to think about manipulating it. "I wear the arm any time I'm out except for bowling, when it gets in the way," he says. "I never wear long-sleeved shirts because I'm proud of it and want to show it off." Like Alice, he uses the prosthesis as he would a natural left arm, to brace or hold things or pick them up. "It's the greatest assist I have," he says. "I want every amputee to be aware of its existence and to try to find out if he or she can use it."

Ken Huff, of Orem, Utah, is a 40-year-old tree expert who used to work for a company that contracted to clear power lines. On the Fourth of July, 1983, he participated in a Civil War reenactment and was swabbing the barrel of a cannon when it went off. Ken lost both hands below the elbows, an eye, and an eardrum. He now has two Otto Bock hands, the right one in a special socket that hooks behind his elbow for better grip. The other socket grips his arm bones.

The Otto Bock Hand

The Otto Bock is the oldest and most reliable powered hand available. Ken's hands pick up myoelectrical signals in his forearms; the wrist extensor muscles

---

*An upper extremity amputee with an active phantom limb often gets better results with a Utah arm. Controlling it with remnant muscles makes the arm feel real; the phantom limb "inhabits" the prosthesis and improves with "use."

open the hooks, and the flexors close them. Like the Utah arm, the hands are modular. The batteries inside the forearms pop out for replacement. The on-off switches are in the palms. The hands can open or close at different speeds, and their controls can be fine tuned. One of Ken's wrists is a ball and socket joint that he can flex in any position and get close to his body; the other is a quick disconnect so that he can substitute the Otto Bock Greifer. The Greifer (German for "terminal device" or "hook"), which looks like a thick claw, is a tool for heavy-duty work.

"Any problems I have now are caused not by my hands but by my eyesight," says Ken, who lost most of the vision in his remaining eye. "If I could see, I could return to work and do just about everything I used to do, including climbing trees. I just have to be careful when I pick up things like beer cans, or I crunch them—the hands have about 22 pounds of pinch.* They make it easier to do most tasks."

This restoration of function makes it seem as if myoelectric hands and arms are here to stay. "Anyone who loses an extremity wants it back very badly," says Yakovich. "Those who have used conventional prostheses for years won't want to relearn natural movement and retrain their muscles to use the Utah arm, but new amputees will want it for psychological reasons. As the state of the art improves, it will be the rule rather than the exception in upper extremity prostheses."

Meanwhile other researchers are taking both theoretical and experimental approaches to the improvement of prosthetic technology.

Skeletal Attachments

Amputees complain unanimously about the socket-harness assemblies that are used to attach their artificial limbs. New types of sockets fit more comfortably; some even preclude the use of harnesses (Larry reports having seen a videotape in which a man with a suction socket on his artificial arm used the prosthesis to lift a 6-foot chain saw). Most amputees, however, must put up with hot, sweaty attachments. Harness systems restrict movement and must be carefully fitted to avoid restricting blood flow as well. Researchers are therefore experimenting with various methods of attaching a prosthesis directly to the bone of the stump. As might be expected, they are starting with the metal and bone cement used in artificial joints and are also investigating materials like fiber-reinforced HA ceramics.

In addition to making amputees more comfortable, direct attachment will spare them certain bizarre experiences, such as that of a 33-year-old San Francisco woman who lost her $9000 artificial arm last year when thieves yanked it from her shoulder along with her purse. The problems with percutaneous attachment are, of course, enormous; the disadvantages include the fact that it

*The average adult male has about 18 pounds per square inch.

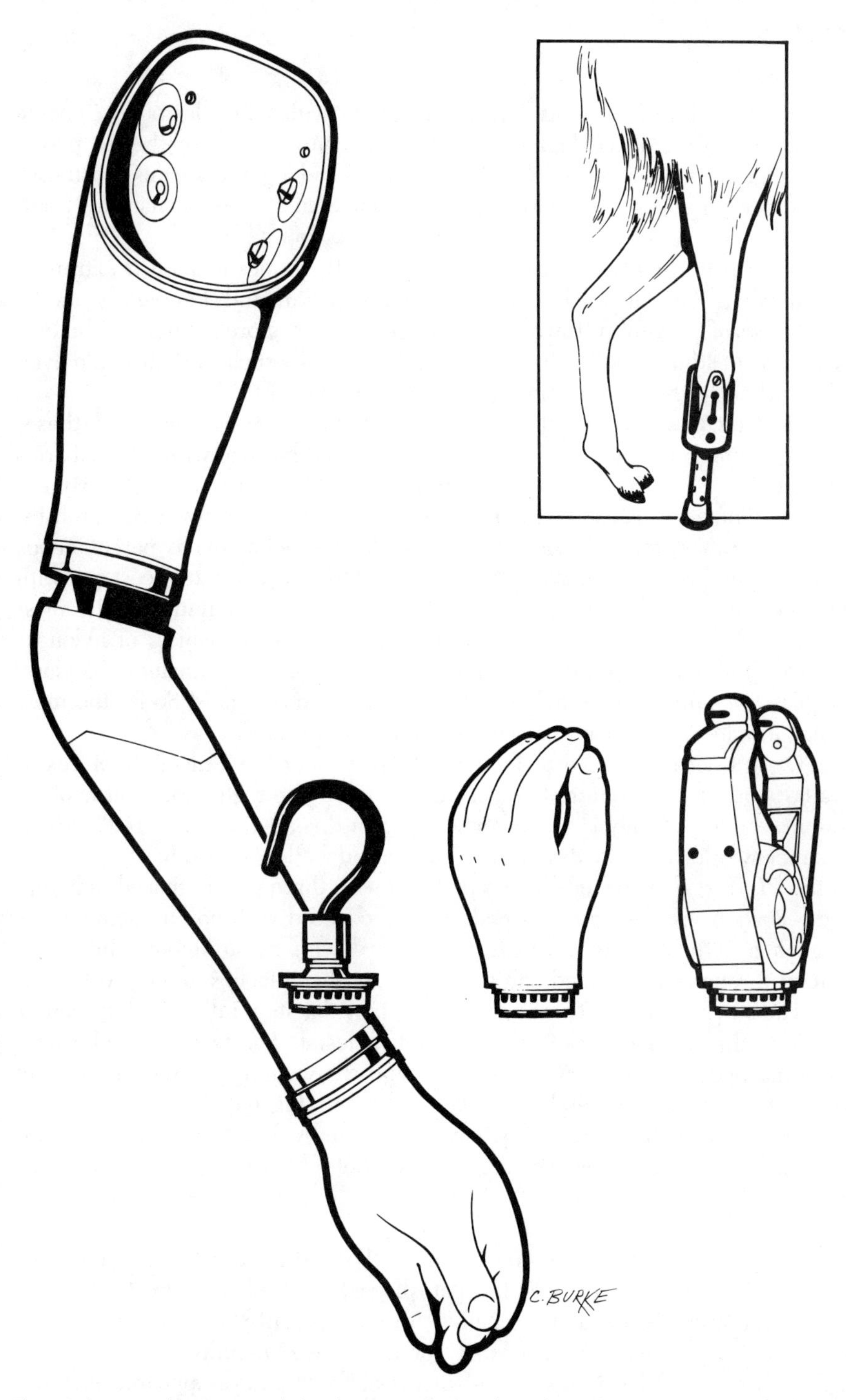

The Utah arm; the Otto Bock hook, cosmetic hand, and Greifer; a percutaneous load-bearing skeletal extension (PLSE) with a prosthesis attached to a goat's leg.

might make a lower extremity prosthesis, in particular, less desirable. "People get along with artificial legs with poor joint control because they're poorly connected to them anyway," says Flowers. "If I had a percutaneous prosthesis, I'd probably be frustrated with most of the knee and ankle mechanisms available because I'd know more of what the leg was doing."

By transmitting body weight directly from the bones to a lower extremity prosthesis, a skeletal attachment would eliminate pain and skin damage on the soft tissues of the stump. It would also provide better, more natural attachment, reception of stimuli, and swing through. Theoretically such a device could even be attached to the remnant muscles for better control.

Investigators have developed attachments for lower extremity prostheses (which are more easily tested in animals) that could be used with artificial arms as well. C. William Hall, M.D., Institute Medical Scientist in the Department of Bioengineering, Electronic Systems Division of the Southwest Research Institute in San Antonio, Texas, refers to skeletal attachments as percutaneous load-bearing skeletal extensions (PLSEs). A PLSE has two interfaces: one with the bone and another with the skin. Researchers have found that the most biocompatible materials for the skin interface are velour fabrics of nylon or Dacron; those most appropriate for the bone attachment include Bioglass, Proplast, or porous polyethylene, alumina, or various metals. So far the most suitable material has been porous sintered titanium mesh.

Over a decade ago Vert Mooney, M.D., now at the University of Texas, Dallas, did the first clinical trial of a PLSE on the right upper arm of an amputee. Dr. Mooney is no longer working in this area, but Dr. Hall's group has meanwhile created a device that has worked well in animal tests. The stem of the PLSE ends in a step-shaped pedestal with three evenly spaced radiating lugs. Portions of the stem and pedestal are covered with porous titanium for bone ingrowth. The lugs, which are covered with nylon velour fabric, will extend through the skin and bolt onto the yoke that attaches to the prosthesis.

The surgeon amputates the shin bone, taps the stem into its hollow center and cuts the lower end to fit the stepped pedestal. The tendons are brought over the pedestal and sutured together, and the wound is closed in layers to cover the implant completely. In 3 weeks the dressing and splint are removed. The skin over the lugs is cut away to expose them; by then tissue has grown into the velour to seal the area. The lugs are then bolted to the yoke and prosthesis. By diverting it at right angles to the bone, the lugs make stress more manageable.

Dr. Hall's group decided to give their PLSEs the ultimate test—they implanted several of the devices in Spanish goats. Fitted with prostheses, the goats walk immediately and normally with no apparent pain. They are turned out to pasture and have lived normally for as long as 14 months. One goat with a PLSE on its hind leg jumped a 6-foot corral fence and disappeared into the annals of animal liberation; occasionally it reappears from the brush to visit its companions but has not been recaptured.

Sensory Feedback

Touching or lifting an object causes nerves in the fingers, hand, and arm to relay sensations to the brain concerning the results of that action. An amputee, who has no such feedback, must instead watch his or her terminal device to make sure that it has the right grip and force. This observation is easier with conventional arms: hooks are more compact, can be brought closer to the body, and can manipulate small objects more precisely than a cosmetic hand; a cable-controlled hand gives more feedback than does a myoelectric one.

Researchers have tried placing pressure gauges inside artificial hands or hooks that relay signals to the stump. A system like this created by Collins also had goniometers that measured elbow and wrist angles, axial wrist rotation, and finger (or clamp) opening so that the amputee could tell where the arm was without looking at it. Despite such efforts, conveying feedback without tying up the amputee's sensory system remains probably the single most serious problem facing researchers developing prosthetic arms.

Nerve Signals

If muscle signals can move prosthetic limbs, it seems logical to take a step closer to their source and instead pick up the nerve signals themselves. Researchers in Canada and Boston are trying to do just this, and Jacobsen's group has discussed with them the possibility of integrating such capability with the Utah arm. The technical problems, however, are formidable. Obviously any percutaneous system lends itself to infection. Researchers have been able to get nerve cuffs to stay around nerves, but since several muscle nerves go down a path encircled by a cuff, it would be difficult to sort out their signals. It is not clear whether more electrodes would pick them up better. Nerve signals would be hard to interpret because they fire muscles rather than performing acts directly themselves. Researchers do not yet fully understand the central and peripheral nervous system integration of feedback information.

Since myoelectrical signals are partly interpretations themselves, investigators may decide that it is just as well to let them do the job.

Movement Therapy

Neville Hogan, Ph.D., Associate Professor of Mechanical Engineering at MIT, believes that movement control is a much deeper problem than has been recognized widely so far. "Much work in robotics and prosthetics has focused on free movement, without a context," he says. "If an artificial limb has environmental forces acting on it, the control problem changes drastically, and it becomes unstable very easily." Hogan's group and other researchers are making mathematical descriptions of the organization and execution of movement in intact upper extremities, and so far experimental results have coincided with their theories. Movements seem to be coordinated globally in terms of overall motion rather than on a joint-by-joint basis; by looking at how an amputee has deviated from this norm, researchers can help him or her correct a movement. The MIT group hopes that their theories can be applied to lower extremities as well.

In the laboratory the group is testing artificial arms that move by simultaneously changing activity groups of "muscles" instead of just one. They hope that an amputee can make such movements to use tools and perform other practical tasks. The group is also trying to simplify the coordination of a prosthesis with the natural limb (to perform two-handed tasks, for instance); if this ability is useful, they can program it into the Utah arm. Other simple control changes might give an artificial arm more speed (with less control) and more springiness (so that the amputee could lean on it while using a plane, for instance). The researchers are also working on a more precise microprocessor-based myoelectrical system that gives 5 times the accuracy of present myoelectrical readings at the same speed.

"The main thrust of all my work," says Hogan, "is to give the amputee the ability to change the system dynamics—as a consequence of that, several other problems like coordination become easier to solve."

Several other problems remain, however, for both upper and lower extremity prostheses. These will be discussed in the next chapter.

7

Bionic Swings, Stances, and Gaits

No one has yet built a two-legged walking machine, and with good reason. Walking is an extremely complicated activity. It is a very marginally stable system: we start from a precisely balanced posture and go into a controlled fall; our legs catch us, and we maintain balance so as not to topple over in any direction. To take one complete stride, the body must ennervate every muscle from the ribs down at least once in both directions, contracting and relaxing. Modeling and understanding walking, and coordinating and repeating it in an artificial system, are no easy tasks.

The arms are involved in fine, dexterous movements, the legs in gross movements. Prosthetic arms and legs are like apples and oranges in terms of the kinds of problems they pose. The upper limb is basically a position-velocity system, the lower limb a force-accleration one. Although the controls for upper and lower limb prostheses can be similar, their environments, task descriptions, levels of force, time scales, and types of activity are very different. Arms are more difficult to calibrate, and more amputees give up on using them. Walking, on the other hand, is more basic to human needs, and amputees will make great efforts to use prosthetic legs. A person needs two legs to walk, so although artificial legs are harder to fit and not yet as sophisticated as the Utah arm, they are much more widely used.

The ratio of lower to upper extremity amputees is about ten to one. Most lower limbs are removed to preclude gangrene as a complication of diabetes or peripheral vascular disease. Here again, advances of technology in polymers and microelectronics have resulted in improved prostheses.

Conventional Prostheses

In their attempts to approximate our natural variety of movement, researchers have created all kinds of artificial legs and feet that can be alternated by amputees like different pairs of eyeglasses. If they wish, below-knee amputees can now get standard lightweight wooden legs and feet with cushioned heels and flexible toes. Women can choose between versions for flat shoes and modified prostheses for high heels. Dee Malchow, who lost a leg below the knee in a boating accident 21 years ago, is now having a cosmetic mold made of her natural leg. "They paint silicone material on my natural leg, turn the material inside out and make the mold for the other leg so it will match," she says. "To me that is long overdue. Women amputees are a definite minority, and most prostheses are made by men, who often don't understand the importance of cosmesis."

Above-knee amputees can have the soft flesh of the thigh fitted with a suction socket that exercises the muscle and makes for less weight and better control. The most sophisticated above-knee prostheses are the heavier, more expensive types with hydraulic knees that operate like power brakes and adjust according to their load. Many of these devices respond to foot and hip movements and to input from the amputee to mimic true gait more closely. They operate by visualizing the weight line with respect to the knee and responding to its position. When the heel hits the ground, the knee flexes, then stabilizes, then swings through.

The foot on a hydraulic prosthesis can accommodate uneven terrain. One model, the Mauch SNS (swing and stance), has an antistumble mechanism designed to engage the hydraulics to let the knee adjust its descent. Other knees have friction locks; when directly loaded, they engage at about 20 degrees to allow a normal gait with swing and stance control. This type of knee allows the amputee to descend stairs step by step.

MIT Knees

Flowers and other researchers at MIT are working on various kinds of passive and powered/passive artificial legs, all of them generically referred to as "MIT knees." A passive leg is one with controlled dampers that can only absorb energy; a powered/passive leg contains motorlike elements that can produce mechanical power. All the MIT knees are capable of passive control functions.

Either type of leg must have an actuator that is noiseless, energy-efficient, lightweight, and reliable. Both must employ controllers that properly damp the swing phase and provide stability during stance so that the leg does not buckle. A load sensor in the shank provides a control signal to assure stability. The circuitry for all these features must be energy efficient and in some cases must handle very large energy flows.

The first amputee-interactive simulator to be created was an electrohydraulically powered device Flowers built as part of his doctoral thesis. It may be the most powerful artificial leg in existence. In laboratory tests this computer-controlled leg has allowed amputees to walk up stairs foot over foot. It operates through echo control by taking its cues from what the natural leg is doing. The kinematics (dynamics) of the sound leg are recorded and played back half a cycle later to the prosthetic leg, which follows suit. (Echo control can be used also with passive legs.) "Transferring the information from one leg to another is easy," says Flowers. "The difficult part is doing the measuring and making the powered leg work outside of a laboratory." Intricate details must be considered, such as implementing control or override if the amputee stops or changes direction in midstride.

In the laboratory tests with this leg, amputees had to learn a different way to walk; the gait was not as smooth as normal, and at first it was difficult to convince subjects that the leg would not collapse if they did not lock it. They nevertheless liked the way the prosthetic knee flexed during stance, as would a natural knee.

The MIT researchers are also investigating a regenerative leg, which is more sophisticated than a powered/passive leg. A simple passive leg dissipates energy; a regenerative leg stores energy to power itself in a later phase of the gait. Storing energy instead of dissipating it each time the leg stopped would mean that the amputee would not have to carry a big battery pack. The principle is like that of jumping on a pogo stick, which sends it energy that the stick gives back as it pushes up. Trying to do this with an artificial leg is like trying to jump on a pogo stick, then make it wait 2 minutes to respond. The group has completed all the hardware for such a leg but is still working on the circuitry to drive it.

MIT now has one of the most sophisticated monitoring laboratory systems in existence—the Telemetered Real-Time Acquisition and Computation of Kinematics (known as TRACK), which gives a three-dimensional analysis of how well a prosthesis is working. The researchers hope to enable an amputee to come in and test a passive or powered/passive leg that could be adjusted in several visits to his or her needs and desires.

"I'm convinced that simple microcomputer-controlled passive legs can provide more comfort and mobility for amputees," says Flowers. "I think that some form of that design will be a commercial product sometime within the next several years, whether it's developed by our group or somewhere else. Powered legs and powered regenerative legs raise much more complex research questions. I don't know whether their increase in complexity has marginal benefit. It's clear that an amputee could do more with a powered leg, which is more complex than a passive leg. I don't think anyone knows whether the increase in complexity is worth what the amputee would gain from it. I don't see an easy way to find that out—we'll just have to wait until amputees can use the devices, then see how they like them."

Myoelectrical Power

The potential use of myoelectrical power is one of the factors distinguishing the design elements of upper extremity prostheses from those of lower extremity prostheses. We have already noted the theoretical difficulties involved in the construction of any type of powered leg. Unlike a conventional prosthesis, which is controlled by the swing of the thigh, a myoelectric leg would be controlled by muscle signals in the amputee's stump. It is harder to measure myoelectrical signals from a leg stump than it is from an arm because the perspiration, movement, and loads make a hostile environment for the instruments and cause the quality of the readings to deteriorate over time. The several muscles that must be read for custom tailoring are deep beneath several fatty layers. "With good EMG control we might convince an amputee to have an implant that might telemeter out the information, but we'd have to be sure first that the system was workable," says Flowers. A leg stump changes from day to day because of volume shifts, swelling, soreness, and fatigue, and the amputee's endurance can vary as well.

Another potential problem raised by a myoelectric leg would be that of training a prosthetist to fit such a device without requiring computer expertise. Its sophisticated electronics might make a myoelectric leg cost as much as $50,000, including fitting (as opposed to $4000 for a conventional leg).

"I think that myoelectric control is not the first thing that should be done to improve a leg, and we've stopped work on it for now, but I suspect that it will come back," says Flowers. "Some interesting things are happening with it, and it's becoming a much better estimate of what the muscle is actually doing."

As we have seen, Utah arm recipients believe that the cosmesis, let alone the natural, effortless movement it provides, makes that myoelectric prosthesis worthwhile. Some researchers believe that amputee response to a successful myoelectric leg would be similar, regardless of its cost. A group at the Moss Rehabilitation Center and Drexel University in Philadelphia is working on prototypes of a regenerative myoelectric leg. In laboratory tests the leg has provided a gait similar to that of the best conventional passive above-knee prostheses. Like microcomputer-controlled passive legs, however, the Moss-Drexel device probably will not be ready for marketing within this decade.*

The Seattle Foot

The Seattle foot was designed for active and sports-minded amputees and for heavy-duty movement, although it may have wider applications because of its energy-storing ability. The foot has a leaf spring inside it made of heavy-duty polypropylene. This design provides a diving board effect—the weight the amputee puts on the foot is absorbed, and he or she is lifted as the material stretches out again. Unlike conventional feet, which are static, hard rubber, the Seattle foot provides feedback, and the restored energy allows amputees to perform fast activities better and longer. It can be added to most standard artificial legs.

The foot was developed by the Prosthetic Research Group at Harbor View Medical Center headed by Ernest Burgess, M.D. Dee Malchow, R.N., a

*Several pages describing the Moss-Drexel leg have been deleted at the request of Gordon D. Moskowitz, Ph.D., Professor of Mechanical Engineering and Mechanics at Drexel University, who believes that only an engineer could write about his group's work accurately.

40-year-old nurse-consultant at the medical center, was the first to get it 4 years ago. "They've perfected it quite nicely since I first tried it," she says, "and they make it from a human foot mold so that it has toes and looks very realistic. Compared to conventional prosthetic feet, it's like putting on slippers after a hard day's work."

Since the foot has no moving parts, it's very reliable and durable. "I can jump rope, skip, and play basketball with it on," says Dee. "It allows for more energy use and makes running more of an option for an amputee whose socket fits precisely." Other recipients of the foot include an amputee from Alaska who runs marathons and Jeff Keith, who wore it on his cross-country run.

Like researchers in upper extremity prostheses, those concentrating on lower extremity devices hope for long-range improvements.

Gait Measurement

A basic research problem is how to quantify what an artificial leg has done for an amputee. A microcomputer in the leg would use mathematical functions to measure the amputee's gait, interpret the measurements, and decide how to drive the leg for normal walking. Gait analysis must therefore be quantitative. "We can recognize an individual's gait by the sound of his or her footsteps or by the shadow of that person's profile, and even a layman can often tell by observation that someone is walking abnormally," says Max Donath, Ph.D., Associate Professor of Mechanical and Biomedical Engineering at the University of Minnesota, Minneapolis. "Clinicians use such subjective evaluation in working with amputees. We think that there are patterns in recognizing normal gait that can be made quantitative." Objective measurements would allow researchers to compare normal to abnormal gaits and also compare a person's gait at different times before and after treatment.

Donath's group is trying to measure walking function by numbers that they can then compare for consistency. By developing equations for a computer, they have enabled it to distinguish normal from abnormal gait in 98% of cases. This program might then become the basis for a computer-controlled multichannel functional electrical stimulation system applied to the muscles normally involved in walking. The result would be a "walking pacemaker" for paraplegic and quadriplegic persons.

Flexible Sockets

One advance that is already being used by amputees is the flexible above-knee socket, which successfully breaks all the rules for lower limb sockets. Because the thigh tissue is soft, it is difficult for an artificial leg to encase the stump properly and transmit the amputee's weight to the floor. The ISNY above-knee socket, which was developed by researchers in Sweden and at New York University, solves this problem quite nicely.

Instead of one hard socket, the ISNY socket has two parts. A thin, pliable translucent polyethylene socket encases the stump. It rests inside a laminated carbon-fiber frame that can be attached to conventional prostheses for weight transmission. The inner socket provides suction, changes shape with weight or pressure, and is cooler and more comfortable for the amputee. Its translucence makes it easy for the prosthetist to check the fit. The flexible socket is simpler to make, modify, and replace than conventional ones.

Larry Campbell, whom we met earlier, has a flexible socket inside his Utah arm that he finds more comfortable than a suction socket. "There's too much flesh left on the stump, so I get a blood blister from a suction socket," he says, "and there's no way that I'm going to let them cut my arm again to fix it."

Despite all these advances in the field, research prosthetics is not often worth working in at this point because funding is so scarce. For the 50,000 to 100,000 upper extremity amputees in this country, for instance, the situation is something like this: most amputees do not use conventional artificial arms because the prostheses are heavy, clumsy, unpleasant, and generally unnecessary for most tasks. Funding agencies therefore assume that the market for artificial arms is not large. Actually both prosthetists and amputees would welcome better prostheses. It takes funding to improve a prosthesis, but to get funding, a researcher must prove that he or she has already made improvements. A better prosthesis would be expensive to produce unless many amputees began purchasing it, but amputees will not purchase prostheses that have not been improved . . . and so forth. The same is true for lower extremity prostheses. "After all these years," says Dee, "I haven't seen enough prosthetic innovation that's long lasting or impressive."

Besides better designed, more reliable devices, the prosthetics industry is due for other changes. Prosthetists, physicians, and amputees must be made aware of what is available in high technology. They must be convinced to accept these devices. Amputees would greatly benefit from nationally organized information and repair networks. The cost of these improvements must be justified in terms of the benefit society would derive from turning such disabled persons into able-bodied ones.

Surgeons who perform amputations are badly in need of education, especially for upper extremity patients. The success of a prosthesis depends enormously on the shape of the stump that remains. In anticipating patient rehabilitation with a prosthesis, orthopedic surgeons are pretty much on the ball, but vascular surgeons, who do most amputations, are disastrously ignorant. Most amputations are performed on elderly people in poor health, and most surgeons are concerned simply with treating the patient and getting him or her home. They seldom think in terms of how that patient is then to get about.

One surgeon confronted with a patient who required an amputation tried to research the proper methods; he found a single reference in the literature, which advised leaving the remnant limb as long as possible and relying on the imaginative prosthetist to fit a device properly. Surgeons cannot be expected to know what kinds of prostheses are available—what they *can* do, however, is consult with qualified prosthetists *before* performing amputations. Prospective amputees should insist on such consultations, which will become easier as prosthetics evolves increasingly away from a craft into a profession requiring suitable education and accreditation.

Despite these difficulties, prosthetists are unanimous in their belief that a renaissance is occurring in the field. Researchers are questioning the old and trying the new, as evidenced by their search for new materials, revolutionary sockets, and myoelectric limbs. While the costs of such innovations are high, everyone agrees that they are totally misleading compared to the quality of life and amount of independence provided to amputees.

None of these researchers expects to duplicate human movement or return everything the amputee has lost. "The natural movement system is one of the most complicated I've come across as an engineer. We must be very careful about what we choose to provide because many benefits like speed and control trade off against each other. We need to identify what is most important to the amputee," says Hogan. "I have a lot of respect for these very tough problems," says Flowers. "I'm sure that prostheses will improve, but it will be a very slow process, with very few *Eurekas*." Nevertheless, any amputee can offer assurances that prosthetic innovations are worth it.

SECTION III

REPLACEMENT ORGANS

8

Starting with the Basics

Before looking at individual artificial organs it makes sense to begin with those elements of the body that—among other things—comprise, nourish, and hold the organs together. Because cells, skin, blood, and vessels are everywhere throughout the body, research on their artificial replacements has particularly broad implications for bionics and medicine in general.

Artificial Cells

In 1956 Thomas Ming Swi Chang, an undergraduate at McGill University in Montreal, discovered to his astonishment that no one was researching artificial cells for medical applications. Chang bought a bottle of collodion solution (a coating for wounds) at the pharmacy, borrowed some hemoglobin, went home to his dormitory room, and began mixing potions. He got quick results—complaints from his dormmates about the smell of ether and enough data to propose artificial cells as his project for an honors B.Sc. degree in physiology. Today Chang is an M.D., Ph.D., and FRCP (C), a Professor of Physiology and Medicine, and the Director of the Artificial Cells and Organs Research Centre at McGill. He is still making artificial cells.

Like their natural counterparts, artificial cells are microscopical balloons full of watery solution containing substances that perform biological functions. They have semipermeable membranes that keep their contents biologically distinct from their surrounding. Groups of them offer an enormous surface area that has all kinds of uses. Whether inoculated into the muscles, skin, peritoneum, or elsewhere in the body in experimental studies, artificial cells, as Dr. Chang has recently written, can function as "extremely compact, efficient and simple artificial organs."

There seems to be no limit to the types of artificial cells that can be made. Literally any substance that can be suspended in a liquid can go into an artificial cell. Dr. Chang has made cells containing enzymes, biological cells and their extracts, bacteria, isotopes, antigens, antibodies, vaccines, hormones, adsorbents—the list goes on and on. Researchers can even put magnetic filings in artificial cells along with chemotherapy or other drugs, then use an external magnet to pull the cells to their intended destination in the body.

Similarly, all kinds of materials can be used to make the membranes for artificial cells. Dr. Chang still uses his original approach but adds new synthetics as they appear. An artificial cell can be a single balloon or a multicompartmental structure. Its membrane can be synthetic: a polymer like cellulose nitrate or nylon, or a liquid hydrocarbon. The cell can have a biological membrane: protein, heparin-complexed, lipid or lipid-complexed, erythrocyte, or polysaccharide. The membrane can have concentric layers of substances like lipids, or it can be bioresorbable so that it disintegrates over time. Its surface properties can be determined in advance, and it can be made biocompatible or permeable to any extent. "One advantage of preparing artificial cells with synthetic membranes, for instance, is that you no longer have the antigens present on the surfaces of the biological cells, which would otherwise cause an immunological reaction and lead to rejection," says Dr. Chang.

Several procedures now exist for making artificial cells, including the one Dr. Chang tried in his dormitory room nearly 30 years ago. This process is based on the principle that oil and water do not mix. If water is added to a beaker of oil and the mixture is shaken vigorously, the water will disperse into tiny droplets. Similarly, the contents of artificial cells, suspended in a watery solution like protein enzyme, can be poured into an oily chemical solution and shaken vigorously until they form microscopical droplets. A quickly applied chemical reacts to form a thin membrane around each of these droplets. Then a very complicated procedure is used to wash the oil from the cells so that they are once more suspended in a watery solution.

"It's a lot of fun," says Dr. Chang, "until you start trying to develop it for use in patients—then it's a lot of work."

As we will note in later chapters, synthetic cells could be part of an artificial kidney, liver, or pancreas. Researchers hope also to apply artificial cells in enzyme replacement and immobilization, tumor therapy, production of monoclonal antibodies and interferon, detoxification, and biotechnology. They can be used as immunosorbents, red blood cell substitutes, and drug carriers—either synthetic or bioresorbable—for slow release and target delivery. New uses for artificial cells are being discovered all the time.

Artificial cells already have routine clinical applications in dialysis and hemoperfusion. A third area that has reached the clinical stage is their use to remove urea, a waste product, from the intestines. The cells custom-made for this purpose contain an enzyme called urease and an adsorbent for ammonia. The patient simply swallows the cells, which proceed to the intestinal tract. The urease inside them changes urea in the intestine to ammonia, which is then removed by the cells' adsorbents. This procedure has worked well in animal experiments at McGill and Batelle Memorial Institute and has been tested on humans at the Mayo Clinic. "It needs further improvement before being tested again," says Dr. Chang.

In his 1972 book on the subject Dr. Chang notes that: " 'Artificial cell' is not a specific physical entity. It is an idea involving the preparation of artificial structures of cellular dimensions for possible replacement or supplement of deficient cell functions. It is clear that different approaches can be used to demonstrate this idea."

The other major approach to making artificial cells is the reverse of Dr. Chang's procedure. This process, created by Garret Ihler, M.D., Ph.D., of Texas A & M University and further developed by Gottfried Schmer, M.D., of the University of Washington, Seattle, results in "overhauled" cells. Rather than enclosing natural materials inside synthetic membranes, these researchers drain the contents of natural cells and refill the biological membranes with chemicals. These cells can then be compressed into a gel and packed into a cartridge to make a hybrid organ. "This approach has been tested for carrying drugs and enzymes but is still in the research stage," says Dr. Chang. "Its applications are different from those of the artificial cell."

The potential uses for artificial cells tax the imagination, and Dr. Chang is the first to point out that their wide range of use is the natural result of their own properties. "Cells are the unit structures of every living thing," he says. "An artificial cell can therefore be applied in many, many different areas. One of the most exciting is that of biotechnology, such as a cell containing a complex enzyme system that can perform complicated functions. We're working very hard on a variety of applications."

As will be discussed, natural cells sulk and refuse to perform normally when isolated from the body; they can be hard to grow in laboratories and take a long time even when they cooperate. Synthetic replacements that do all that human cells can do, and more, will have plenty of work awaiting them. The same can be said of synthetic skin.

Bioresorbable Skin

The skin is the remarkable supertissue that keeps each of us in and the rest of the world out. Not only must the body's fluid and chemical levels be kept perfectly balanced, it also must be protected against the bacteria that swarm through the environment. Part of this protection is afforded by the two layers of skin: the epidermis or outer layer, which is constantly sloughed off and regenerated, and the dermis or inner layer. The dermis performs an indispensable role in mechanically reinforcing the weaker epidermis; when destroyed, it does not regenerate but instead leaves a scar.

Because of the skin's protective function, burns and other extensive skin wounds can be devastating injuries to the entire body. First-degree burns destroy the epidermis. Second-degree burns (and the wounds left from autografts taken to cover them) destroy both layers but leave residual cells from each because the epidermis lines the hair follicles. These cells will regenerate into skin that is scarred and shriveled but at least *there*. Third-degree burns destroy all skin cells and expose the underlying tissue, fat, and muscle.

Third-degree burns are among the worst possible types of injuries. In a painful process called debridement or early primary excision, burned skin must be ripped off as soon as possible, before it can breed bacteria. The body's metabolism soars into high gear as it mobilizes its forces against infection. Fluids oozing from the exposed wounds cause dehydration, while the respiratory and circulatory systems weaken under the strain. Patients who survive such trauma face scarring that is a physical as well as a social handicap. A healing burn or wound is covered with unspecialized, constricted tissue that if extended over a large enough area can constrict movement or cause deformity. It is estimated that 15% of the 100,000 or so Americans hospitalized each year for burns are injured to this extent. To save their lives, then rehabilitate them, researchers have tried various methods to cover burns promptly and cosmetically.

A logical choice for burn protection is a bandage, and two types of "artificial skin" are essentially that. Op-Site, a clear polyurethane film marketed by Acme United Corp., is an oxygen-permeable covering that keeps skin moist and is suitable for first-degree burns. Biobrane, a silicone rubber–nylon compound coated with chemicals extracted from collagen, has been developed by Hall-Woodroof, Inc., for more serious burns. Both products can be left on burns for up to 2 months, but then a permanent covering is needed to minimize scarring.

A preferred covering for burns is an autograft, a section of skin removed from elsewhere on the body. A severely burned patient, however, might not have enough skin left to donate. The autograft is meshed (slit and spread out) before being placed on the burn, which gives a poor cosmetic effect. Rather than regenerating completely into new skin, it leaves some scar tissue. While it is possible to use animal or cadaver skin instead, the body eventually rejects these transplants.

To increase the amount of autograft available, researchers in Massachusetts and New York have harvested natural skin cells and grown them in laboratories. The cells are combined with other materials to help them grow more quickly and in a better configuration. Sheets of this "test-tube skin" have been tested on both animals and humans with good results. This method is more appropriate for cosmetic surgical procedures or other nonemergency situations, however, because the cells take several weeks to grow.

The ideal artificial skin would be simple and cheap to make quickly. It would flexibly conform to body contours without sticking too tightly or limiting range of movement. While containing moisture and blood, it would allow oxygen to reach the wound and have controlled permeability to water. It would not promote bacterial growth; if infection occurred, it could be infused with antibiotics. The body would accept it, and it would accept the body rather than shriveling into scars. It could be used for different purposes: to cover burns and accident wounds, to correct disfigurement caused by cancer or birth defects, and to improve the effects of cosmetic surgical procedures.

To approximate this ideal, researchers have experimented with various materials, especially collagen and polysaccharide. Collagen, a protein, is the major component of connective tissues like skin and cartilage; it is also found in bones. Polysaccharide is a generic term for several families of sugars. These happen to be the two basic materials of the most successful artificial skin developed so far. Created by Ioannis V. Yannas, Ph.D., Professor of Polymer Science and Engineering at MIT, and John F. Burke, M.D., Head of Trauma Services at Massachusetts General Hospital, the artificial skin has two layers like its natural prototype. It presently exists in two forms: stage one and stage two.

Stage one skin has an inner layer composed of collagen taken from cowhide. Yannas did his doctoral thesis on collagen, which is very useful for making rubbers, plastics, fibers, membranes, or gels. Unlike other substances used in these materials, collagen is nontoxic and bioresorbable. In the artificial skin, it is bonded to polysaccharide glycosaminoglycan (GAG), which is extracted from its most convenient source—shark cartilage. The GAG helps slow down the rate at which the collagen is degraded by the body.

The two substances are arranged in a precise architecture of chemistry and porosity. The inner layer of the skin is an extremely lightweight, translucent, white material that feels like thin foam rubber. The outer layer is very thin, clear silicone rubber. When bonded together and wetted, this two-layer polymer membrane feels like a wet paper towel and can be draped nicely over wounds.

When the skin is applied, it induces the growth of a neodermis into its inner layer. This neodermis differs from the original dermis in that it lacks some specialized tiny skin organs like hair follicles and sweat glands. It does have mechanical strength, basic protective tissue components, and nerve cells, and it is very well vascularized. "It is not a full dermis," says Yannas, "but it is physiologically adequate for the patient to walk around and do as he or she likes." Regeneration of the dermis even to this extent is a major breakthrough. Meanwhile, the collagen-GAG layer has broken down enzymatically. Most of this process occurs within 2 to 3 weeks, although it will be a few months before the dermis recovers its original strength and the artificial layer is completely gone.

After 2 weeks the surgeon removes the silicone upper layer of the skin, which is very weakly bonded to the neodermis. Then the surgeon takes autografts the size of postage stamps from other areas of the patient's epidermis and places them on the neodermis. These grafts regenerate and spread out to cover the whole area.

The skin, then, is not really "artificial." "I called it 'artificial skin' 14 years ago because that misnomer was the best I could think of at the time," says Yannas. "Actually it's a device—a temporary biotemplate. The artificial membranes last only until the natural skin has regenerated."

Stage two of the skin takes this procedure a step further by regenerating the epidermis as well. Before it is applied, the researchers take a quarter-sized biopsy of the patient's skin. They remove the basal cells, those at the innermost layer of the epidermis, which are youngest and divide most quickly. Using a special centrifugation process invented at MIT, they implant the basal cells into the collagen-GAG membrane. In animal tests this procedure takes less than 2 hours.

This skin is then placed on the wound as was stage one. A second operation, however, is no longer necessary. In 2 to 3 weeks, when the silicone layer is peeled off, the patient will have a neodermis and a neoepidermis—original skin layers in their correct anatomical relationship.

The new epidermis seems to be the critical factor in preventing contracture, the tightening in scar tissue that occurs in the dermal layer. "This is important because severe burn victims otherwise leave the hospital looking terrible," says Yannas. "They may have contracture of the whole neck, for instance, so that the head is fixed to the chest. One never sees these people; they work at night and don't walk around during the day."

Developed in 1980, stage two skin has been used on animals, and Yannas expects to begin clinical studies any time now. Meanwhile, Dr. Burke has treated about 60 patients with the stage one skin. The patients ranged in age from very young children to people in their 80s. The earliest have had their new skin for 4 years. They are definitely less disfigured than they otherwise would have been, since the stage one skin has cosmetic results comparable to those of an autograft. It causes little or none of the intense itching that usually follows skin grafts.

Some of the patients were burned over more than half their bodies, and apparently the amount of surface that can be covered with the bioresorbable skin is not limited. It is especially useful for third-degree and may be applied to second-degree burns as well. It is not clear yet what will be the effects of having no sweat glands or other skin elements. "We're not sure about the limits of this biotemplate," says Yannas. "What's clear now is that we've saved some lives and restored the ability of these people to function as socially acceptable, physiologically competent human beings."

The researchers hope to apply their principle to other organs and tissues, using different configurations for specific purposes. "Our main purpose here is to find technology to change the way the surgery of the future will be done," says Yannas. "We believe that by a combination of thought and chance we've discovered a new principle for treating a wounded site: rather than regarding it as a risk to be closed and forgotten as soon as possible, it's an opportune area on which to grow back a new tissue or organ."

A Blood Substitute

Like skin, blood is usually needed and often unavailable for emergencies. Large losses of blood surpass the body's ability to replace it before the organs suffer from lack of nutrients and especially oxygen. It is the interruption of the oxygen-rich blood supply that causes the devastating effects of heart attack or stroke. Like other substances needed for transplant, however, whole blood is expensive and not always easy to get in a matching type. It can cause allergic reactions, and the AIDS epidemic in particular has alerted people that it can transmit disease.* It is very difficult, however, to make a synthetic oxygen carrier as complex as whole blood.

For years researchers have tried to make or preserve hemoglobin, the part of the red blood cell that carries oxygen. Hemoglobin can be removed and either encapsulated for suspension in synthetic liquid or freeze-dried for indefinite storage. Once isolated, however, hemoglobin may not release its oxygen freely or remain in the bloodstream for very long. Scientists are working on chemically modifying the hemoglobin molecule to improve its ability to get oxygen to tissues quickly. "The hardest part of this whole approach is to find material that is really well-defined and that will stay in circulation for a long time," says Robert P. Geyer, Ph.D., Professor of Nutrition at the Harvard School of Public Health. "There are many different patents on artificial blood but no FDA-approved clinical products at all yet."

Using different methods, however, Geyer and his group developed a blood substitute that had exciting broader applications. The preparation is a milky white substance that is very simple compared to natural blood. It is made of liquid fluorocarbons, which replace the red cells in carrying oxygen. Other elements include hydroxy ethyl starch, which performs the blood protein function of keeping water in the vascular system to maintain pressure and preclude shock.

Such a blood substitute could play an important role as a transfusion for Jehovah's Witnesses and other persons who cannot receive natural blood. Other patients could simply be resuscitated with the preparation, then given whole blood, plasma or serum. "It's possible that past a certain critical time, the patient can repair his or her own blood," says Geyer. Even with 60% of the blood replaced by the artificial material, a patient's body could sustain itself: if the liver is intact, it will start making blood proteins at a very fast clip. If necessary, the patient could be given supplementary platelets or proteins. Geyer's group is intensely studying the question of how the material affects the replacement of missing blood cells.

Various preparations of such perfluorochemical products could prove much cheaper than natural blood and could be manufactured and ready for immediate use in large supplies. They cause no allergic reactions, require no matching of blood types, and transmit no diseases. Working with them can also help researchers better understand natural blood.

The most exciting thing about these products, however, is their potential use in cancer therapy and cardiovascular medicine. They can be unique therapy enhancers because of their remarkable ability to carry oxygen. In natural

*Screening procedures now eliminate donated blood containing AIDS antibodies.

blood, oxygen is attached to hemoglobin; it comes off only when the external concentration of the gas drops and reattaches when it rises. In an artificial preparation, however, the oxygen is just dissolved. The more oxygen the patient breathes, the more oxygen it carries and releases easily. It can therefore enhance any type of oxygen therapy by literally pushing the gas into the surrounding area and presumably into the cells.

The particles in these preparations are much smaller than red blood cells, so they can flow through capillaries constricted by burns or shock. They can also reach areas that are inaccessible to red cells because a heart attack has damaged them or because they are cancerous and poorly vascularized. Injecting a perfluorochemical product intravenously or with an implanted pump and having the patient breathe pure oxygen can suffuse a target area with the gas. Oxygen added to the heart after an infarction (or the brain after a stroke) can reverse existing and preclude further damage. Oxygenating the center of a tumor greatly increases the efficacy of radiation, even at a much lower dosage. If researchers find good methods for controlling oxygen toxicity, these artificial oxygen carriers can help such therapy come into its own.

Combined with oxygen therapy, then, different preparations of perfluorochemicals could be used to treat heart attacks or strokes, enhance cancer therapy, or reverse carbon monoxide poisoning (since that gas cannot block the fluid from carrying oxygen). Each of these preparations would be mixed differently according to its particular purpose and added to normal blood levels. The product for cancer therapy, for instance, would contain water, fluorocarbons, emulsifier, salts, and substances to maintain its pH. With luck, the various mixes could be prepared to remain in the body for different periods—a blood substitute should circulate as long as possible, while a radiation supplement should be excreted quickly.

"These preparations are actually more like drugs than like artificial blood," says Geyer, "but they are unusual drugs. Some researchers prefer to call these perfluorochemicals 'devices' because essentially they are materials that carry oxygen."

Geyer's group began working on perfluorochemicals in 1967 as a means of perfusing organs. "We were the first to show that an animal could be kept alive in this way; this developed the concept of total body replacement, in which all organs are being perfused," he says. This work had an immediate result. Having heard the media describe the artificial materials as liquid Teflon (as they often still do today), the individuals holding the patents on that product called Geyer to ask where he got the substances and how they worked. "Teflon has nothing to do with these products at all," he says.

Tests in animals went so well that the team began seriously to consider these blood substitutes for use in emergencies, shortages, and transfusions for people who refused whole blood. Although their first grant application had been turned down by reviewers who refused to believe it possible, they worked their way up to maintaining completely bloodless mice and rats on preparations containing perfluorochemicals. These animals need no booster doses of the materials, for they quickly replace missing blood components. The hydroxy ethyl starch (which is FDA-approved for use separately) is metabolized; the fluorocarbons are exhaled, and the emulsifier is excreted in the urine.

The Green Cross Co. in Osaka, Japan, meanwhile consulted Geyer about the preparations and began to develop similar products themselves. Since Harvard did not allow Geyer's group to do so, Green Cross was able to patent the preparations. The Japanese researchers formulated Fluosol-DA, a perfluorochemical product, for clinical use that could supplement treatment of various diseases. Because of restrictions imposed by their food and drug administration, initial clinical work had to be done in Germany on brain dead patients. Once that series of tests was completed, the Japanese received permission to use Fluosol-DA on volunteers, starting with the president and other employees recruited from Green Cross. They then received permission for further tests in humans and have used Fluosol-DA on many Jehovah's Witnesses.

The American FDA has decided that although Fluosol-DA looks safe, more work is necessary to show that it does, in fact, benefit patients. Presently Green Cross manufactures Fluosol-DA, and Alpha Therapeutics Corp. in California distributes it to half a dozen or so American groups with FDA approval to test it. Its use in Japan is similar, although they seem to be testing it in a larger number of different patient populations rather than targeting a few groups. "I would guess that the Japanese have tested Fluosol on about 800 patients, including anemia, ischemia, and stroke victims, and we've used it on a few hundred," says Geyer. "In these large groups there certainly have been few serious problems." For the foreseeable future, the product will be tied up in clinical trials while groups such as Geyer's seek to make improved products of different kinds. "We can do clinical work, but we want enough animal data first," he says. "This research has to be done correctly, ethically and scientifically, so that for everyone, including the patient, it makes sense."

Fluosol-DA has many exciting possibilities, but each of its variants would be a product for a specific use rather than a replacement for whole blood. In its final form as a blood substitute, Fluosol-DA will still lack antibodies, clotting factors, and other essential elements. As such it will supplement rather than replace existing blood supplies, and it may make a big difference in areas that suffer from shortages. Meanwhile, as we are about to see, steady supplies of natural blood cause other researchers problems of their own.

Synthetic Vessels

Cardiovascular surgeons are often called upon to construct a bypass for a congenitally defective, damaged, or obstructed blood vessel. Whether it is a large, debris-choked major artery or a small peripheral or coronary one affected by atherosclerosis, diabetes, smoking, or trauma, the method of choice is to take a natural replacement from elsewhere in the body. Usually the surgeon will borrow one of the saphenous veins from the leg, which like other vessels have inner linings that preclude their becoming clotted with blood.

Natural vessels, however, are not easy to harvest. The patient simply may not have spares—some people have such advanced atherosclerosis that most of their vessels need replacement. Up to 25% of the time recovered vessels cannot be used, often because they are in no better condition than the worn-out ones. If expected to provide the heavy-duty service of an artery, a small leg vein is likely to clot or just give up. The leg surgery is longer, more troublesome, and more likely to cause postoperative pain and complications than the coronary surgery itself.

Since the 1950s surgeons have instead been able to use artificial vascular grafts to replace larger vessels. Usually made of Dacron or expanded Teflon, the grafts come in various lengths and sizes; the Dacron can be woven, knit, or velour with smooth or textured finishes. Once these grafts are implanted, their inner surfaces become lined with fibrous material, making them biocompatible but also narrowing their diameters. For a patient whose vessels are stiff and clogged from atherosclerosis, the risk of blockage in a synthetic graft is just a tradeoff. When grafts replace vessels smaller than ¼ inch in diameter (as 75% of them are), however, the thick lining poses a crucial risk. An estimated 300,000 Americans, especially those with poor peripheral circulation, could benefit each year by receiving better small blood vessels.

Researchers have approached this problem by trying to line artificial vessels with materials that preclude clotting. Scientists at Tufts University and Indiana University have seeded Dacron grafts with endothelial cells taken from the linings of patient's blood vessels. These cells form a reasonably smooth lining and produce compounds that help regulate clotting. Meanwhile researchers at the Cleveland Clinic Foundation are developing cardiac vessels about ⅛ inch wide out of a microporous structure of polyurethane and silicone that has been biolized.

Brown University researchers are likewise trying to sidestep the issue of biocompatibility by creating bioresorbable blood vessels. Because these vessels would eventually disappear, leaving natural replacements, clotting would be much less of a problem. The natural vessels would consist mostly of collagen; it is uncertain at this point whether such scar tissue would be pliable enough for this purpose. "Natural blood vessels lose their flexibility with disease or aging, but they still function," says Dr. Sasken. "We really don't know yet what kind of leeway we have with their pliability."

When their experiments with dogs indicated that the grafts were deteriorating too soon, the Brown group switched from complete grafts to bioresorbable coatings on Dacron and Teflon vessels. "The test results on both types could be more promising than they look right now," says Dr. Sasken, "but those in mice show that we have the ability to prolong resorption time."

Presently the most viable alternative to using standard synthetic grafts is

Lyman's method of making new clot-repellent surfaces. Lyman began by asking himself why blood clots and how to design polymers to control that process. He noticed that blood platelets do not stick to surfaces coated with the protein albumin. By manipulating single molecules or atoms, Lyman rearranged the chemical groups on polymer surfaces to attract the albumin and make it stick, thereby preventing clotting. Using a new process of their own, Lyman's group made a copolyurethane material into a white, spongy, elastic blood vessel. The polymer technology is so involved that Lyman had to form his own company, Vascular International, to make vessels for use in human trials. Because he is worried about competition and prematurely raised expectations, Lyman tries to say as little as possible about his work; he does note, however, that his group has made vessels with diameters of 2 millimeters—slightly more than $1/16$ inch. They have even had some success implanting vessels the diameter of a paper clip wire.

The grafts have surfaces ranging from smooth to textured. "This has nothing to do with their effectiveness," says Lyman. "What's important is the surface structure of the polymer and the chemical groups that are sticking up and exposed." Textures that can be felt are measured in microns; in contrast, these surfaces are measured in Ångströms—millionths of an inch. On that scale even smooth surfaces look like mountain ranges.

The surface characteristics of the polymer had to be balanced with others required to make the material elastic. Each juncture between a natural and a synthetic vessel is a trouble spot: the suture line should be angled, the thread a monofilament, and the stitches uniform. If the graft is not flexible enough, it will not pulse like the natural vessel to which it is attached. This discrepancy damages the juncture, causing clots to form. "It is the same problem as clotting anywhere else," says Lyman. "Thickening at the juncture is tolerable in large diameter vessels because they have room. With smaller diameters, it becomes critical—below 6 millimeters, it's a no man's land where synthetic grafts have not worked well." Lyman's group, however, had made synthetic vessels so pliable that they pulse just like natural ones. "The new polymers we've developed also have tissue ingrowth into the juncture and throughout the graft," he says, "so unlike the old Dacron grafts, they don't depend on the suture line."

The grafts have performed well in animal tests. To obtain data they needed for the FDA, the team had to sacrifice several animals in which they had implanted grafts 15 to 34 months previously. The vessels were still working beautifully.

The FDA has approved the polymer vessels for human tests. Prospective candidates undergoing reconstructive surgical procedures must have no natural vessels suitable for transplantation and must otherwise face amputation. Lyman's team is analyzing the grafts' mechanical properties and elasticity as well as the healing process, which was different in each species of animal tested. "So far we're encouraged by the results of the human trials, which are consistent with what we've seen in animals," he says.

Less encouraging are his chances of getting enough funding. Like Geyer,

Lyman has in the past had a grant application denied because the reviewers believed that he was proposing the impossible. Initially he had found farsighted sources willing to fund a broad effort ranging from basic research like surface chemistry and polymer synthesis to surgical implantation in animals. The situation has become more difficult, however.

This lack of funding is especially unfortunate because Lyman's work, like that of Geyer, Yannas, and Dr. Chang, has widespread applications. Theoretically his approach could result in an infinite number of new biomaterials, many of which Dr. Chang could then use to make cells. The principle for making blood vessels could also be applied to the fallopian tubes, bile duct, ureters, bladder, and intestines and might improve existing prostheses for the trachea and esophagus. Although these vessels are not blocked by life-threatening substances like blood clots, they pose formidable challenges in polymer chemistry. "It seems, though, that the blood vessel studies that took us 15 years are beginning to accelerate the other areas we're looking at," says Lyman.

As we have noted, some of these other areas include better understanding of cell growth and more sophisticated diagnostic techniques afforded by custom-made material surfaces. The ability to grow specific cells would be a major step toward regenerating whole organs. Meanwhile, we can investigate some of the ways in which organs, other body structures, and their functions are being replaced now.

9

Bionic Breathing and Speech

Many of us have trouble breathing due to asthma or allergies; fewer know what it is like to live with only one weak lung or to have a windpipe suddenly become too narrow to transport sufficient air. Similarly, the sore throat or laryngitis that may strike us temporarily voiceless is not nearly as frightening as the prospect of never being able to speak again. Fortunately, the development and use of prostheses for breathing and speaking are decreasing the number of people in either predicament.

Artificial Voice Production

The larynx is a hollow cartilage at the top of the trachea that contains the vocal cords. It acts primarily as a protective mechanism for the trachea (windpipe) that permits swallowing without choking. The larynx is also a valve through which air moves in and out of the lungs; by allowing sufficient air pressure to build up in the chest, it enables us to lift heavy objects.

Finally, the larynx is the source of voice production. When a person exhales to speak, the vocal cords close and vibrate, making sound that the mouth shapes into words.

If cancer or some other condition requires a total laryngectomy, the larynx, including the vocal cords, is removed. The trachea is diverted to the neck for breathing through a stoma, completely separating it from the esophagus. "With a stoma, you can breathe, but you can't blow your nose or cool your soup," says Lou Catalani, a 58-year-old driver for Purolator who lives in Stow, Ohio. "You also lose much of your sense of smell." Deprived of the filtering effects of the nose, many laryngectomy patients need to humidify the air in their environments. They also need to select an alternative means of voice production.

"There is no way to restore laryngeal function completely following total laryngectomy," says Melinda Harrison, M.A., CCC-SP, Coordinator of Speech-Language Pathology in the Department of Otolaryngology and Communicative Disorders at the Cleveland Clinic Foundation. An alternative location for voice production can be generated, however, by the hourglass-shaped muscle system called the pharyngoesophageal (PE) segment, which is located in the lower throat and upper esophagus.

Like the vocal cords, the PE segment requires air flow to vibrate. Until a few years ago the conventional method of delivering air to the PE segment, called esophageal speech, was the sole means of voice generation. The laryngectomy patient would take the small amount of air that could still be inspired through the mouth, and by pressurizing the lips or pushing the tongue against the palate, inject the air into the esophagus, thus creating a vibration, or voice.

Only about 50% to 60% of laryngectomized patients are able to learn esophageal speech, and only about 10% to 15% of them develop excellent voices. Scarring from the surgery can cause tightness in the esophagus that often interferes with the ability to produce voice. Other physiological factors can likewise interfere with esophageal speech acquisition. "Motivation is a critical factor to successful voice restoration, but even many well-motivated patients find esophageal speech extremely difficult to master and often don't achieve optimal voice. The speech may be understandable, but it tends to be husky, rough and deep sounding, with little variation in pitch or volume," says Harrison. "My one thought after my laryngectomy was to develop the best speech I could," says Lou, "but even though I practiced, the only esophageal word I learned was 'Scotch.'"

VOICE PROSTHESES

For many years, researchers have been working on various surgical procedures to help patients acquire esophageal speech. All these procedures attempted to establish a means by which lung air would be the activating power for the procedure. Such attempts were typically fraught with complications. A breakthrough came 5 or 6 years ago, when Mark I. Singer, M.D., then at the Indiana University School of Medicine, and Eric D. Blom, Ph.D., previously of the VA Medical Center in Indianapolis, developed a relatively simple surgical approach that used an equally simple prosthesis. The surgical procedure involves puncturing the back wall of the trachea to make a tract into the esophagus. This is a blind opening into which a one-way valved prosthesis can be placed. The device allows air to be diverted into the esophagus; at the same time it prevents food, liquids or saliva from entering the trachea and keeps the tract patent.

The basic Blom-Singer prosthesis is a small T-shaped Silastic device. The crosspiece is flat; the stem is a small tube slit at the far end. The tube has a hole in it just beneath the crosspiece. The T is placed horizontally through the stoma and tracheoesophageal tract with the hole facing down. When the wearer covers the stoma with a finger, exhaled air enters the tube through the hole and is diverted into the esophagus through the slit, which acts as a one-way valve. Since the slit opens like a duck's bill as the air is sent through, the device is often called a "duckbill prosthesis."

Blom and Dr. Singer, now of Head and Neck Associates in Indianapolis, have continued to modify their surgical procedure and voice prosthesis. Early critics of the device pointed out that the finger occlusion was not an ideal method for producing speech, so Blom and Dr. Singer developed a tracheostomal breathing valve that is placed over the voice prosthesis in the stoma. The

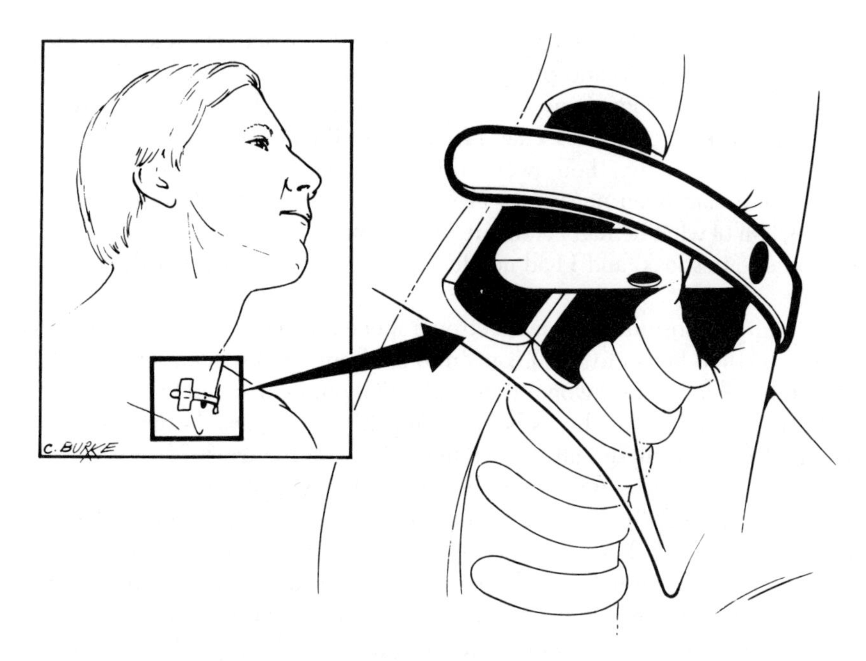

The Blom-Singer voice prosthesis. A breathing valve can be placed over the stoma once the prosthesis is in place.

valve contains an inverted rubber diaphragm that closes during exhalation to divert the air through the voice prosthesis. Those who are able to wear the valve successfully find that their manner looks more normal without finger occlusion and speech is often easier to produce.

The medical literature on voice prostheses is quite repetitive, since most models follow the basic concept of the Blom-Singer device. Alterations are designed to help retain the prosthesis in the tract and maximize voice production.

The Blom-Singer prosthesis, which costs approximately $15 and lasts a few months, is available in several lengths, sizes, and variations to suit individual patient needs. It requires little maintenance. "I replace the valve every day but need to clean the prosthesis only twice a week," says Lou. "It takes only a few minutes. The more I heal, and the better my stoma gets, the less cleaning I have to do."

The operation for voice restoration is done at a few institutions immediately at the time of the laryngectomy but at most facilities a few months later. "I think it would be better to wait until the stoma begins to heal," says Lou. Once they undergo the voice restoration procedure and the prosthesis is in place, patients are often told that they sound much as they used to. "This is probably because this method of voice production enables the patient to speak with more natural inflection, stress and phrasing," says Harrison.

Lou had his laryngectomy 3½ years ago because of cancer. (He used to smoke four packs of cigarettes a day.) He entered the hospital expecting to have a partial laryngectomy, then awoke from the operation to learn that a total laryngectomy had been necessary. "I wrote the surgeon a note telling him to put back half of what he had taken out," says Lou. "I was totally unprepared for being unable to speak, and I had no idea at the time that voice prostheses were available."

Lou left the hospital with an electronic speech device, underwent radiation treatments, then began to learn conventional esophageal speech. Once he had completely healed from the operation and radiation, he decided to try the voice prosthesis method. "I've always been a big gabber," he says. "I drive my wife crazy, and I talk to myself all day while I'm driving the truck."

Using his prosthesis with finger occlusion, Lou was able to speak well, but he was not satisfied. "I'm a perfectionist, so I noticed when people wondered why I pressed my neck to talk," he says. "And as an Italian, I found it hard to talk without using both hands." He therefore elected to be fitted with the tracheostomal valve, which has given him excellent results for the past year.

"The valve has its problems, but I wanted it badly enough to overcome them," Lou says. "It doesn't want to stick in very hot, humid weather—it kept coming loose when I took a vacation to Atlantic City in September, but everyone out there is so noisy that I couldn't talk much anyhow. The biggest problem I have with the valve is that sometimes I have to sneeze and don't open it fast enough, so the seal breaks."

Lou believes that he has made the best choice for his voice restoration. "I can do much more than even a really good esophageal speaker," he says. "I can talk longer and vary my voice much more. The valve allows me to talk better than I can with my finger; I can blow my nose a bit, and I've even started to whistle. If they come up with something better, I'll be the first to try it, but meanwhile I love what I've got and expect to use it for a long, long time."

The surgical-prosthetic method of voice restoration has relative disadvantages as well. These problems include the need to have further surgical procedures, use finger occlusion if a stomal valve cannot be worn, and buy and maintain a prosthesis and wear it continuously to prevent the tract from closing. Some laryngectomy patients do, in fact, prefer the conventional method of esophageal speech. "I think there's some resistance to the procedure, especially by good esophogeal speakers, who don't need it. There's also something of a protective attitude on the part of people who relearned to speak on their own, without needing a crutch," says Harrison. "A lot of them don't want to go through that extra little step of surgery after the big operation," says Lou.

If they know about the devices and choose them, most of the 10,000 or so patients who have laryngectomies each year can benefit from voice prostheses. The results of the surgical-prosthetic method are so excellent that it is considered a primary procedure for those patients who want it. "Our success rate at the Cleveland Clinic has been about 85%," says Harrison. "There are few contraindications to the procedure, and we've had very few complications. The procedure has changed my whole attitude toward working with laryngectomy patients because I know that the ultimate results of their treatment will be so much better than they used to be. Nothing is like a normal voice, and we'll probably never replace it, but we're getting much closer."

"I'm personally tickled silly that voice prostheses came out just in time for me," says Lou. "I think my device is fantastic; I tell people about it, and I want someday to go to Indianapolis to show Blom and Dr. Singer what they've done for me."

ARTIFICIAL LARYNGEAL DEVICES

In addition to prostheses used in surgical voice restoration, laryngectomy patients can use artificial larynges that are recommended early on in rehabilitation. Most patients use either an electronic or a pneumatic device within days or weeks after surgery. The electronic larynges are either cervical (held against the neck) or intraoral (positioned to produce sound through the mouth), the type Lou took home with him. Intraoral pneumatic models require the patient to blow through the stoma to vibrate a rubber diaphragm; the sound is put in the mouth through a tube.

"It's great to know that right after surgery you can make some kind of noise and even use the phone, because not being able to speak really works on the mind," says Lou. Some patients continue to use artificial larynges instead of switching to voice prostheses; their voices, however, sound mechanical. "The devices definitely serve a good purpose, but I always make sure that a patient is well educated about the alternatives before I'm satisfied that he or she really doesn't want to undergo the surgical-prosthetic method," says Harrison.

FUTURE DEVICES

More exotic electronic voice prostheses and larynges are either in progress or in the planning stages. Medtronic, Inc. is developing a digitally synthesized speech source consisting of a tiny speaker implanted between the vocal cords and the spine, which receives its power and radio signal from a pocket-sized transmitter. The patient uses muscles in the mouth, tongue, and throat to produce the speech, which may be tuned to imitate the qualities of his or her natural voice.

Another type of electronic larynx, a battery-powered gadget the size of a half dollar, is placed against the palate. The patient speaks by flicking the tongue against moisture-sensitive electrodes.

A proposed model that will take years to develop consists of electrodes implanted in the throat muscles and connected to a pocket-sized computer and speaker. The electrodes would read the muscle signals that precede speech production. Of course, scientists would first have to map and interpret these signals themselves to program the device.

Another type of artificial larynx, now being tested at the Cleveland Clinic, works through functional neuromuscular stimulation (FNS), which we will discuss in detail later. The device is not for laryngectomy patients. "A certain class of patients with cancer or scar tissue from surgery suffer paralysis of the nerves and muscles that activate the vocal cords," says Jacobs. "They either can't breathe at all or feel very uncomfortable when they do, so they need a permanent stoma and can't talk." In cooperation with the Department of Otolaryngology and Communicative Disorders, Jacobs' group is developing a device the size of a thumb that will be implanted in the neck. It will make the vocal cords open during inspiration and close during exhalation, as they do normally.

While implanting the device, the surgeon takes two nerve muscle pedicles, places an electrode cuff around each, and transplants them to the muscle that governs the vocal cords. Once healed, the transplanted nerves can control the muscle and, indirectly, the cords. The group has devised two different methods for stimulating the nerve cuffs so that vocal cord opening can be synchronized with breathing. One uses a strain gauge, the other implanted electrodes to measure transthoracic impedance. Two wires will extend to either the gauge on the trachea or electrodes in the intercostal muscles, and two output wires will connect to the nerve cuffs.

"We've implanted the device on one side of dogs, using the denervated side as a control," says Jacobs. "The next step is to denervate both vocal cords, do two nerve transplants with an electrode cuff on each, stimulate the cords on both sides and see—or rather hear—what happens."

The Artificial Trachea

The trachea is not a highly popular body structure on which to operate: it can be problematic, and its proper treatment requires an intricate knowledge of all its surrounding structures. In general, physicians do not usually suspect that a tumor in the trachea, like one in the larynx, could be the reason that a patient finds it difficult to breathe.

That condition became clear, however, to William E. Neville, M.D., Director of Thoracic Surgery at the University of Medicine and Dentistry, when he was practicing at Hines VA Hospital in Chicago. As a thoracic surgeon Dr. Neville found himself always maneuvering around a patient's trachea. One of the hospitals at the Chicago medical center was exclusively for patients with lung problems. "I saw many patients whose lungs or tracheas could not be removed because of cancer," says Dr. Neville. "Others couldn't breathe after they were taken off respirators. Manufacturers didn't make endotracheal tubes as well as they do now, and after removal of a tube, stenosis [narrowing] of the trachea often occurred." These problems prompted Dr. Neville to begin work on an artificial trachea for use when the natural tube could not be repaired surgically.

The Neville Tracheal and Tracheo-Bronchial Prostheses, the only artificial tracheas on the market today, have been used all over the world. Each prosthesis is made of silicone copolymer fashioned into either a straight or a bifurcated tube. It is resilient, yet rigid enough not to collapse under the pressures of respiration. The tube is airtight, nonreactive, smooth inside, and impervious to tissue ingrowth and bacteria. Sewing rings at either end, like those in heart valves, are made of thin silicone tubing wrapped in Dacron and bonded to recesses in each end of the tube. The tube diameters narrow at both ends to fit inside the natural trachea. The prostheses come in different lengths and widths. They can be trimmed and sutured through the silicone or attached end to end. Once implanted, a tracheal prosthesis is rapidly incorporated within the surrounding tissue.

Dr. Neville implants the artificial trachea as a last resort, usually for patients with severe narrowing of the natural trachea due to a benign or malignant condition. He has used it on about 75 patients and estimates that about that many individuals need such a device each year. One patient that he treated in Chicago has had an artificial trachea for 14 years. "Her doctor was a thoracic surgeon," says Dr. Neville. "He diagnosed her as having a benign tumor at the bifurcation of the natural trachea and flew her to my hospital in his private plane. We did her surgery as an emergency procedure." Dr. Neville reports that surgeons in Poland use the prostheses for patients with inoperable cancer who cannot breathe—the surgeons simply implant the devices right through the tumors to establish the airways.

Although artificial tracheas are not in great demand, their use does dramatically save lives. "I just wish that we had these devices 20 years ago, when we were unable to repair tracheas and people would die," says Dr. Neville. "It was a very difficult situation."

This would be an understatement applied to the case of Lillian Petrocelli, a 44-year-old former nurse from Lincroft, New Jersey, who has survived with an artificial trachea for 6 years. Lillian divides her life into two stages: Before and After Europe. In October 1978 she went to Italy with her mother and a friend. "I've had asthma all my life that developed into bronchiectasis and emphysema," she says. "In Italy, the air conditioning was off, the heat hadn't come on yet, and the Pope's coronation was taking place. Rome was full of the exhaust of thousands of little cars."

Lillian had an asthma attack and was taken to a local hospital unable to speak. Her companions did not communicate well with the physician, whose incorrect treatment sent her into a coma. "I was intubated with an endotracheal tube, which should be left down for only 24 to 48 hours," Lillian says. "They left mine in for 6 weeks." The intubation caused an endotracheal-esophageal fistula (a hole connecting the trachea and esophagus) that was not discovered for several months. Meanwhile Lillian was being given intravenous medications but no caloric fluids; a physician later told her that she came within 6 days of starving to death. She also developed a chronic *Pseudomonas* infection in one lung.

Shortly after Thanksgiving Lillian and her life support systems were flown to Holy Name Hospital in Teaneck, New Jersey, where for the second time she underwent an operation for stress bleeding caused by all her medications. Her endotracheal tube had finally been replaced by a regular tracheotomy. About 2 weeks later Lillian awoke from her coma to find herself on a respirator, unable to speak, and paralyzed from the neck down.

At the end of February 1979, as Lillian gradually recovered, her physicians removed her tracheotomy tube. She was able to eat but could not drink liquids without choking. A barium x-ray examination finally revealed the fistula that was allowing fluids to seep directly into her lungs. The healed scar tissue surrounding the fistula was too massive for surgical reconstruction, so Dr. Neville was called in to remove the trachea and implant the straight prosthesis. "I awoke from the surgery and thought, 'Well, I can swallow; I can breathe, and there's no respirator, so I guess I made it,'" Lillian says. "The prosthesis has been in ever since, and I've had no problems with it."

Lillian notices the tube if she tries to bend back her head or swallow several big pills at a time. "I don't have the mobility that people normally do for swallowing," she says. "Other than that, there's no pain or discomfort from the device. It seems to have no effect on my asthma, but obviously I can never have another tracheotomy." It took Lillian a good 18 months to relearn to walk with a cane and a walker, but now she is able to function relatively well when her *Pseudomonas* infection is repressed. Fortunately the chronic infection has not spread to the prosthesis site.

"I'm just thankful that Dr. Neville's procedure worked and I'm alive," Lillian says. "Depending on circumstances, I would definitely recommend the prosthesis to others. It saved my life and has been fantastic for me. Thank God for Dr. Neville."

Artificial Lungs

The best known function of the natural lung is that of exchanging the carbon dioxide in the blood for oxygen. Each lung contains tiny air sacs suspended in a net of capillaries so narrow that only one red blood cell can squeeze through at a time. Every 4 minutes, 6 quarts of blood containing 25 trillion red cells pass through the lungs. Each cell excretes carbon dioxide and absorbs oxygen through the delicate sac membranes. If these membranes were spread out, they would blanket almost 85 square yards.

THE HEART-LUNG MACHINE

During open-heart surgery a heart-lung machine consisting of a blood pump and an oxygenator takes over the heart's pumping action, allowing it to lie still and bloodless. The oxygenator does less work than the natural lung because it feeds pure oxygen to a patient whose body functions have been suppressed by anesthesia. In surgical procedures the machine most commonly used is the bubble oxygenator, which bubbles oxygen into the blood, removes carbon dioxide, defoams the blood, and returns it to the body. Because bubbling damages blood, a membrane oxygenator is more appropriate for prolonged periods of healing and recovery. This device contains a thin membrane through which the blood exchanges its gases. More natural than bubbling, the procedure causes fewer complications.

The spiral flow or Kolobow oxygenator, marketed by Sci-Med Life Systems, Inc., was invented by Theodore Kolobow, M.D., a former associate of George Clowes, M.D., who had introduced the membrane oxygenator in clinical practice in America during the mid-1950s. Now at the National Heart Institute, Dr. Kolobow originated the concept of extracorporeal carbon dioxide removal and was involved in the revival of assisted respiration that occurred in the mid-1970s.

THE RESPIRATOR

Another form of lung support is the respirator, a bedside tank that supplies oxygen-rich gas through a tube into the patient's throat. A person who has lost a lung and is left with a weak one may spend the rest of his or her life in this condition. "The number of lung patients chained to beds is not large, but they are not being served, and we'd like to get them ambulatory," says Richardson. To do this, the Brown group is trying to develop an implantable artificial lung, which presents a formidable challenge. "We need as much gas transfer on as small a surface as possible," says Dr. Galletti. "It's a tall order."

THE BOOSTER LUNG

Dr. Galletti is interested in this project partly because he was a resident during the epidemic of silicosis that hit Switzerland after World War II. While in Zurich, he saw respiratory patients sleep out on a porch in midwinter because they suffocated in heated rooms. "Compared to cardiac or renal failure, terminal respiratory disease is in some ways the most frightening form of death," he says.

The Brown researchers imagine the artificial lung as a substitute for a removed lung and a support for the remaining natural one. It would be implanted for 6 months to a year, during which time the natural lung would provide residual biochemical activity and presumably show its gratitude by regaining health. This concept is based on the lack of scientific evidence that a person could survive with a prosthetic replacement and no lungs. "The lung does much more than exchange gases," says Dr. Galletti. "It may have as many as 40 different types of cells and seems to be as complex as the liver. We don't fully understand many of its functions."

The team has tested the booster lung in laboratory animals. At first they used so-called miniature swine, which weigh about 175 pounds, are highly intelligent, and can be nasty to people they do not trust. When approached, the pigs indicated loudly that their natural lungs worked just fine, and they put up such a fight that the laboratory assistants refused to go near them. Finally the researchers switched to sheep, which like pigs are human size but unlike them are much more docile.*

One might hope that these pigs, along with the one that escaped from Lyman's laboratory and went speeding down the obstetrical-gynecological floor of the Utah hospital, have gone on to be role models in the animal liberation movement.

Surgical procedure. As remarkable as the device itself, the surgical procedure used for the artificial lung involves two steps, preparation and implantation. The sheep's right lung is removed, and the cavity is lined with a split-thickness skin graft taken from its abdomen. "Split-thickness" skin goes to a depth just above the hair follicles, which otherise would fill the chest cavity with a furry mess. The skin is meshed so that when pulled, it spreads out to cover about three times as much area as the initial graft. The skin heals, fills in the spaces, and somehow vascularizes. Its purpose is to provide a biocompatible lining for the lung. "We found out by serendipity that it works better than we had any right to expect," says Richardson.

Respiration is provided by a stoma cut between the ribs, so the artificial lung must be connected only to the circulatory system. The pulmonary artery and vein are clipped and connected by a long shunt that creates flow resistance so that enough blood goes to the remaining lung for oxygenation. The shunt is bonded to a helical Teflon spring that keeps it from being crushed by the pressure in the cavity created by sebum from the skin graft.

The second half of the procedure occurs in 3 weeks. The surgeons open the chest between the ribs for access to the midcavity. When the graft has taken well, the skin has grown up into the vessel-shunt connections to seal them and keep out bacteria. It has also stiffened the mediastinum (a membrane separating the two halves of the chest) so that the left lung and the heart stay put. The surgeons remove any lumps of sebum, select an implant of the proper size, and connect it to Y ports in the artery-vein shunt. The major length of the shunt itself is removed; the implanted lung now provides enough flow resistance to keep an adequate blood supply moving through the contralateral normal lung. One lobe of the artificial lung points down toward the diaphragm, the other up toward the head. Because it will not get soggy and sag like the natural lung, the prosthesis is more balanced and predictable.

Lung design. The booster lung itself consists of wide-pore tubing resembling white macramé cord that is coiled inside a plastic bag serving as a pleural sac. To approximate the blood flow resistance of the remaining lung while saving

*Sheep can also be conscientious. One animal that received an early version of the artificial lung was placed under 24-hour surveillance. A researcher visited the laboratory at 4 AM to find the technician sound asleep and the sheep attentively watching *him*.

space, the tubing has been woven in a serpentine pattern with continual reversals of curvature that can be tightly looped. Because all the tubing connections are at one end of this pattern, the prosthesis can be shaped into a lobe that can itself be looped. It still has a central connection to the shunt that imitates the natural lung structure.

This pattern requires tubing that does not bend or kink when sharply curved. The researchers use a specially prepared Teflon tubing, the walls of which are like Styrofoam structured out of small fibrils; Dr. Kazuo Tanishita, a former Brown student now teaching at Keio University in Japan, weaves the material and sends it to them. The tubing length can be compressed very easily and tightly; when it is coiled, the inner side compresses evenly and does not buckle. It is so pliable that it can easily be knotted and untied with no effect on its shape. Blood flows through it freely.

The lung design can cause difficulties. The tube membrane, with its microfibrillar structure of long, thin slits, must selectively filter the right substances. As used in the lung, the simplest double-serpent pattern has four ends to it—two inlets and two outlets—that must be combined into common ports off the artery-vein shunt. Blood platelets may very well race to form clots at these junctures. "Over the long term, tissue ingrowth would be more of a problem than clotting, but I have something up my sleeve to tackle that," says Richardson.

Fluid is a potential problem. It gets into the natural lung from body fluids in the blood vessels; when not automatically drained, it can cause pulmonary edema or waterlogged tissue that interferes with gas exchange. In the artificial lung, however, fluid cannot easily get through the tubing wall into the gas exchange area. A hole would have to be made in the tubing (which the platelets would temporarily plug up, conveniently enough), or water vapor passing through it would have to reach the dewpoint and condense. If the lung were adequately ventilated, and its owner did not breath cold gases, fluid would not collect inside the plastic bag; if it does, it can be drained through a tube inserted into the stoma.

Fluid that collects outside the bag is "artificial"—it comes from the skin used to line the cavity. The Brown researchers have been using Scottish porridge to test drainage systems that might be used to remove it.

The ECMO trial. The main problem posed by the artificial lung is the common one of inadequate funding. Lung research, which is out of vogue, received a major setback when the National Heart, Lung, and Blood Institute (one of the National Institutes of Health) conducted a nationwide test during the mid-1970s. The ECMO (extracorporeal membrane oxygenation) trial tested patients with end-stage respiratory disease; those placed on artificial lungs lived a week longer than those in the control group. "The study was a good example of a clinical trial that was ill-conceived and ill-timed," says Dr. Galletti. "The trial concluded that using artificial lungs for people with end-stage respiratory disease prolonged their lives and allowed their lungs to progress to a new, yet unknown form of terminal pathology. Those involved felt that the first priority should be to direct efforts toward understanding the cell biology of the lung to interpret this process. ECMO *should* have concluded that it is possible to support the lungs artificially, and that we should simplify our mechanisms for doing so and find out what caused the deaths."

Two foreign groups decided that ECMO was a typical example of the American tendency to draw the wrong conclusion from the right facts. By questioning traditional approaches, these groups have had remarkable long-term results with existing artificial lungs, which are designed only for short-term support.

Joel D. Cooper, M.D., Chief of Thoracic Surgery at Toronto General Hospital, recently had an American patient poisoned by paraquat, an herbicide that is very specifically toxic to the lungs. Dr. Cooper placed the man on an artificial lung, with hemofiltration to remove the toxin, for 6 days. Then he transplanted a new right lung. After a week the new lung had become poisoned by toxic residue, so Dr. Cooper put the patient back on the artificial lung for 6 days and performed a second, successful transplant. A year later the man is alive and well.

A group in Milan meanwhile decided that patients in the ECMO trials had been killed by inappropriate respirator treatment. They decided to lengthen inspiration and shorten expiration (both reversals of the standard procedure) and adjust their own extracorporeal gas exchange systems to emphasize carbon dioxide removal rather than oxygen supply. The group kept patients alive for 3 weeks on artificial lungs, and some of those patients survived.

We can conclude, then, that artificial lungs could be very useful indeed. So too can devices that are neither totally artificial nor true "organs."

10

Regular or All-Purpose Organs?

Although it refers specifically to completely man-made devices, "artificial organs" is also a generic term covering quite a diversity of machines and gadgets. An artificial organ can be a small plastic capsule filled with natural tissues, a bulky machine hooked to a patient's circulatory system, or a special structure dissolved by the body once its purpose is fulfilled. We have mentioned some examples and will be seeing others, so it is appropriate to explain a few of their principles here.

Hybrid Organs

A hybrid organ combines artificial materials with natural tissues that behave like the organ from which they were taken by biopsy. Researchers might try to create a hybrid organ when they have not completely identified or are unable artificially to duplicate a natural organ's functions.

We will see that hybrid organs like the liver and pancreas have consisted of live cells from those organs seeded on one side of a selectively permeable membrane that protects them from blood flowing on the other side. The cells can cling to the outside of a vessel shunt or be enclosed in a membrane capsule and implanted in a well-perfused area. Cell products can diffuse through the membrane in response to blood stimulus, but larger immunological agents in the blood cannot cross the barriers to attack the cells themselves.

In a sense, hybrid organs combine the advantages of both transplants and prostheses while minimizing the disadvantages of each. "For the time being, we have gone as far as we can go with pure technology," says Dr. Galletti. "To make significant progress now we need brand-new technologies, or materials, or in some cases concepts. For organs that perform complex metabolic functions, the hybrid approach is presently more workable than the artificial." It is also equally as fraught with difficulties and, since hybrids must be grown in laboratories, rather expensive.

Bioresorbable Organs

Like the "artificial" skin and blood vessels described earlier, a bioresorbable organ acts as a scaffold for the regrowth of a missing tissue or organ. This framework, which has been seeded with natural cells perhaps treated with drugs to enhance their growth, slowly resorbs as the organ grows back. The procedure resembles that of using an artificial organ to maintain a patient awaiting a transplant, except that it is precisely finite. The trick is to find a material that is both bioresorbable and capable of supporting cell growth, then time its disappearance with the growth rate of the natural tissue.

"We could use this principle for a cardiac patient with a myocardial infarction, for instance," says Dr. Galletti. "We could take a biopsy of his myocardium and grow the tissue on a little sponge attached to a tube. We would then implant the sponge, place the tube in the aorta, and allow both to disappear as the patient grows a new myocardium. None of the technical problems with this procedure is insoluble."

It is easy to imagine the hybrid and eventually the bioresorbable organ approach superseding that of purely artificial organs. Until more breakthroughs occur in microbiology and immunology, however, artificial organs will maintain the lead in the medical future. Some major examples are the subject of this chapter.

Artificial and Hybrid Livers

In a sense the liver is an appropriate target for replacement with an artificial organ. It is vastly overdesigned: about one tenth of it could perform the functions necessary for survival. Over half the liver can be removed, yet it will cooperatively regenerate if the disease has been caught in time, before there has been too much scarring. An artificial liver, then, could last a patient until either his or her own liver regenerates or a donor organ becomes available for transplant. It could also support a patient's metabolism in about half the cases of transplant surgery, during which the recipient has to function without a liver.

In another sense the liver is the worst organ to try to imitate with an artificial device, even for a finite period such as in acute hepatic coma. It is both large and enormously complex. The liver cells have thousands of functions divided zonally by their distance from the arterial supply and temporally by their varying oxygen contents. "Unfortunately we know very little about what the liver does," says Dr. Nosé. "Anyone who claims to know all about liver function is fooling himself or herself."

Basically the liver seems to perform three functions: regulation of carbohydrates, lipids, clotting factors, and protein metabolism, among other things; synthesis of all kinds of important chemicals; and detoxification. It seems likely then that the most sensible way to replace the liver would be with a hybrid rather than an artificial organ. It would take much natural liver tissue, however, to perform the crucial functions in the time that would be available. Either the patient's blood would have to be perfused through the liver cells, or they would have to be injected into the body.

Unfortunately, liver cells last only a few days in culture and lose some of their biochemical functions in the meantime. "Someone who figures out a way to maintain functioning liver cells in culture will have done so by chance, no matter what he or she claims," says Richardson. Not only are the cells difficult to maintain, they also require a specific geometry to their organization because they need contact and like to influence each others' function. "Liver cells are very fussy about what they touch," says Dr. Galletti. "They sense the micro-geometry of the material, and they respond by changing their shapes and becoming either very stupid and passive, or very intelligent, actively exchanging complex molecules. We don't know what characteristics of polymers give them those signals." Overcoming these difficulties and testing hybrid livers in laboratory animals would raise another issue: there are good animal models for diseases like diabetes but no comparable models on a suitable scale for acute hepatic failure.

A replacement for long-term rather than acute liver disease would complicate things further. The liver manufactures proteins, including clotting proteins in the blood. With the hybrid principle it would be difficult to send such large molecules into the bloodstream without using a membrane so porous that immunological agents could sneak through and devour the cells. "It's the cells themselves, rather than their products, that the body considers foreign," says Richardson. "We might need to develop an artificial lymphatic system that would enable us to take materials from a region, in this case extravascular, and deliver it someplace where it might be picked up by the circulation without exciting an immune response."

THE HYBRID LIVER

All this suggests that an artificial liver would be confined to use as a research tool rather than in clinical application. Some hardy souls, however, have tried to make a practical one. Dr. Chang hopes someday to combine artificial cells that detoxify wastes with others capable of performing metabolic functions. In 1957 researchers at the Cleveland Clinic Foundation took liver tissue from dogs and attempted to isolate and maintain the cells. They perfused the liver slices with plasma, then sent the plasma through an animal with liver failure. The technique worked for a few hours, but the project was shelved because of its difficulty.

In the early 1970s three groups of researchers were constructing hybrid organs using the same basic principle with different cells. Researchers led by Richard A. Knazek, M.D., Senior Investigator at the NIH, grew cancer cells in a hybrid device to study hormonal production. As discussed in more detail later, a group at Brown University and some of their colleagues made a hybrid pancreas using live beta-cells and have likewise proposed a model for a hybrid liver. Dr. Wolf worked on a hybrid device at the New York Blood Center, where he had access to liver cells. "My approach was to identify the engineering problems and solve them one by one as we encountered them," he says.

"We ran into generic problems, so it was really irrelevant what kinds of cells we used." Dr. Wolf has elected not to renew his most recent NIH grant for the project, in part because of the technical difficulties he encountered in carrying out further experiments.

At this point, then, it is impossible for prostheses to synthesize or regulate as the liver does. Artificial liver research does not have high priority—and for once it is the researchers rather than government agencies or manufacturers who are not interested. "I think that most certainly it will be possible one day to make a hybrid or artificial liver, although definitely not in the immediate future," says Dr. Wolf. "The technology we attempted to apply will be developed by biotechnology companies for other purposes. The artificial organ will reemerge as a spin-off of the commercial ventures where they're trying to make specific proteins for a variety of reasons."

There is far greater hope, however, of mimicking the detoxifying function of the liver. Researchers at the Cleveland Clinic Foundation and elsewhere are using a system that can not only support liver patients but also treat, and perhaps arrest or reverse, many other diseases. This approach, in fact, may change the entire philosophy of artificial organ technology.

PLASMAPHERESIS

Plasmapheresis uses a procedure similar to that of kidney dialysis, except that it removes large toxic molecules rather than small ones. It detoxifies blood plasma in a two-step procedure of plasma separation and filtration.

The patient's blood is first drawn into a centrifugal or membrane device to separate a portion of the plasma from the whole blood. The plasma may be discarded and replaced or treated on-line and returned to the patient. One form of plasma treatment, for instance, is by perfusion over sorbents. Activated charcoal is used to remove various amino acids, mercaptans, and other sulfur-containing materials that a failed liver cannot process. Ion-exchange resins remove bilirubin, a breakdown product of red cells normally excreted in the bile, that can build up in tissues and cause jaundice. The cleansed plasma is then remixed with the blood cells and returned to the patient.

Using membrane plasmapheresis, the Cleveland researchers have temporarily revived comatose acute liver failure patients, some of whom have survived. They have also considerably reduced itching in chronic liver disease patients. "I know that we can do even the most difficult hepatic assist," says Dr. Nosé, "because once we have the plasma, we can do anything the liver can do if we know the specific function of the liver metabolism. Currently we can treat the moderate liver insufficiencies and those created by known metabolites."

They have also saved the life of 37-year-old "Ted Stevens" (who requested anonymity). A technician and part-time model who lives in a Cleveland suburb, Ted found out 8 years ago that he had abnormally high levels of enzymes, bilirubin, and cholesterol. He was eventually diagnosed as having primary sclerosing cholangitis and secondary biliary cirrhosis. Both these diseases cause inflammation, scarring, and hardening of the bile ducts so that Ted's body is unable to excrete waste materials normally processed by the liver. Symptoms usually appear in middle age after some type of infection; Ted's condition may have been aggravated by a particularly traumatic divorce.

Ted's earliest symptoms included what he calls "hideous" itching, "hot flashes," and weight loss. He had a T tube implanted in his liver to divert the bile through a stoma, but the procedure did not work. After the surgery his strength began to fail; one night while dancing at a disco, he lifted his partner to dip her and surprised them both by dropping her instead. He felt generally ill and feverish and began sleeping heavily. Having modeled for Barbizon in Paris, Ted was most upset by his appearance: his skin turned sallow and his eyes yellow; his hair fell out, and he broke out in xanthoma, a rash of hard pimples that obliterated his mustache and eyebrows.

After exhausting other medical and surgical possibilities, Ted went to the Cleveland Clinic Foundation for further tests. The staff took one look at his cholesterol level—2500 milligrams per deciliter—and told him to check in immediately. He was dying.

Plasmapheresis had been used at the clinic for about a year when Ted began his treatments 4 days per week. "My first treatments were hard on the machine because pumping in my plasma was like throwing mud into a car's oil filter," he says. "Every half hour someone had to come in and put in a clean filter." Within a few days, however, the rash on Ted's face began to clear up.

Ted had the 2½-hour treatments 3 or 4 days each week for a few years. Over the next 4 years he was cut down to twice weekly, then once a week, and now he goes to the clinic biweekly. "I've been on this schedule for 4 months," he says, "and my cholesterol level doesn't go as high as it did when I first started. After treatment, it dips to about 175, and within 2 weeks it rises to 400—the level it used to reach in 3 days. If I watched my diet a little more carefully, it would probably go even lower."

This slight evidence that Ted's liver is improving agrees with the stability of his condition and the way he feels. "I'm much better since going on the plasmapheresis," he says. "I haven't recovered all my strength, but if I don't have to do any strenous work, I hardly even know I'm ill. I still party and disco—without lifting anybody, though—and I ran for office last year. Before the treatment, I could hardly walk." Although most patients have to stay in bed for a day after treatment, Ted has been able immediately to return to work or go campaigning door to door.

The plasmapheresis may enable Ted to fulfill his long-range plans to move to New York and resume his modeling career. "Attitude is really important," he says. "Sometimes I get depressed because I have no choice but to come in regularly for treatment. Without these diseases, I wouldn't even be in Cleveland, and we're hoping to decrease my schedule to once a month so that I can get out of here." Eventually Ted wants to try a liver transplant. "I haven't deteriorated one iota in 6 years," he says. "If this detoxification can keep me stable for several more years, by that time liver transplants might be more successful."

Meanwhile, he is grateful for his good luck. "Thank God for the Cleveland Clinic and this treatment—no question about it," he says. "I definitely had one foot in the grave when I went in there. I'll testify to any doctor that plasmapheresis is wonderful."

The detoxifier could be combined with other liver support systems like those currently used in intensive care units. Acute liver failure could be treated daily for as long as necessary. Patients with chronic liver failure could be treated weekly or monthly. "If we could identify what needs to be removed, some kind of liver growth factor could be added that might enormously help cells grow better or improve metabolic status," says James Smith, M.D., Ph.D., Staff Physician in the Department of Artificial Organs and the Blood Bank at the Cleveland Clinic.

Membrane plasmapheresis is a definite step toward an artificial liver. "We have many key components available now, but the technology hasn't yet been optimized," says Dr. Smith. "How fast that happens will depend on how much interest and research this work generates. We could build a practical device, but we would need to do a large trial to get it accepted by the medical community. Along with funding, acceptance is our biggest roadblock."

CRYOFILTRATION

By taking the filtration-sorbent system a step further, the Cleveland group has come up with an all-purpose treatment that is still difficult to do with liver disease but could alleviate an enormous number of other serious disorders.

The treatment, called cryofiltration or cryogelpheresis, recalls the old idea that the body has "humours" that must be kept balanced to ensure good health. Like kidney failure, many diseases (including those of old age) are characterized by abnormal blood chemistry levels, particularly excesses of certain solutes. Researchers are not sure whether these solutes are just by-products of diseases or are actually involved in their processes, but in many cases removing them from blood plasma has alleviated the symptoms. There are drastic ways to do this: plasma exchange alone (in which plasma is replaced with albumin) has been tried on over 50 diseases.

Cryofiltration is unique in that it removes the pathological molecules from the plasma specifically so that the rest can be returned to the patient. Many of these molecules are large proteins that will precipitate from solution into a gel when the plasma is cooled. Working with Malchesky, Dr. Nosé has developed a special cooling system with a macromolecule filter that removes the chilled gel.

This procedure holds particular promise for autoimmune diseases like myasthenia gravis and lupus. An autoimmune disease is one in which the body's protective mechanism turns traitor and produces autoantibodies: substances that attack its own tissues. These can form immune complexes (combinations of antigens and antibodies) that circulate in the blood and seem to function in certain diseases. The levels of these molecules stop rebounding and remain low after several cryofiltration treatments.

The Cleveland team has tested cryofiltration on class 2 and 3 rheumatoid arthritis patients for whom all other forms of treatment have failed. Ten cryofiltration treatments have resulted in a 77% response in patients who seemed past therapeutic help. There were over 50% improvements in subjective evaluations like pain reduction, morning stiffness, walking rate, and strength. These effects have lasted from a month to years, and generally the treated patients have responded better to medication. "We've had eight patients continue on maintenance treatments from 1 to 4 years, many of them after a few years of intermittent therapy once a month or so," says Dr. Smith. "The disease seems to burn out—they go longer between treatments or are closer to remission."

One of these patients is Marjean Wallover, whom we met earlier talking her physicians into replacing both her knee joints. Marjean's arthritis is not confined to her knees. "It travels," she says. "It can go from my wrist to my shoulder to my back—it's a miserable, miserable disease. Only someone who has had it can ever understand the torture we go through."

For 20 years Marjean tried different kinds of drugs and gold shots. As her arthritis worsened, she even made trips to Canada and Santo Domingo to stock up on an effective liquid medication; later she learned that it was just a highly concentrated form of the cortisone that she was already taking. "Arthritis victims are suckers for all kinds of quackery," she says. "We're in so much pain that we'll do anything." In addition to her knee surgery, Marjean had a laminectomy on her spine, an operation that caused complications because years of taking steroids had destroyed her surrounding tissues.

In August 1982 Marjean began cryofiltration treatments at the Cleveland Clinic as part of an FDA trial. Each patient was scheduled for two treatments per week for 5 weeks. Marjean had her first treatment, drove home to Pennsylvania, and began having an unusual, severe reaction. "The joints flared all over my body," she says. "My husband had to call an ambulance to take me to the hospital." From there she called Dr. Smith and informed him that although she wanted to continue in the program, she also wanted to make sure that she did not again experience such a reaction. Marjean was admitted to the Cleveland Clinic and supervised during her next few treatments, which caused no harm and quickly made her feel better.

After completing the series for the FDA trial, the patients began stretching treatments out to once each week, once biweekly, and so on. Marjean is up to one treatment every 3 months. "If I feel bad before then," she says, "I just call and arrange to go in sooner. Usually it makes me feel better right away." The length of each treatment is determined by the patient's weight and the amount of plasma to be filtered. Marjean, who is an average height and weight, is treated for about 3 hours.

"I would recommend that an arthritis patient investigate cryofiltration if other treatments are not working," says Marjean. "But people should be prepared for some nuisances." For one thing, the patient must find a rheumatologist who accepts the procedure and can provide careful supervision. The FDA has approved the machine used for cryofiltration, but the AMA has yet to approve the process. "They want it more highly monitored, or something," says Marjean. "I can't tell them what to do, but I know that it has made me feel much better."

The patient seeking this treatment needs to have a positive attitude. He or she will have to make frequent trips, at least at first, to the treatment center. A person ill enough to try cryofiltration probably has a damaged vascular system and will need a fistula: a juncture in the arm (or leg) at which the end of a vein is joined to the side of an artery to increase the blood flow to the vein and enlarge it. "That means making sure that no one ever takes your pressure or does other blood work on that arm, unless you want it to burst," says Marjean. "My first fistula clotted just as I left the hospital, and when I found out that they'd have to replace it, I cried all the way home." Patients beginning cryofiltration should expect not a cure but rather a slowing down of the disease process.

Within the limitations of her disease, Marjean still tries to stay active—keeping house, cooking, driving, shopping, swimming, and doing all she can. "There's no question that the cryofiltration has been successful and a big help for me," she says. "I have less pain and stiffness, and I can reduce my usual effective dosage of medication by nearly 20% and still feel fine. I've lived more normally for the past few years than I have for quite a while."

So far the Cleveland team has treated 19 patients with rheumatoid arthritis and another 20 or so with other kinds of diseases. Cryofiltration systems manufactured according to their specifications have been tested at other centers nationwide and abroad. Dr. Nosé wrote recently that "overall improvement of 90% with cryofiltration treatment is achieved when all published results up to April 1983 are reviewed."

The researchers plan to apply cryofiltration and other plasma purification systems to atherosclerosis and other diseases, concentrating for the time being on the detoxification system. "Like old machines, our bodies rust and clog up," says Dr. Nosé. "Instead of just taking drugs when different organs fail, we need to overhaul and clean overselves up. We have no organ to get rid of these waste products that we gradually accumulate, because we're supposed to die when our productive time is over. With artificial organ technology, however, we can compensate for this missing part." A simple blood test can show whether someone with an inherited predisposition to an autoimmune disease, for instance, has high levels of cryomolecules that may cause more specific symptoms later on. Used as preventive medicine early in the disease process, cryofiltration might arrest or some day even reverse terminal diseases like kidney failure or cardiomyopathy. "But it's very difficult and will take much education to convince people to use artificial organs as preventive medicine," says Malchesky.

In some respects cryofiltration can compensate for the medical community's ignorance of many vital organ functions. Although the artificial kidney has been available for 40 years, researchers still argue about how to define uremic toxins. Blood plasma contains *thousands* of compounds; by using selective filtration, researchers can find out what roles these various agents play in disease processes. While doing so, they can somehow thwart pathological problems to maintain organs for regeneration or transplantation. In the future cryofiltration may be used as widely as blood transfusions.

Plasmapheresis and cryofiltration illustrate that looking at illnesses from a strictly physiological viewpoint does not always work. "No organ works by itself," says Malchesky. "Heart disease is not caused by the heart. The objective of the artificial kidney has been not to replace that organ but to remove uremic toxins. This approach is more practical than that of trying to replace an organ."

It also creates a semantic difficulty. Such metabolic and immunological assist devices perform the functions of one, two, or even several organs. "We're puzzled as to what to call these systems," says Malchesky. "Are they artificial immunological systems? Although they assist or preserve specific organs, they involve the whole body."

Perhaps these new systems, whatever they are named, signal a whole new understanding of artificial organ technology. "By crossing into other areas of medicine, we may be diffusing—or just confusing—the artificial organ concept," says Malchesky. "This is bad in that the idea lacks an entity—our journal *Artificial Organs* may go out of business—but it's good in that it gets integrated into other levels of medicine."

From machines that act as "all-purpose organs," we turn to devices that try to perform a minor function of just one.

11

Pumps and Pancreata

From 3% to 5% of Americans have diabetes and know it; another 2% to 3% have the disease and are unaware of it. The third leading cause of death by disease in this country, diabetes monopolizes more patient years than heart disease.

Less than 5% of the human pancreas performs the important function of constantly regulating the glucose (sugar) level in the blood. After each meal or snack, the blood glucose level rises and must be restored to normal. Clustered in groups called islets of Langerhans, endocrine pancreatic cells begin this process by increasing their production of insulin, which opens the glucose receptors in body cells. An adequate insulin supply generally assures that the cells will absorb enough glucose to meet their energy needs while the blood level is maintained between 80 and 120 milligrams per deciliter.

If the pancreas does not supply enough insulin or the body cannot use the hormone effectively, the tissues will be unable to absorb glucose. They instead break down fats for energy, a process that nauseates the diabetic patient and may lead to coma and death. Meanwhile glucose floods the bloodstream and is excreted in the urine, causing dehydration. Chronic glucose imbalance or some other abnormality associated with diabetes seems to break down small blood vessels and contribute to long-term complications: eye and kidney failure, circulatory and nerve problems, heart disease, and loss of limbs.

A device that could duplicate the immediate, constant pancreatic function of supplying insulin might eliminate some of these problems. Since disability and premature death from diabetes cost the United States over $10 billion each year, the artificial pancreas enjoys high priority among researchers in bionics.

Work toward development of an artificial pancreas is another example of how artificial organ research can make a unique contribution to a better understanding of disease processes. In many ways, diabetes is still a mystery. Researchers do not know, for example, why beta-cells like to cuddle in islets. More important, they do not know exactly what optimal glucose control might be or what it could contribute to the diabetic patient's health. "The complications of diabetes are probably related to both glucose regulation and the disease process itself. We think that eye and kidney complications are closely related to glucose control, but we're not sure why diabetic patients die of coronaries, which involve big blood vessels," says O. Peter Schumacher, M.D., Chairman of the Department of Endocrinology at the Cleveland Clinic

Foundation. "It's important to control diabetes well, but high percentages of careful diabetic patients still develop complications. The controversy is what controlling it well means," says Edmund P. Chute, M.D., Medical Fellow and Resident in the Department of Surgery at the University of Minnesota, Minneapolis. "Nature often knows things that we haven't worked out yet. The pancreas responds to glucose production very quickly; then the insulin level drops; then it climbs again, then drifts down. Ideally we should mimic this pattern," says Richardson.

Treatment of diabetes depends on the severity of the disease. Some patients can stabilize their glucose levels with managed diet and exercise. Since the 1920s, more seriously ill patients have taken insulin injections. Conventional insulin therapy of one or two injections each day has for some patients given way to multiple daily injections (MDIs) with intensive home glucose monitoring. More recently diabetic patients have begun using subcutaneous and even implantable insulin pumps to deliver a more steady—and with some models, a more natural—supply of insulin. Still in the laboratory is the coup d'état—a hybrid pancreas that uses live beta-cells to do what they do best.

The Pride of Utah: The SPAD

The large majority of insulin-dependent diabetic patients still rely on standard injections for their daily supply. Their task has been made easier by gadgets like button infusers or portable pumps that require less frequent needle sticks. The problem with these devices (and with the injections themselves) is that the insulin is shot into the leg, arm, or abdomen and circulates systemically. To work most effectively and do the least harm, however, insulin should be delivered into the portal circulation.

A good example of biological optimization, portal circulation refers to blood flow from the capillaries of the gastrointestinal tract and spleen through those of the liver before entering the hepatic vein. The liver is unique in having two blood sources (portal and arterial) and only one blood outflow vessel. The pancreas drains into the portal circulation first; the liver uses about half the insulin it receives to throttle blood glucose production. The rest of the insulin then circulates through the body. If there is too much peripheral insulin and not enough in the portal circulation, the liver goes haywire producing glucose, and the cells gobble it up. Injecting enough insulin to satisfy the liver oversaturates the rest of the body. This situaton would cause abnormalities even if the blood sugar level were normal; it makes a diabetic patient feel sick.

Created by Dr. Stephen and Jacobsen, the subcutaneous peritoneal access device (SPAD) is an attempt to solve this problem. Once implanted, the device allows the diabetic patient to inject insulin directly into the peritoneum, where it can be swept into the portal circulation.

The SPAD resembles a large mushroom whose hollow stem is mounted on a flat disk with a hole through the middle. This mushroom is turned sideways and implanted into the linea alba, a tendonlike structure of abdominal muscles. The "cap" of the mushroom, made of white sponge, rests just under the skin with the stem and disk extending into the mesenteric peritoneum, an upper section whose blood supply drains into the portal vein.

The diabetic patient injects insulin into the spongy cap. Diluted insulin runs through the SPAD quickly, providing a bolus effect—the extra jolt of insulin needed to compensate for the rise in blood sugar that occurs at mealtimes. More concentrated insulin sinks to the lower side of the stem and is drawn out by circulating peritoneal fluid, creating a basal infusion, a steady trickle of insulin that controls blood sugar most of the time.

The SPAD must last in a particularly hostile environment ruled by an immunological system called the omentum. The abdomen can fall prey to various kinds of bacterial infections resulting from a ruptured appendix, a perforated ulcer, or the like. Its defense mechanism is highly vascularized tissue that migrates to the inflammation site and walls it off in order to capture the invaders. Unfortunately, the omentum reacts similarly to any helpful implant in the abdomen. The Utah researchers have protected the SPAD from encapsulation by making it out of a special copolymer developed in their laboratory, a material with a soapy outer film that does not become coated.

Since August 1980, 28 patients at Utah have received SPADs. One of them is 29-year-old Paul MacDonald, a Utah undergraduate earning a degree in computer science. A diabetic patient since age 7, Paul needs four to five insulin shots each day. Eighteen months ago he joined the SPAD program, first as a patient, then as a computer analyst.

"I got on the program just as my kidneys were starting to show damage from the diabetes," Paul says. "We've since been able to flatten off my kidney function line, and I'd like it to stay that way." Thanks to the combined basal-bolus effect of the SPAD, the curves between Paul's blood glucose and insulin have been matching up nicely. "As a student, I have schedule changes every day, and it's difficult for me to maintain a reasonable blood sugar level with multiple injections," he says. "With the SPAD, I can drag out one injection for 8 to 10 hours if I don't eat anything, and my sugar level stays constant overnight."

Like any other therapy, the SPAD has its drawbacks. Aside from possible infection or encapsulation, the device itself can fail. "The 28 of us have had something like 51 buttons—everyone's had to be replaced at some point," Paul says. "My first button had an exit port that was too small, so it blocked, then unplugged all at once while it held about 70 units of insulin."

Other patients have had the sort of problems one can expect with a gadget most people know nothing about. One patient, Cordelia, tried to inject her insulin in a Parisian restaurant. The maitre d' threw her out. Cordelia anxiously went searching for another restaurant, made it to the end of the block, and collapsed from her insulin overdose. She woke up in an emergency room to see a very worried French intern staring in horror at the scar and slight protuberance on her stomach.

Despite such difficulties, initial results with the SPAD look good. "Many diabetic patients worry that low glucose levels will knock them out while they're driving a car, for instance, or that high levels will cause complications," says Jacobsen. "All of our patients seem more immune to hypoglycemia and worry about it less with the SPAD. It smooths out glucose levels, and in one set of studies of diabetic nephropathy it's been shown to arrest kidney failure processes."

"The results of our human trials have been very, very encouraging indeed," says Dr. Stephen. "We treated only those patients who had complications and could hardly maintain decent metabolic control. They have responded beautifully. While any diabetic patient, even a type 1, who is doing well with injections should be left alone, this class of patients might find the SPAD worthwhile." "I wouldn't trade the SPAD for anything," says Paul.

If cryofiltration is the pride of the Cleveland Clinic, the SPAD holds that status at Utah. The button might also be used for chemotherapy or thalassemia (Mediterranean anemia), whose victims must regularly flush their peritoneal cavities to control iron levels. "Except for the materials problem, the SPAD is probably the simplest device that has ever come out of our laboratory," says Jacobsen. "It has the potential to help more people than anything the Center for Biomedical Design has ever done."

The SPAD is unique in providing access to the portal circulation, but researchers suspect that a precise basal-bolus infusion of insulin offers the real advantages in diabetes management. As a result, many diabetic patients are experimenting with "artificial beta-cells" in the form of insulin pumps.

External Insulin Pumps

The artificial beta-cell, a "closed loop" system, is a computerized bedside device that draws blood from the patient, measures glucose levels continuously, and administers appropriate insulin doses in response. Many of these machines have been studied in different laboratories, although they are obviously impractical for general use. More recently diabetic patients have used small, portable "open loop" systems to deliver insulin continuously. The term "open loop" is a misnomer; patients with these external insulin pumps actually close the loop themselves by periodically measuring their glucose levels and adjusting their insulin dosages accordingly.

At least a dozen types of external insulin pumps are available, some of them as small as credit cards. They infuse basal levels of insulin automatically and can be set to deliver a bolus before each meal. Battery powered, they are worn on belts and connected by tubing to needles stuck in the patients' skin. Patients change the tubing and move the needle every few days. They can remove the pump and temporarily clamp off the tubing to shower or swim.

The pumps offer several advantages over MDIs. They help maintain tight glucose control, so their users feel more energetic and optimistic. (Newer models may enable patients to preprogram a bolus for the "dawn phenomenon," the extra shot of insulin that is sometimes needed immediately in the morning.) Pump patients always have their insulin handy without having to fuss with injections. They can schedule meals more flexibly—and some patients are tempted to celebrate with food binges. An unexpected advantage is that the cholesterol levels of many diabetic patients seem to decrease rapidly when they start using external pumps.

Pump use does, however, cause problems. The pump is *always there;* its batteries, tubing, tape, blood tests, and button pushing are constant reminders that the patient has diabetes. Its cost ranges from $1500 to $2500; supplies, which must be mail ordered, can run from $300 to $500 per year. Many physicians as well as hospital and ambulance staffs do not know about these devices and would not think, for instance, to drop out the pump battery in an emergency to stop a patient's insulin reaction. Insulin deteriorates at body temperature, so the infusion is less effective as the pump's reservoir empties. Researchers report anecdotally that established retinal problems seem to worsen when pumps are used and that diabetic acidosis occurs more often.

The external pump is best described as a helpful hell of a nuisance. A recent study of external pumps in *The New England Journal of Medicine* showed that about half their users gave them up after a year. Early in 1982 *Diabetes Forecast* published a notice requesting feedback on their use from insulin pump owners across the country. The results of 120 letters appeared in an article later that year in which patients descibed feelings ranging from "undying devotion, to grudging tolerance, to utter contempt," to a mixture of all these. Ruth Lundstrom, of Worcester, Massachusetts, for instance, was nearly arrested in a coffee shop when a waitress mistook her pump for a bomb and called the police. Pumps have been mistaken for radios, pagers, and garage door openers; the article mentions one little boy who saw a pump owner at a zoo and asked her if she used the device to feed the animals. Ronald Poinsett, of Aurora, Colorado, wrote to describe how when sleeping he tries to avoid rolling over on his pump and his cat. Yvonne Fischer, a 28-year-old teacher from Boulder, Colorado, perhaps wished she *had* rolled over on *her* cat the night it chewed through the tubing while she slept. One writer offered suggestions for fashionably dressing the pump; another listed several similarities between an external pump and a newborn baby.

Nancy De Long, a 34-year-old secretary-reporter for a small local newspaper in Wakeman, Ohio, nearly lost an interview when the subject mistook her pump for a tape recorder, but she would never give it up. She has been diabetic since age 4, insulin dependent since 1955, and on two shots per day since 1980.

Nancy is a brittle type 1 diabetic patient whose blood sugar can fluctuate from 600 to 20 milligrams per deciliter within an hour. She is well accustomed to having insulin reactions and losing consciousness. Her husband Michael has spent years sleeping lightly: touching her at night to see if she was sweating, waking her to check her blood glucose, or, worst of all, finding her unconscious and reviving her by spooning sugar into her mouth. When Michael went on business trips, Nancy would take a bottle of soda to bed and set the alarm so that she could measure her glucose level midway through the night. One morning she recovered consciousness to find her older daughter, 12-year-old Julie, standing by her bed in tears. "Mom," she said, "I'm going to miss the school bus, but how can I leave when you're sick like this?"

Over the past 2 years Nancy's diabetes worsened until she became too tired to work. After consulting Dr. Schumacher, she decided to try an external pump manufactured by Cardiac Pacemakers, Inc. "We chose my pump type because I'm so brittle," she says. "A malfunctioning pump with a large reservoir would be worse for me than one that holds less insulin. And once I program and lock my pump, pushing the buttons won't change the insulin supply. During my insulin reactions, I'm disoriented; I reprogram the clock radio and look for other buttons to push."

Shortly after getting her pump, Nancy found her health and life-style dramatically improved thanks to the constant basal infusion. "I've had many fewer insulin reactions," she says. "We threw one party when I went a week without a reaction and another when I stopped taking daytime naps." Her family needed to adjust: it took Michael about 6 weeks to learn how to sleep soundly again, and Nancy's niece announced that her aunt would resemble one of the school geeks who wear calculators on their belts. Her husband and daughters are thrilled, however, that Nancy can now plan activities with them.

"I would recommend the pump to patients having trouble controlling their sugar with injections," Nancy says. "You really need to have problems to appreciate the time and expense of the pump. I myself dread the day when the pump malfunctions and I have to go back to injections, even temporarily. The flexibility I have now is just wonderful!"

The Implantable Insulin Pump

The nuisance factors of the external pump all but disappear with the implantable insulin pump, invented in 1969 by researchers at the University of Minnesota, Minneapolis, led by Henry Buchwald, M.D., Professor of Surgery and Biomedical Engineering. The group has implanted the pump in 30 diabetic patients, types 1 and 2, with remarkable success.

The only pump currently approved by the FDA for implantation (although not yet for insulin), the Minnesota pump is marketed by Intermedics Infusaid Inc. It is made of titanium in the size and shape of a hockey puck. The construction is quite simple, since the only moving part is a bellows separating two chambers. The upper chamber holds about 50 milliliters of a glycerol solution with insulin, which is injected through a septum in the top of the pump. The lower chamber holds a liquid similar to Freon. Body heat turns this liquid into gas, which expands and forces the insulin in the upper chamber out through a capillary tube at the side. The flow varies from 0.5 to 5 milliliters per day depending on the pump (they are individually made), temperature, altitude (height increases it), and diameter of the capillary tube. Once implanted for a few months, the pump flows with precise regularity.

The pump is implanted below the skin of the upper torso with the catheter tip extending into the superior vena cava. The high blood flow in that vein keeps the catheter free of blood clots. "We like the idea of portal delivery of insulin, but access to the portal system is not yet perfected in adults, and chronic portal cannulation has not been tested in human beings," says Dr. Chute, one of the team members. "Intravenous insulin delivery is the next best thing because the insulin is rapidly delivered to the liver."

The third patient to receive the Infusaid pump for insulin delivery was Dianne Nelson, a 39-year-old housewife from Minneapolis. Dianne's diabetes was diagnosed late in 1980. She went on insulin immediately and for 4 months experienced the unstable glucose levels of a brittle diabetic patient. Dianne tried an external pump and disliked it intensely but was very excited by the idea of an implantable device. "I hoped that it would stabilize my sugar, and most of all, that the steady basal infusion would help prevent complications," she says.

Now Dianne's glucose levels are close to normal and stay that way. "Between using this pump and regulating my diet and exercise, I really have to cheat to get abnormal sugars," she says. Dianne hardly notices the pump. "It's implanted in a pouch they created in the upper left side of my chest," she says. "It's half a pound of metal, but I don't feel it at all. I golf a lot; I dive off diving boards when I'm swimming; I just forget that it's there."

Dianne's pump is refilled once a week. "My husband Gordy pushed me into getting the pump because he was devastated when he heard I had diabetes," she says. "But when they told him that *he* would be refilling it for me, he said, 'No way am I going to do that—I'm a banker, not a doctor!'" The next thing Gordy knew, he was slipping on surgical gloves and unfolding a drape while his wife swabbed herself with Betadine. In a 20-minute procedure Gordy removes the old insulin, measures it, and injects a fresh supply.

In June 1984 Dianne received her second pump after spending a month feeling ill and finally recognizing that the first one had clogged. Insulin can gum up inside a pump. The Minnesota group therefore adds a glycerol solution to the insulin, but other problems can occur. "Dianne's pump didn't fail because of the insulin consistency or a fault in the mechanism," says Dr. Chute. "Our preliminary conclusion was that the insulin precipitated at a point in the catheter where a suture was too tight." Whatever the cause, Dianne had no doubts whatsoever about getting a new pump. "When I gave myself the first insulin shot after we figured out the pump problem, I cried," she says. Her second surgery required a hospital stay of 2 days (rather than the 2 weeks of the first implantation), and she has had no problems since. "I can't see why anyone wouldn't want this pump once they get all these little problems worked out," says Dianne. "I wish it were available to more people."

Even more than the SPAD or an external device, the Infusaid pump would bewilder emergency personnel in cases of insulin reaction or malfunction. It has established a history of reliability since its FDA approval for cancer drugs, heparin, and morphine, however. So far it has delivered insulin for 3½ years without any maintenance whatsoever. Although the pump does not give a bolus, the intravenous insulin delivery seems greatly to lower hypoglycemia, hypoinsulinemia, and especially diabetic acidosis among its users as compared to either external pumps or injections. It also seems to improve residual pancreatic function, enabling that organ perhaps to kick in a bolus when needed. "Our glucose control is better than conventional therapy, and our patients are happier," says Dr. Chute. "An independent psychologist from the University of New Mexico recently interviewed 23 randomly selected Infusaid pump patients, 9 of whom were from our group. She found that all 23 would ask for the pump again if they had the decision back, mainly because of their relative freedom from life-style restrictions and hypoglycemic reactions."

The simplicity of the Infusaid pump suits it admirably for some uses while making it safe, reliable, and less expensive than one might expect. Dr. Buchwald feels, however, that eventually patients will have available a family of devices ranging from simple to complex depending on their proposed use. "Probably a multiple flow, programmable, monitoring and recording instrument will be useful in treating diabetics with certain types of brittle, type 1 disease," he has written recently.

Glucose Sensors

Two implantable variable infusion pumps are in the works, one at Johns Hopkins University, as discussed in Chapter 26, the other at the University of New Mexico, but the big breakthrough would be a pump that could regulate itself with its own glucose sensor. "A glucose sensor would take care of 98% of all that ever had to be done to treat diabetes," says J. Stuart Soeldner, M.D., Head of the Section of Clinical Physiology at the Elliott P. Joslin Research Laboratory and Associate Professor of Medicine at the Harvard Medical School. "It's an extraordinarily difficult device to invent because it would have to act like the control room of an ocean liner, recording what everything is doing and making adjustments accordingly." Such a device would have to be accurate, drift-free, capable of being calibrated, quick to respond, and biocompatible so that the body would not swallow it. It would have to be relatively cheap and use small amounts of blood or tissue fluids.

Dr. Soeldner and his colleagues have invented a glucose sensor consisting of a Teflon-bonded platinum electrode that has unique properties. It is run as a potentiostat—a three-electrode system that allows researchers to control the voltage between the platinum "glucose electrode" and a reference electrode. "The invention part of it is that using a reference electrode, we discovered that glucose in particular seems to be readily oxidizable when that voltage is at a certain value," says Dr. Soeldner. In other words, the electrode emits a current when it touches glucose. Since other substances are oxidized at other voltages, a similar device could conceivably be used to monitor metabolic changes in critical care patients with other types of disease states.

Dr. Soeldner's group is starting a series of animal implantations with the electrochemical sensor. Their big concerns now are what animal models to use, what size and shape to make the sensor, what else to measure to get FDA approval for human tests, what other materials the sensor may catalyze, and whether these materials might become toxic or carcinogenic. Their technical problems include finding the best geometry for the sensor itself and the most suitable diffusion membranes to cover it. "When we started, there were too few biocompatible diffusion membranes—now there are too many! Every polymer chemist has one, and none of them has proved biocompatible for more than 5 years," says Dr. Soeldner.

The group hopes eventually to perform human trials in stages. Implanted near the navel, the sensor will at first relay glucose levels to a calculator-sized display unit so that the patient can monitor MDIs. A later version will control an implantable pump whose flow rate will be adjusted by a microcomputer worn on the belt. Finally, the sensor, pump, and computer will all be implanted.

Another sensor that operates on different principles has been invented by Dr. Bessman and his associates. In 1977 the group performed animal tests with a complete sensor-pump-feedback system that uses a unique sensor and pump constructed in their laboratory. The sensor consists of two electrodes that measure oxygen. One electrode is covered with a layer of glucose oxidase (an enzyme that oxidizes only glucose) in plastic form. In the absence of glucose, both electrodes read the same. When glucose is present, the enzyme consumes oxygen to oxidize it, so the reading of the enzyme-covered electrode drops below that of the other electrode. The difference between the two readings is interpreted by a complex formula that gives a measure of glucose concentration. The development of this sensor has proceeded slowly because the tissue concentrations of oxygen are very low, requiring a precise measurement, so the electrodes must be extremely sensitive.

The pump is based on a patented principle using a piezoelectric crystal (the same kind that produces the beep in alarm clocks). It is the smallest pump ever made; the entire system, including the power supply, takes up less space than three packs of Lifesavers. Seven more successful implantations were reported in 1981, and the unit is now under development for total implantation in humans.

Once perfected, a glucose sensor could make an enormous difference in the treatment and life-styles of diabetes victims. "A sensor will probably be more valuable for non-insulin-dependent diabetic patients," says Dr. Soeldner. "It could also be useful for high-risk persons who want to diagnose and treat their diabetes as early as possible."

The Hybrid Pancreas

Unfortunately, there remains another glitch in all these research advances, namely, that the rate of insulin production is not linear with glucose concentration. Apparently only nature knows why. The ideal solution to diabetes, then, would be a pancreas transplant; the next best a device that in some way could use beta-cells to produce the correct amount of insulin in response to glucose levels.

As mentioned previously, a hybrid pancreas has been developed by a New England research team including, among others, Dr. Galletti and Richardson of Brown University, Clark K. Colton, Ph.D., Professor of Chemical Engineering at MIT, and William Chick, M.D., now Professor of Biochemistry at the University of Massachusetts, Worcester. Basically the system involves beta-cells separated from the bloodstream by a biocompatible artificial membrane with selective permeability. Glucose molecules diffuse through the membrane to the beta-cells, which respond by producing insulin that passes back through the membrane into the bloodstream. Larger immune agents, however, cannot squeeze through the membrane's small pores to sense and attack the foreign beta-cells. "The hybrid pancreas represents the enslavement of cells—we imprison and protect them so they can perform their function blindly but to good purpose," says Richardson.

One type of hybrid pancreas consists of a cigar-sized plastic jacket surrounding a coil of semipermeable plastic tubing. One port is a blood inlet connecting this tubing to an artery, the other a blood outlet connecting it to a vein. Two other ports on the shell are used to charge it with insulin-secreting beta-cells. Carried in a fluid that is injected through a syringe, the cells attach to the outside of the tubing.

This device is designed to be implanted in the body or attached externally to the bloodstream much like an artificial kidney. Blood flows through the tubing, which has a thin inner layer that filters the blood and a spongy outer layer that holds the cells in place. Pressure factors as well as diffusion make the glucose pass into the surrounding fluid at the jacket's arterial end and the insulin pass back into the tubing at the venous end to be carried through the bloodstream.

The whole structure has been meticulously designed in terms of priming volume, diffusion rate, and pressure to reduce the response time of the beta-cells as much as possible without fracturing and collapsing the tubing. The tubing is fit tightly inside the chamber, for example, because too much fluid washing past will dilute the insulin.

Hybrid pancreata seeded with rat beta-cells have been tested on beagles. A prototype has been mounted on a dog's back with an artificial kidney shunt hooking it to the iliac artery and the contralateral iliac vein. Dr. Jean-Jacques Altman, a Parisian colleague of Dr. Galletti's, has temporarily attached a similar device to a diabetic volunteer, whose blood glucose became normal within 2 hours. Many technical problems still must be overcome. The device must have tubes wide enough to minimize clot blockage yet narrow enough to have thin walls. The thicker the tube walls, the less oxygen can get through to maintain the cells and the less effective the diffusion and glucose response.

Other problems include the molecular weight limit of the membrane, currently set at 50,000 daltons. "The beta-cells might slough off low–molecular weight antigens that could pass through the membrane and get into the bloodstream, where the body would create antibodies to it," says Colton, "or the system could stimulate circulating immune complexes." The membrane surface area has to be large enough to support the cells, which must spread evenly and work well enough to do their job. Just getting enough beta-cells is a problem: normal beta-cells can be maintained but do not proliferate. Modified cells (like naturally occurring or induced insulinomas) will grow but may lose their regulatory ability and just produce insulin constantly. The solution might be a hybridoma that would respond to glucose but also grow in a laboratory. Yet even if all these problems were solved, researchers would still wonder whether the glucose control would be good enough to justify the risk of implanting such a device.

Microencapsulation

Other investigators have also experimented with a hybrid pancreas turned inside out: they placed the beta-cells inside the membrane with the blood circulating outside them. They have also enclosed insulin-secreting tissue in pieces of tubing closed at both ends and placed these capsules in the abdominal cavities of diabetic animals, with remarkably successful results. This idea has also been tested by Franklin Lim, M.D., of the Medical College of Virginia, who combined Dr. Chang's method of creating artificial cells with his own islet-friendly membrane. Dr. Lim enclosed each islet in a semipermeable microcapsule and injected the capsules into laboratory animals. Over a year later the enclosed cells were found to be alive and well inside the animals. Whether humans would need regular booster shots of them to correct diabetes is not yet known. The capsules could be injected inside or outside the bloodstream and might thrive in the human peritoneum.

Dr. Lim is now working with Anthony M. Sun, M.D., of Connaught Laboratories in Canada; Dr. Chang spent 15 years persuading them to take on this project. Connaught Laboratories and Damon Biotech in the United States have created a new joint venture company, Vivotech, to make and test the capsules. Human trials are expected in about 2 years. If the system works, the microencapsulation principle could be applied to other chronic conditions like immune system deficiencies and pituitary and thyroid diseases.

These are some of the approaches and problems involved in trying to imitate the small portion of the pancreas that produces insulin. Like other body parts, the pancreas is challenging because of ignorance about disease processes as well as technical difficulties involved in arresting them. Researchers agree that although they do not know how much glucose control is needed, any diabetic patient who is doing well with conventional therapy ought to continue with that. For others, different approaches to the artificial pancreas may be suboptimal but can definitely help restore metabolic equilibrium.

"We can't guess now which pancreas substitute will work best. Research on all these systems is justified because each has its problems and strengths," says Colton. "I'm absolutely convinced that my successors will treat diabetes differently from the way I do. Personally I think that for the average patient, some variation of the insulin pump is going to become viable," says Dr. Schumacher. "I think that the hybrid pancreas offers the most hope of all by a long shot, although there will be real problems if the membrane is walled off," says Dr. Stephen.

Regardless of which version of the pancreas is most successful, it will probably create new tasks for diabetes specialists. "Although it's the most viable one that we can see today, the bioengineering approach to diabetes will be hard for people to accept," says Dr. Soeldner. "Diabetic patients fantasize that they'll get well or they'll get transplants, and that just isn't going to happen. Our task will be to educate patients that these devices may be the way to optimal health." We are about to meet the group of patients with the longest history of having learned just that.

12

Dialysis in Wonderland—and Other Processes Elsewhere

The Department of Biomedical Engineering at the University of Utah has its own modern building at the foot of the Wasatch Mountains that offers a view across the valley to the distant Salt Lake. Outside a third-floor entrance, a bed of pink, white, and lavender petunias blossoms in what looks like an enormous rusty-red planter. Leaving the building with a visitor, Dr. Kolff pauses and points to the colorful display. "Do you know what that is?" he asks.

"Petunias," the visitor suggests.

"That," says Dr. Kolff, "is the nose cone of a rocket. We once used it to make an artificial kidney."

It is logical that the dialysis machine was the first artificial organ, since the natural kidneys play a crucial role in good health. The kidneys maintain the delicate equilibrium of dozens of chemicals; control the pressure, volume, and acidity of blood; produce hormones; and filter the blood to remove excess fluid and waste products. Together the kidneys contain 2 million nephrons that process all the blood every 2 minutes. Inside each nephron is a glomerulus, which removes fluid from blood plasma. Tubules of the glomeruli filter the fluid to draw out waste products, then return necessary substances to the blood flowing through a web of capillaries that enclose them. The waste fluid leaves the kidneys through ureters and is stored in the bladder.

Like the liver, the kidneys are greatly overdesigned. A person can survive with only one kidney, and this organ will function on only a fraction of its capacity. But the kidneys can be fatally damaged by any of several birth defects, diseases, infections, obstructions, toxins, trauma, or shock. When the kidneys fail, the body swells with water and begins accumulating wastes and poisons. The blood pressure rises. The victim feels weak and nauseous, sinks into a coma, and eventually dies.

People can survive acute kidney failure if fluid and toxins are removed from their bodies until their kidneys can regenerate. Those with chronic kidney failure can likewise survive if they undergo such removal at closely spaced, regular intervals. This removal, hemodialysis, is the purpose of the artificial kidney.

The process used by the kidneys to remove wastes and water is a complex one, and according to Dr. K6lff, understanding it is no help in mimicking it. "It's a misconception to think that we need to understand everything about organ function before we can replace it," he says. "We can cheat nature to get a practical result. Dialysis is different from natural kidney processes, but it's very effective."

Hemodialysis

Hemodialysis (or simply "dialysis") is actually two simultaneous processes: diffusion, which rids the body of toxins, and ultrafiltration, which removes excess fluid.

We have already encountered diffusion in the artificial pancreas; it is based on the principle that chemicals will equalize their concentrations on either side of a semipermeable membrane. In the artificial kidney the membrane can take the form of coiled tubes, sheets, or hollow fibers. Blood runs on one side, a clean fluid called dialysate on the other. The dialysate contains chemicals the body needs and lacks those that need to be removed. As these chemicals diffuse across the membrane and equalize their concentrations, the used dialysate is replaced with a fresh mixture. Repeated changes of dialysate result in the removal of urea, creatinine, and other toxins and the addition of sodium, chlorine, and other missing chemicals. This passive diffusion of chemicals is less efficient than the active transport of the natural kidney, which pushes the molecules across its tubular membranes until concentrations are higher on the far side.

Because it can impose suction on the dialysate side and positive pressure on the blood side of the membrane, the artificial kidney can pull water across it. This waste fluid is called ultrafiltrate. Patients can lose 5 to 10 pounds of water in a 5-hour hemodialysis treatment. The resulting drop in blood volume and rise in its colloid osmotic pressure cause water from both extracellular and intracellular spaces to move into the blood compartment. The decreases in blood volume and pressure are two reasons that some patients do not feel very well during and after hemodialysis.

Hemodialysis machines have included some strange and rather unsophisticated contraptions, a few examples of which will be discussed here.

THE FIRST ARTIFICIAL KIDNEY

Had it not started the science of bionics, the first artificial kidney would still be a remarkable device and a tremendous credit to Dr. Kolff, who for all practical purposes is its creator. "I am not really the inventor of the artificial kidney," he stated in a recent lecture. "The artificial kidney was described by Abel, Rowntree, and Turner, three Americans, in the City of Groningen, The Netherlands, during a Physiological Congress in 1913. I was 2 years old at that time, and I do not remember it, but in that same city I began to work on artificial kidneys in 1939."

Dr. Kolff claims that all his motivations to build artificial body parts have been based on what he noticed or knew about his patients. As a young physician in Groningen, Dr. Kolff once had to tell a young man's mother that her only son was dying of kidney failure. "What struck me about this patient's condition," he says, "was that I saw a way of doing something about it." Dr. Kolff began experimenting with cellophane tubing used to make sausage skins; he poured in some blood and urea and swished the tubing around in a saline bath. After 5 minutes most of the urea had diffused through the cellophane into the saline. The next step was to make a machine that could filter a patient's blood in the same way.

Dr. Kolff's work was interrupted when the Germans invaded the Netherlands in 1940. The following year a National Socialist was named to head the Department of Internal Medicine at the University of Groningen. Dr. Kolff resigned in protest and moved to a hospital in Kampen. He convinced an industrialist, H. Th. J. Berk, to donate materials and build the first clinically useful artificial kidney. This "rotating model" kidney consisted of cellophane tubing wrapped around a drum that was partially immersed in saline solution. The device had no pump; instead the patient's blood was drawn into a burette that was repeatedly raised to the ceiling on a pulley. From there the blood drained into the rotating drum, where it circulated through the tubing. During the war, Dr. Kolff made eight of these kidneys, scrounging the materials or buying them himself.

In 1943 Dr. Kolff treated the first of 17 local patients with the artificial kidney. "I was very tired and desperate at times," he says. "We had to work during the day and dialyze at night. It took at least 6 hours, so often I didn't get home until dawn. On one occasion I was so overtired that I cried and cried." Although the patients were too ill with renal and other diseases to survive with the device, Dr. Kolff noticed improvement in each, and one patient recovered consciousness long enough to write his will. "His son, a physician, stayed with him until he died early one morning," says Dr. Kolff. "When I came in that day, I found a note from the son saying, 'The method is good, but we were too late.' "

In 1945 Dr. Kolff saved the life of his seventeenth kidney patient, a 67-year-old woman who lived many years after her recovery from acute renal failure. Five years later he emigrated to America and continued work on artificial organs at the Cleveland Clinic Foundation. The science of bionics had begun.

Ironically, the surviving kidney patient was a National Socialist.

THE MAYTAG KIDNEY

An interesting successor to the rotating drum kidney was the Maytag kidney, the result of Dr. Nosé's interest in dialysis at home rather than in a hospital. In 1960 Japan did not have artificial kidneys, so researchers had to make their own. "I thought that since the kidney was the washing machine of the body, why not use a washing machine to make one?" says Dr. Nosé. So he cut out a window screen, wound it into a coil, and dropped it into a washing machine. It was easy to make, it was cheap, and it worked very well.

Two years later Dr. Nosé came to America to work with Dr. Kolff. The latter feared the possible legal consequences of home dialysis (an issue to which the Cleveland Clinic was very sensitive), but within a few years the concept had caught on. Although Dr. Kolff was intrigued, he still balked at the idea of building artificial kidneys out of washing machines. "So I waited until he left the office one day, then called several washing machine companies and told them that if they would donate a dozen machines for dialysis, they'd become famous," says Dr. Nosé. Maytag was among the companies that responded, so Dr. Nosé built a Maytag kidney.

Dr. Kolff had a change of heart and began promoting the device. His group sent 28 patients home with Maytag dialyzers, some of whom used them for several years. The cost of the dialyzer and 2-months' worth of supplies was $265, compared to the $10,000 cost of other artificial kidneys for home use at the time.

Then a Maytag lawyer showed up in Dr. Kolff's office. He explained that if a patient had an accident while using one of the dialyzers, the nuisance value of a lawsuit against his company would be at least $22,000. Maytag was going to stop donating machines and inform its dealers that they sold them for dialysis at their own risk. "It just shows you how incredibly narrow-minded people can be," says Dr. Kolff. "From that time on, I had to purchase Maytag washing machines on the black market."

The Maytag kidney is associated with Dr. Kolff rather than with Dr. Nosé. "That's fine," says the latter. "Dr. Kolff was my teacher, and I worked under him, so any project I developed at that time is really his."

THE ORANGE JUICE KIDNEY

At some point Dr. Kolff enlisted the help of Bobbie Brooks, an apparel company in Cleveland, to make the twin coil disposable dialyzer, popularly known as the orange juice kidney. In his basement Dr. Kolff made a contraption of seven orange juice cans, around which he wrapped a 10-meter roll of window screening and cellophane tubing that had been stitched together at Bobbie Brooks. "For the next 20 years or so," he says, "all coil kidneys were made around a core the size of a fruit juice can." His choice of that material arose from necessity. "Beer cans," he says, "were too small."

DIALYSIS TODAY

There are now over 300 different types of dialyzers available worldwide. An estimated 250,000 patients are on dialysis; they are joined by 20,000 newcomers each year. By supporting patients awaiting transplants, the artificial kidney has saved the lives of 80,000 recipients who no longer need it. Tens of thousands of other people have survived acute kidney failure or have had successful surgery as a result of the heart-lung machine, which developed directly out of Dr. Kolff's kidney research.

Victims of accidental poisoning or drug overdose have been saved by procedures derived from the artificial kidney. One of these, hemofiltration, uses a much more porous membrane to squeeze gallons of water and large molecules out of the body. In hemoperfusion the blood is run through a sorbent, like charcoal, to remove large amounts of wastes and toxins. The charcoal is usually encapsulated in semipermeable membranes to form artificial cells. "The use of artificial cell devices like ACAC, Hemosorba, Dialaid, and Hemacol is routine treatment for acute poisoning in medical centers around the world," says Dr. Chang.

Encapsulated charcoal cells packed inside a disposable cartridge are attached to standard dialysis machines in many centers to increase efficiency and reduce cost. Membranes enclosing the cells are 100 times thinner—and their diffusion rate that much faster—than those in artificial kidneys. The cells do not remove water, electrolytes, or urea, so Dr. Chang hopes someday to combine his cartridge with a small ultrafiltrator and figure out another way to remove the urea.

DISADVANTAGES OF DIALYSIS

Dialysis is performed on most patients three times each week for 5 to 6 hours per treatment. The patient must go to a dialysis center and be hooked up to a large machine, while the country pays the crippling expense—as much as $25,000 per patient per year.

Like patients undergoing cryofiltration or similar procedures, a dialysis patient must have a fistula to use the machine. Two huge needles attached to tubes are stuck in the fistula during each dialysis to carry blood to and from the artificial kidney. New arteriovenous shunts with access buttons may simplify this process, but maintenance of any type of access site is going to be complicated by blood clots and infections.

Time consuming and rigid as it is, the dialysis treatment schedule is far from ideal. People do not urinate three times per week for 5 to 6 hours; they do it throughout each day. "Hemodialysis is just a great wrenching around of blood chemistries at set times during the week," says Dr. Stephen. The procedure can cause side effects: cramps, itching, fatigue, nausea, dizziness, impotence, blood cell damage, bleeding, low blood pressure, or fever. These effects result partly from the shock to the body of being suddenly cleaned out and dehydrated and partly from the diffusion across a cellulose membrane, which is not completely thorough in removing or retaining substances. Patients receiving dialysis must follow strict dietary rules and restrict their intake of fluid, protein, potassium, salt, and other substances.

The psychological complications of dialysis can be devastating. Tied to their machines, patients feel dependent and lose their self-esteem. They can become withdrawn and hostile toward their families and medical personnel. Some experts have observed that dialysis patients have a suicide rate higher than that of the general population.

Naturally, researchers are exploring alternatives to center dialysis. Medicare is conducting an 8-year study in three states to see if home dialysis is a reasonable alternative. It is by far the least expensive treatment available and less of a hassle for the patient. But not everyone can use it, and those patients who do find their families under terrible strain. "We worry a lot about power failures," says Leola Burr, who acts as technician for her husband's dialysis at their home in Salt Lake City. "And my husband has a problem with blood clots. But two things they warned us about haven't happened: we're not divorced yet, and I haven't become an alcoholic."

An ideal alternative to standard dialysis would be an implantable artificial kidney. The next best would be a portable one, and here some exciting advances have been made.

THE "WEARABLE" ARTIFICIAL KIDNEY THAT IS NOT WORN

The two portable artificial kidneys heard of most often are the Downstate kidney and the wearable artificial kidney (WAK). The kidney developed at the Downstate Medical Center in Brooklyn, New York, is the size of an attaché case and weighs about 22 pounds; its tank holds 21 liters. It must be plugged into a power outlet and has an adapter for European travel.

The quality of several hundred patients' lives has been temporarily improved by the completely portable WAK, which was developed at the University of Utah. WAK should really be PAK, for portable artificial kidney. "It was never intended to be wearable, and it's not even a basic scientific breakthrough," says Jacobsen. "It's an engineering breakthrough—a well-designed standard kidney that allows patients to avoid hospitals and have some mobility. It's simple, safe, and easy to use, and it can someday be applied for other related purposes."

The WAK has three parts, two of which are used at a time. The kidney itself weighs about 7 pounds. A short line of tubing draws in the patient's blood, which is dialyzed using the popular hollow fiber cartridge. Like standard dialyzers, the WAK sets temperature and pressure, pumps blood and dialysate, checks functions, and shuts down during emergencies like air or blood leaks.

Tubing about 25 feet long connects the WAK to either a mixer or a 20-liter tank. The mixer, which hooks up to a purification system, mixes dialysate concentrate with water. The tank, which holds the mixed dialysate, is basically a carrying case that will fit under an airplane seat.

The WAK enables its user to dialyze each day. While dialyzing, the patient can get up and walk as far as the dialysate tubing allows to answer the door or the telephone, let the dog out, check the roast, or whatever. An older version could be completely detached and worn for as long as the cleansing fluid and the small bag collecting ultrafiltrate held out. Because most people do not feel like strolling around the block while dialyzing, the emphasis of the WAK was changed.

The main advantage of the WAK is that it enables kidney patients to travel farther than a day's journey from their dialysis centers. To do so now, a patient must try to reserve time at a center near his or her destination. Some centers do not have space; many hesitate to treat seriously ill patients they do not know; others are just incompetent with strangers. "I had a bad experience when we visited my uncle in California," says Rian Peek, who lives in Murray, Utah. "I have a great fistula, but the technicians at a private dialysis center in San Diego stuck me nine times and couldn't find it. Although I was only 12, they wouldn't let my parents stay. I sat there crying for 4 hours with a sore arm and no dialysis while my family was out having fun."

The FDA approved the WAK in 1983, but the Utah researchers could not find an American company willing to manufacture it. "It is like the automobile," says Dr. Kolff. "Until you import a small one from Japan, the established companies are not interested." Junken Company, Ltd., in Japan has made a slightly smaller model of the device that will be imported by WAK Utah. Meanwhile, the 30 or so WAKs in existence are leased out to patients by Wonderland Travel in Salt Lake City for Dialysis in Wonderland trips.

DIALYSIS IN WONDERLAND

Dialysis in Wonderland is a program of vacation trips for kidney patients that enables them to dialyze while away. The program, which began in 1975 with a trip to the outdoor Shakespeare festival in southern Utah, includes vacations at several national parks, monuments, and recreation areas as well as in Hawaii and the U.S. territories. Patients have dialyzed outdoors, on houseboat decks, and in tents or cabins and have enjoyed hiking, boating, skiing, fishing, white-water rafting, sightseeing, and lazing around.

Participants in the trips have ranged from age 11 to 76 and have included blind, diabetic, and wheelchair as well as dialysis patients. They are accompanied by family or friends and a full medical staff. As of July 1985, 531 people (including 238 different patients) had undergone 1338 dialyses on 70 trips. They came from 35 states and Canada. Most of the trips range from 5 to 7 days and cost from $400 to $600; Medicare began covering dialysis costs in 1982. (This was broad-minded of them: when the Medicare and Medicaid inspectors took a trip, the patients waited until they were in their bathing suits, then started a mud fight with them.) The staffs found that patients' morale, health, and appetite improved considerably during their vacations.

The patients themselves are happy to attest to this. Loren Burr, a 63-year-old retired partner in a Salt Lake City accounting firm, took the 8-day trip to Yellowstone National Park in July 1982. While there, he got to pose for a Canadian magazine that wanted a photo of someone dialyzing in front of Old Faithful. "We'd be dialyzing outdoors, and when the geyser was due to erupt, they'd load me into the van and rush me up to do it." he says.

Loren has been unable to take another trip because of health problems but would like to at some point. "I would definitely recommend Dialysis in Wonderland," he says. "My impression was that the participants on our trip were all very pleased." Loren's wife, Leola, whom we met earlier, also enjoyed herself even though she was on crutches because of a broken hip. "They took as good care of me as they did of Loren," she says, "and I had the best seat in the sight-seeing van. I would highly recommend their trips to anyone who is incapacitated in any way."

Lee Butterfield, a 27-year-old employee of Industries for the Blind in Salt Lake City, has been on 15 Dialysis in Wonderland trips in 2 years. His motivation resulted from diabetes, which damaged his kidneys and left him legally blind. "I started going on trips because my eyes and my kidneys started going all at once, and I was kind of bent out of shape and depressed," he says. "I had just started to deal with it, and I think the trips had a positive effect. For some people it's like a switch turned on—they suddenly realize how much they can do."

Lee's favorite trip was white-water rafting on the Colorado River. "The Salmon River is okay too," he says. "but you don't get as wet. On the Colorado, you get tossed around and soaked." Like Loren, he believes that all those vacationing were satisfied. "We had a few people get uptight during their first treatment because they were dialyzing in bathing suits, but once they tried it, they were fine," he says. "After the trip, no one asked about dialysis; they asked about other trips."

The real expert on Dialysis in Wonderland, however, is Rian Peek. At age 23, Rian has survived enough medical catastrophes to kill a mule. Her kidneys failed as the result of a strep infection that she had when she was 11. She has had two failed kidney transplants. An accident during a surgical biopsy caused her to suffer a stroke at age 18; a few years later she went into cardiac arrest during surgery on her fistula. Now she does volunteer work, runs around constantly, and can be difficult to track down.

Rian has been on 22 Dialysis in Wonderland trips. "Next year I have to go to Yellowstone," she says, "I haven't been on that one yet. I go because it's fun; I don't have to arrange for dialysis; it's easy and safe; I like to go alone rather than with my parents, and I love dialyzing on the WAK because time goes by so fast."

Rian recognizes that some participants are nervous about taking the trips, and she enjoys watching them change. "The trips prove that you can lead a normal life," she says. "You escape from the hospital and your problems, and you feel physically and mentally healthier because your attitude has changed. I'm still partially paralyzed, but I water and snow ski and hike and do about everything anyone else does. I got my confidence and my energy from these trips."

The consensus among Wonderland alumni is that Lake Powell is the best trip because of its low cost and variety. Participants can swim, fish, hike, water

The compact wearable artificial kidney (WAK) enables people with chronic kidney failure to dialyze while on vacation trips.

ski, ride in canoes, sightsee at the Indian ruins, or do absolutely nothing. They also agree that Dialysis in Wonderland should become a nationwide service. Since Wonderland Travel is losing money on the trips, this is a distinct possibility. The program may also be broadened to include all persons with handicaps or need for medical supervision. Meanwhile, the WAK will continue to offer many patients an enriching, rewarding experience.

CONTINUOUS AMBULATORY PERITONEAL DIALYSIS

The WAK is an important step toward an implantable kidney. In a sense, however, each of us already has an implanted "artificial kidney" (namely, the peritoneum) that has been used in some patients since the early 1900s. The peritoneum is an extremely well-perfused, semipermeable membrane that can filter solutes. There are two types of peritoneal dialysis: intermittent, which hardly works and is not often used, and continuous ambulatory peritoneal dialysis (CAPD), which effectively compensates for its inefficiency by going on 24 hours a day.

To use CAPD, a kidney patient must have a catheter implanted in the abdomen. He or she attaches the tubing from a bag of dialysate to the catheter, hangs the bag up high, and releases the tubing clamp so that the fluid drains into the abdomen. The patient then reclamps the tube and folds the bag into a pocket. To change the fluid, he or she drops the bag to the floor and releases the clamp so that the abdomen can drain. The patient then switches to a fresh bag of dialysate and repeats the process. This change occurs four times each day, the last just before bedtime so that the patient can sleep through the night.

CAPD has a twin called CCPD, for continuous cycler peritoneal dialysis. CCPD is done at night using a cycler, a machine that drains fluid into and lets it out of the abdomen. The cycler does four or five exchanges each night, with the last one left in the belly all day. It is just a mirror image of CAPD without the bags.

CAPD was approved for adult use in the late 1970s and for children a few years later. Although the treatment of choice for children is a kidney transplant, many researchers consider CAPD an ideal way to buy them time.

Until the recent past, a newborn infant with kidney failure was unable to be dialyzed and too small for a transplant, so it was essentially left to die. Now CAPD is saving the lives of such babies, since it is the only dialysis technique that works well for them. About 5 years ago, researchers in Houston taught mothers to do CAPD on their infants, interspering eight exchanges each day with supplemental nasogastric feedings. Some of the patients grew almost to normal size at 1 year, making them candidates for transplants. "We've used CAPD successfully on premature infants all the way up to adults," says Robert J. Cunningham, M.D., Head of the Section of Pediatric Hypertension and Nephrology in the Department of Pediatric and Adolescent Medicine at the Cleveland Clinic Foundation. Children older than 8 or 9 years can do CAPD on themselves very well.

Like hemodialysis, CAPD is an imperfect treatment with many disadvantages. Because the dialysis must be continuous, the abdomen is always full of fluid. Large amounts of glucose must be used to pull the water off. The burden of the dialysis process is shifted from a medical staff to the patient and family. Brian Adkins, a 15-year-old tenth-grade student from Waynesburg, Ohio, who wants to be a nephrologist, sees this as a distinct disadvantage. "If I'm out playing with my friends, I always have to stop and come in to change bags," he says. "I also have to change my dressing twice a day, weigh myself to see if I'm holding fluid, and keep records of my weight; the fluid color, volume, and amount; any medication I add; the time; and so on."

Despite the hassle, many children like CAPD so much that they create another problem for their physicians. "If they do really well on CAPD, they resist having transplants," says Dr. Cunningham. Brian is one of these patients, although he is giving the medical profession another chance. "I feel better on CAPD than I did with my transplant before it was rejected," he says. "I'd pick CAPD over a transplant for the rest of my life, but I'm going to try another kidney to see if their new techniques will keep me from rejecting it."

The biggest problem with CAPD has been peritonitis. Brian, for instance, has had five catheters in the 3 years he has been on CAPD. "The cuff on my catheter keeps working its way out and I get a terrible infection—my stomach hurts so bad that I can't believe it," he says. More commonly, infections occur simply because switching bags involves placing a spike inside the catheter in what must be a sterile exchange. "CAPD has five to ten times the hospitalization rate for complications of hemodialysis, and almost all of these complications are due to infections," says Dr. Stephen. For this reason, treatment of CAPD patients often costs more per year than those on an outpatient dialysis unit even though the latter treatment is much more expensive.

Peritonitis in CAPD is coming under control, however, especially in children. "Our pediatric patients do better than our adults do," says Dr. Cunningham. "They average about one bout of peritonitis every 12 to 14 months. This may be because they're more motivated, or because the parents are more meticulous with their children than they would be with themselves. These possibilities would also explain why there is much less peritonitis in the home than there is in the hospital."

The advantages of CAPD, especially for children, far outweigh the difficulties. CAPD is less expensive than hemodialysis; it is completely portable, and it can be done anywhere. "It's easy for me to go on trips—all I need is my supplies," says Brian. The bag exchanges, which can be adjusted to fit a school schedule and normal social life, are far less restrictive than standing appointments at a dialysis center. Children on CAPD can play sports, even basketball and touch football, and go swimming or camping. They stay out of hospital settings and seem to develop greater independence and self-esteem. They also avoid the side effects of hemodialysis, which very young children might interpret as a punishment. The abdominal fluid does not seem to bother patients; they grow used to it and forget it is there.

The biggest advantage of CAPD for children is that the continual dialysis keeps their condition more stable and minimizes dietary restrictions. Children on dialysis do not grow well simply because they get disgusted with their strict diets and do not eat. "CAPD allows patients to eat a better diet and grow better," says Dr. Cunningham. "If they eat too much salt or protein, for example, they just change the concentration of their solution and pull it right off. They can even go to McDonald's and have junk food with the other kids." When Brian started on CAPD at age 12, he was 3 feet, 11½ inches tall. Three years later he is 4 feet, 4 inches and weighs about 70 pounds. "I get to eat more on CAPD, and I do, because I like everything!" he says.

The WAK and CAPD are improvements on a standard therapy that has worked for years. Some researchers feel, however, that the time has come to treat kidney failure using a different approach.

Continuous Renal Replacement Therapy: An Alternative to Dialysis

The patient is a middle-aged diabetic woman. She developed low blood pressure during heart surgery and was placed on an intra-aortic balloon pump (IAB). Then her kidneys failed, and she gained 26 pounds of water. Now she needs antibiotics and hyperalimentation, which require adding much more fluid to her swollen body, yet she is too ill to tolerate dialysis.

Fortunately, an alternative is available. A simple, self-contained device is removing 14 liters of waste fluid from the patient's body each day. Meanwhile she is receiving clean fluid, nourishment, heart and blood pressure medicines, antibiotics, and other good substances. Soon the device will be adjusted to draw off her extra fluid at the rate of only 2 to 4 pounds per day, gently returning her to normal.

This patient is an example of clinical necessity being the mother of invention. The Cleveland Clinic Foundation has an unusually large number of open-heart cases, some of whom develop kidney problems after surgery. "We'd be called in to dialyze these patients, and we'd end up giving more fluid than we removed because their blood pressure would go down," says Satoru Nakamoto, M.D., Medical Director of the Hospital Dialysis Unit in the Department of Hypertension and Nephrology at the clinic. "In the meantime the cardiologists would have to give them TPN [total parenteral nutrition] and medicine and so on, so they would add 5 to 6 liters a day. And many patients couldn't tolerate dialysis at all. We decided that we had to do something."

What the researchers did was to become the first Americans to use a system introduced by a West German physician in 1977. Instead of intermittent dialysis, they tried slow continuous ultrafiltration (SCUF), which can remove fluids constantly.

The kidney glomerulus is one of the few systems in which blood goes through an artery into capillaries and back into an artery to maintain high pressure. The SCUF system is essentially a glomerulus stretched into a straightforward cylindrical design with an artery at one end and a vein at the other. It is still a high-pressure system, with a highly permeable membrane inside an inexpensive hollow fiber filter the size of a flashlight. The system can be attached to an arteriovenous shunt in the forearm and held in place with Ace bandages. (A similar device being developed at Brown University uses a single coiled tube; prototypes have been tested externally on dogs.)

Two forms of therapy are available from this one circuit: SCUF, which removes fluid, and CAVH (continuous arteriovenous hemofiltration), which removes fluid and toxins. SCUF is appropriate for patients like the woman in our example, whose kidneys have failed but who needs hyperalimentation. The amount of fluid given determines how much will very slowly be taken off, that is, an equal or greater amount. "In dialysis we might try to remove 5 pounds of fluid in 4 hours," says Dr. Nakamoto. "With SCUF it would take 24 hours. This makes it ideal for postoperative patients, especially cardiac patients who are acutely ill."

CAVH is a slightly faster system that takes off as much waste fluid as possible. It is the opposite of SCUF in that the waste amount determines how much clean fluid will be returned to the patient. Because large volumes of water take solutes and waste products with them, it is similar to a plasma-water exchange.

Both SCUF and CAVH leave the patient very stable; in fact, without blood pressure changes to tell them when to stop, physicians must be careful not to take off too much fluid. Like CAPD, CAVH is less efficient than dialysis but can run continuously. "The system is totally different from hemodialysis," says Dr. Paganini. "It's like an artificial glomerulus. Since we don't have a tubule yet, we take off fluid, discard it, and pretend that the fluid we make up to return to the patient is what the tubules would have retained. So we've broken away from dialysis to mimic the natural kidney more closely."

The researchers plan to try CAVH initially with an external filter along with blood and water tubes. Ideally the water tubes could eventually be channeled to the bladder, giving the patient the psychological lift afforded by a restored function. SCUF would cut costs by alternating with and extending dialysis treatments. An experimental version of the system is cigar sized. "We're applying it now to acute and hope to apply it to chronic renal failure," says Dr. Paganini. "It will be low-efficiency continual therapy that may have application as an implantable artificial kidney down the road. But we can't rush in and say we have a kidney that's going to replace dialysis. It will take 10 years to develop it." The development will be pushed by Dr. Nakamoto, who wants to see continuous therapy used for patients with chronic disease before he retires.

Technical problems still to be dealt with include those of how to obtain blood access, add heparin as needed, and compensate for the system's inability to remove potassium. The researchers must figure out a way to clean up the fluid that is removed, control it, recycle it, and remove all the wastes. A filter miniaturized for implantation will still need to have its membrane changed.

Dr. Paganini is optimistic that these problems can be overcome. "Miniaturizing a regular dialysis machine is a good idea, but it may be time to rethink how we want to control end-stage renal disease," he says. "The divine prototype is ultrafiltration that plays with fluid. Diffusion works, but we want to mimic the natural kidney's methods of doing it. I think that a new direction is probably the best way to get an implantable kidney, and I'm really excited about the way we're going."

Which kidney failure treatment is appropriate depends on the patient's physical condition, motivation, life-style, family situation, time constraints, and other factors like previous surgery. As circumstances change, patients can switch from one treatment to another. It is somewhat difficult to compare treatments for kidney failure because they are used on patients from various age groups who have other diseases. If these differences are factored out, hemodialysis is still the standard, but the government pipeline for such treatment is emptying out. Meanwhile, funds for researching alternatives are scarce precisely because artificial kidneys are considered standardized treatments. By the time the devices decrease in size and possibly become implantable, most of the problems of kidney transplantation may be solved.

From the rotating drum to SCUF, the artificial kidney seems to have come full circle—from a device that turned its back on nature to one that tries to mimic it. Throughout this development, however, it has maintained a distinct bottom line: the artificial kidney has several defects, but a great many people would never have survived without it.

SECTION

IV

ASSISTED HEARTS

13

Heart Helpers

The heart-lung machine, an eventual result of Dr. Kolff's research in dialysis, has in turn made possible other bionic devices. By enabling surgeons to improve the techniques of cardiac surgery for repairing and revascularizing the heart, the heart-lung machine has contributed to the implantation of heart valves, antiarrhythmia devices, and various blood pumps, including artificial hearts. Discussion of these devices in this and succeeding chapters should be preceded by a brief description of how the natural heart works.

The Natural Heart

The heart has four chambers and four valves. The two atria, or upper chambers, collect blood being returned to the heart from other parts of the body. Two one-way valves allow blood to flow from the atria to the ventricles; two others allow it to be pumped from the ventricles into the large arteries that carry it away from the heart.

The heart can be thought of as a pair of pumps. They work simultaneously in two cycles: during diastole, blood enters the ventricles, and during systole, the ventricles pump it out of the heart.

The right atrium receives blood from every part of the body except the lungs. This blood drains into the right ventricle, which contracts to push it through the pulmonary artery into the lungs. There it releases waste gases and picks up oxygen. The blood then returns to the left atrium, flows into the left ventricle, and is pumped through the aorta to vessels all over the body. It releases nutrients and oxygen and picks up waste products (most of which are removed by the kidneys) and gases. Veins in the upper body carry it back to the heart through the superior vena cava; those from the lower body return it through the inferior vena cava. The blood flows to the right atrium and ventricle and is again circulated through the lungs.

The heart beats about 40 million times per year. During the average lifetime it pumps 5 million liters of blood. The left and right sides of the heart pump approximately equal volumes that change according to the body's demands. Strenuous exercise raises the volume and pressure of blood returning to the right atrium, increasing the heart's output by making it beat faster and pump more blood with each systole.

Heart Valves

The four heart valves keep blood moving in one direction by closing so that it cannot flow backward. Each valve has a different anatomical design to carry out this function. The atrioventricular valves, the two between the heart's upper and lower chambers, squeeze shut when the ventricles contract so that no blood is forced back into the atria. The two semilunar valves close to keep blood in the aorta and pulmonary artery so that it will not flow back into the ventricles during diastole.

Congenital conditions, aging processes, or disease can result in defective valves. If a valve fails to open easily or shut completely, the blood regurgitates. To compensate, the heart pumps harder and faster, causing shortness of breath, dizziness, or chest pain. These and other symptoms indicate that a person should have a "valve job," or replacement of a natural valve with an artificial or hybrid substitute.

In addition to not causing clots, artificial valves must allow free blood flow yet close tightly and very quickly, within a few hundredths of a second. They must be durable, quiet, and easy to implant. The first heart valve implantation was performed in 1952 by Charles M. Hufnagel, M.D., at the Georgetown University Hospital. The hundreds of valve models that have appeared since then can be divided into two general categories: biological valves and mechanical valves.

Biological valves are natural ones taken from pigs; they are sewn to Dacron-covered suture rings, then sterilized and tanned. These valves work very much like their human counterparts: blood pressure forces their tissue to expand so that blood can flow through its sections. They cause less clotting than do mechanical valves, but they do not last as long.

Biological valves can also be made of other natural materials like dura mater (a tissue surrounding the brain) or pericardium (a fibrous sac surrounding the heart) from cows. Some of the blood pumps being developed at the Cleveland Clinic Foundation have dura mater tissue valves. For clinical tests the researchers eventually plan to use valves made of pericardium treated with glutaraldehyde; these may not require the use of anticoagulants.

Mechanical valves come in three types. All three styles have suture rings whose components include metal or plastic coated with Dacron, Teflon, or Biolite carbon. Some valves can rotate within their rings to provide longer wear. Recipients of any type of mechanical valve must take anticoagulants.

The oldest type of mechanical valve is the ball-in-a-cage design. This valve has struts that look like a cage containing a hollow ball. Usually the cage is made of titanium, the ball of silicone rubber or Pyrolite-coated graphite. The bottom of the cage is an open ring smaller in diameter than the ball. Blood flowing through the valve lifts the ball to the top of the cage; as it back flows, it pushes the ball down against the ring to close the valve. The disadvantages of this design are the space problems it can cause and the audible clicking of the ball against the struts.

Bileaflet mechanical valves have two flaps like swinging doors that open and close.

Tilting-disk valves open and shut as blood flows through. Various models of these valves have different pivoting mechanisms, housing fits, guide struts, shapes, and opening angles. The two best-known tilting-disk valves are the Bjork-Shiley and the Medtronic-Hall (or Hall-Kaster) valves, both of which were in the news during the first clinical implantation of the Jarvik-7 artificial heart.

The Bjork-Shiley valve, made by the Shiley division of Pfizer, Inc., was the first effective tilting-disk valve. Although Shiley has come out with a new stronger model, their valves seem to be less durable than those made by Medtronic, Inc. "Over the years we've had fractures of the struts holding the disks of the Bjork-Shiley valves, and recently we've been using Medtronic-Hall valves," says Dr. Olsen. "We think that the latter are superior because in our artificial hearts we've never broken a disk or the valve struts."

At the time of Barney Clark's implantation in December 1982, Dr. Jarvik wanted to use the Bjork-Shiley valves in the Jarvik-7 artificial heart because they were the model with which the team had the most experience. Dr. DeVries wanted to use the Medtronic-Hall valves and stated during the surgery that he hoped that the Bjork-Shiley valves would not break. His statement was prophetic: the mitral (left atrioventricular) valve in Barney's artificial heart broke and had to be replaced. "At the time, we knew of the Bjork-Shiley struts breaking but did not have sufficient animal data to substantiate a change to the Medtronic-Hall valves," says Dr. Olsen. "Since then I've had enough cases to say definitely that the Medtronic-Hall valves are superior in the artificial heart."

To confuse matters further, Medtronic came out with a new valve that proved to be inferior. "Their new Convexo-Concave Valve called the D16 did fracture, and they asked physicians to stop using them just before Barney Clark's implantation," says Dr. Olsen. The valves have, however, been used in many humans with no problem. "The changes in the D16 involved minor modifications in the shape of some of the struts to streamline the flow and reduce blood damage," says Dr. Jarvik. "Those modifications stress the disk more when it closes against the strut. That's probably irrelevant in humans."

It is important to distinguish the implantation of artificial valves in humans from their use in animal and clinical tests of artificial hearts. Since valves can cost about $2000 apiece, the Utah researchers use them in different animal tests of artificial hearts until they break. "A design that is not selected for an artificial heart is not necessarily bad for human use," says Dr. Jarvik. "In an artificial heart powered by our big console, the rise in ventricular pressure over time is 1½ to 2 times what it is in the normal heart," says Dr. Olsen. "This means that after the disk opens, the heart bangs the valve shut very hard. It was this action of the heart console that caused Barney Clark's heart failure on day 13 of his implantation. The Heimes Heart Driver and the electrohydraulic heart are both slower, almost identical to the rate in humans. By changing to other hearts and drive systems and using the Medtronic-Hall valves, I think we'll downplay these problems."

Dr. Olsen's experience is that neither Shiley nor Medtronic is negligent in quality control. "All the companies we've dealt with have been most cooperative and helpful in resolving our common problems," he says. A completely different attitude has been expressed by two former employees of Shiley. In August 1984 George Sherry, a tool design engineer, charged that there were welding defects in the company's 60-Degree Convexo-Concave Valves that cause the struts to break. About a month later, Larry G. Hamilton, a quality control manager, said that Shiley had rejected his suggestions to improve the quality of the valves. Shiley had apparently introduced a new quality control test in February 1982 that had reduced the incidence of fractures, but according to a letter they sent to physicians in February 1983, their third in less than a year, the valves were still breaking. Of the 80,000 valves manufactured, 57,000 have been implanted; news reports state that estimates on the number of deaths they have caused range from 64 to five times that many. FDA inspectors found serious problems in Shiley's manufacture of heart valves; informed of these findings, the Public Citizens Health Research Group (a Washington, D.C., organization founded by Ralph Nader) requested the agency in June 1985 to withdraw the valves from the market.

Harry Zimmerman, a 65-year-old chemical engineering consultant from Baton Rouge, Louisiana, need not worry about this particular problem; his artificial heart valve is made by Hemex Scientific, Inc. Harry has a heart condition that is either congenital or the result of childhood rheumatic fever; it causes his aortic valve to leak. As he grew older, Harry noticed that he was becoming increasingly short of breath. When he developed aortic fibrillation about 8 years ago, he traveled to Houston to be examined at the Texas Heart Institute. "The cardiologist told me that with the leakage, my heart was working at about 75% of efficiency," says Harry. "But my arteries are in better condition than those of the average person, and I seem to have less chance of heart problems than most people my age. The tissue valves they had available weren't much better than my natural valve, so I was advised to keep going as long as possible in case a better valve appeared on the market."

Over the next several years Harry became more short-winded, and his family physician became increasingly concerned about him. By the time he was examined in Houston in 1984, Harry needed a valve job immediately. Fortunately, St. Luke's Episcopal Hospital had FDA permission for clinical trials of a new mechanical valve. Made of graphite encased in nylon, the new valve is thought to last longer than a tissue valve. Harry had it implanted in May 1984.

"The graphite has remarkable properties from an engineering standpoint and is an ideal material for this use," says Harry. "The valve has no flex fatigue and no hysteresis, and it poses less risk of clotting than other mechanical valves.

The heart tissue grows in over the nylon pad, so it becomes like part of the body. It's supposed to be good for at least 12 years."

Harry takes an anticoagulant but has had no problems with his valve over the past year. Instead he is rediscovering activities he had discontinued several years ago. "I used to use up all my oxygen so quickly that I couldn't walk a mile because I'd be panting so hard," he says. "Now I chop logs, dig in our big garden, and play golf without having to ride in a cart." Although technically retired, Harry works as a consultant full-time thanks to his renewed stamina. "I wouldn't do anything else but recommend this valve to other people," he says. "And if I had to get the valve over again, I'd go straight to Houston and have it done."

Bradycardia Pacemakers

The blood is not all that flows through the heart; proper heart function depends also on the flow of electrochemical impulses that regulate the heartbeat.

The heart's natural pacemaker is the sinoatrial (SA) node, a spot of atypical muscle fibers at the junction of the superior vena cava and the right atrium. Electrochemical impulses from the SA node travel through specialized groups of fibers throughout the heart to signal and time its contractions.

Congenital defects, disease, or trauma can affect the functioning of the SA node or the ability of the heart to respond to its signals, resulting in partial or complete heart block. In this condition few or no electrical signals reach the ventricles, which may tire of waiting for them and begin contracting arbitrarily. Heart block can make the heartbeat slow or irregular; it may cause erratic beating or cardiac arrest, either of which can be fatal. The effects of heart block can be minimized, however, by the use of an artificial pacemaker.

A pacemaker system has a pulse generator consisting of circuitry and a power source connected to an insulated wire lead with a tiny electrode at its tip. The pacemaker is implanted beneath the skin, most often in the upper chest near the shoulder. The electrode wire is threaded through a vein to the heart and directed to the right ventricle or to the atrium in the case of a single-chamber device (dual-chamber pacemakers have electrodes implanted in both chambers to sense and pace in each). A pacemaker may otherwise be placed in the abdominal wall with the lead attached on the outside of the heart. The wires carry electrical signals from the pulse generator to the heart muscle. The rhythmical contractions that result are similar to those caused by the SA node.

Since the first clinical implantation by two Swedish researchers in 1958, pacemakers have become the most successful electronic implant ever used in humans. They have evolved from fixed-rate models to demand pacemakers (which send out signals only when needed) to two-speed devices to physiological pacemakers that can be programmed from outside the body. Now that pacemakers contain either microprocessors or custom-programmed chips, physicians can check their operations via telemetry, that is, over the telephone.

Through modem hookups, they can monitor pacemakers from patients' homes by telephone and have ECG strips printed out at their communication centers. Less versatile pacemaker "programmers" now exist as hand-held battery-operated portables that can query, display, and alter up to 40 parameters.

Marcel Malden, M.D., a 61-year-old neurologist from Tacoma, Washington, has had a pacemaker since January 1984. Dr. Malden first noticed symptoms of his heart condition in July 1983, when he became dizzy while climbing 9600 feet up Mount Rainier. Episodes like this began occurring more often in the fall. One morning in November, Dr. Malden ran up the stairs in his home, became dizzy, and discovered that his pulse was irregular. When he consulted a cardiologist a few days later, Dr. Malden learned that he had sick sinus syndrome, a complex arrhythmia that can consist of slow heartbeat alone or alternating with too rapid heartbeat or atrioventricular block. "My heart was beating too slowly," he says. "With exertion it accelerated erratically, but sometimes at the height of my effort it would slow down, causing the dizziness, then pick up speed again. With rapid short-term effort, like running up stairs, my heart sometimes did not accelerate fast enough." Having been instructed by his cardiologist to think about getting a pacemaker, Dr. Malden requested the local representatives of major pacemaker companies to send him information on their products.

Dr. Malden had already chosen a pacemaker by the following January, when he felt ill while relaxing at home one Sunday and discovered that his pulse rate was 36. By the time he had the device implanted the following afternoon, the rate had dropped to 26.

His knowledge of physiology and his particular heart problem led Dr. Malden to select an Intermedics Cosmos pacemaker. "I wanted a sequentially paced, two-lead DDD pacemaker that sensed and paced in both the atrium and the ventricle," he says. "The surgeon and I decided on the Cosmos because it had those capabilities and could store and print out a great deal of information. It was the most technologically advanced model, still investigational at the time."

Dr. Malden wanted a demand pacemaker so that his heart could work on its own as much as possible. Since his pacemaker senses separately in the atrium and ventricle, it paces each chamber only when necessary. Sequential pacing, in which the atria contract first, allows the ventricles time to fill with 20% to 25% more blood.

Although he feels well and functions normally with his pacemaker, Dr. Malden still experiences discomfort from time to time. He continues all his activities like jogging, mountain climbing, and cross-country skiing. If he starts his exercise slowly, his heartbeat climbs to 110 or so on its own, and he can keep going and feel excellent. Sometimes if he stops, however, his heart rate can drop suddenly to the paced rate, causing a momentary feeling of instability. Slowing down gradually helps, but there can be an interim area in which his exertion is not enough to make his heart beat quickly but too great for his paced rate.

The Cosmos pacemaker can be fine-tuned through reprogramming to alleviate this problem. Meanwhile Dr. Malden has been enormously pleased with his selection. "My pacemaker is incredibly clever," he says. "It remembers every beat of my heart up to 16 million, both the natural beats and those that were paced in the atrium, in the ventricle, or in both chambers. Every 3 months or so, we use telemetry to obtain and print out a mind-boggling amount of information. It even tells us the state of the battery and the resistance to the passage of electricity between the lead and the myocardial wall. My physician can program the rate of pacing, refractory periods, voltage and amperage of the stimulus, and impulse delays—I found that reprogramming the delay by five hundredths of a second made me perform and feel better."

Dr. Malden had the opportunity to visit Intermedics, Inc. and tour the factory, where he was extremely impressed with the quality control. "The company has the name of every single employee who worked on each pacemaker," he says. "During my tour, the people who built my pacemaker wore red hearts on their tunics so that I could identify them." He is likewise impressed with the field representatives he has dealt with. "Their knowledge of the physiology of the heart's conduction system and the technology of the pacemakers is extremely comprehensive," he says. "Cardiologists can feel very much at ease in cooperating with them."

Dr. Malden does not think of his pacemaker as an implant, a "foreign body." "My machine is very much a part of me," he says. "It's a good friend—and I like it!"

Antitachycardia Devices

Tachycardia is a condition in which rapid bursts of electrical impulses cause one heart chamber to beat faster than the other. This occurs when the SA node starts sending erratic impulses or an impulse elsewhere in the heart detours on a different conduction path and starts a heartbeat prematurely. In sustained tachycardia such irregular beats can last from seconds to hours, reaching rates of 300 beats per minute (normal rate is well under 100).

Supraventricular tachycardia (SVT) occurs in the atria and ventricular tachycardia (VT) in the ventricles. Ventricular tachycardia causes most of the 600,000 heart attacks that occur in America each year. In this condition the ventricles beat too fast to fill with sufficient amounts of blood. Sustained VT can degenerate into ventricular fibrillation (VF): useless, disorganized quivering of the heart muscle that causes nearly 350,000 sudden cardiac deaths in this country each year.

About 15% of the 1.5 million Americans with cardiac arrhythmia have SVT, which like VT can diminish blood flow. The condition is not life threatening, but its symptoms (weakness, dizziness, faintness, and shortness of breath) can be uncomfortable and disabling.

Drugs are generally the treatment of choice for both types of tachycardia.

Nearly 30% of patients, however, find that drugs are ineffective or intolerable because of their extensive side effects. Researchers have developed implantable antitachycardia devices to correct the problem, but most of the early models had to be triggered by the patient or physician or relied on a single detector. Newer devices are much more sophisticated.

The first totally automatic implantable antitachycardia device approved by the FDA for commercial distribution is the Intermedics CyberTach-60. The device, which senses and automatically stops SVT, is the size of a matchbox, weighs 3 ounces, and is implanted like a pacemaker. If it senses tachycardia for eight consecutive heartbeats, the CyberTach-60 delivers preprogrammed mild electrical impulses to interrupt the conduction path of the SVT signal. The device also records its detection and treatment of the tachycardia. Like a pacemaker, it can provide bradycardia (slow heartbeat) pacing at 60 to 80 beats per minute.

Intermedics is now clinically testing the second-generation Intertach antitachycardia device. The Intertach can distinguish between tachycardia and an exercise heartbeat. If jamming the SVT pathways with electrical impulses does not work, Intertach starts a preprogrammed sequence of alternative tactics. Like CyberTach, it records the number of SVT occurrences and the responses to each; it uses the most recently successful tachycardia treatment when the next episode occurs. Like a pacemaker, Intertach can initiate heartbeats to correct bradycardia; it can be externally programmed and checked through telemetry over the telephone.

The manufacturer is now testing devices to control VT and hopes at some point to develop an implant that could treat SVT, VT, and VF. At least one other device is slightly closer to this goal.

The Automatic Implantable Defibrillator

We have noted that if not corrected, VT can degenerate into VF, which cuts off circulation and blood flow to the brain, killing its victim very quickly.

Like those taken for other arrhythmias, drugs prescribed for VF are often ineffective, cause side effects, can aggravate rather than suppress erratic heartbeats, and do not necessarily reduce mortality. Drugs also do not act immediately on VF. The only way to stop fibrillation quickly is to have medical personnel administer strong jolts of electricity to the heart.

Along with his colleagues, Michel Mirowski, M.D., Director of the Coronary Care Unit at the Sinai Hospital in Baltimore and Associate Professor of Medicine at the Johns Hopkins University School of Medicine, began work in the late 1960s on a device that could monitor and immediately correct VF. Overcoming much initial skepticism on the part of their peers and working without the benefit of government grants, the group developed an implantable defibrillator now manufactured by Cardiac Pacemakers, Inc. Successfully tested in over 700 patients at 35 centers, the device is expected to receive FDA approval for marketing very shortly.

Early versions of the automatic implantable defibrillator (AID) looked like early pacemakers. While pacemakers treat asystole and bradycardia, however, the AID corrects malignant ventricular tachyarrhythmias. To do so requires a much more complex arrhythmia detection system and electrical pulses a million times stronger than those of a pacemaker. While antitachycardia devices correct SVT, the AID can stop VT, often the immediate forerunner of fibrillation. It is more properly referred to, then, as an implantable cardioverter-defibrillator, a device that can treat the whole range of ventricular tachyarrhythmias.

The AID, which weighs about 10 ounces, is implanted in the abdomen. One lead with an electrode in its tip is threaded into the vena cava; another with an electrode patch is placed outside the pericardium of the lower heart. The device also has a bipolar electrode: either two screw-in epicardial electrodes placed on the exterior of the left ventricle or a bipolar electrode catheter threaded into the right ventricle.

"I usually forget that it's there," says 63-year-old Jack Stubbart of Haleiwa, Hawaii, who has had an AID since August 1982. "It makes a slight lump in my abdomen, and it gets in my way a little bit when I'm trying to put on a sock or tie a shoe on my left foot. But I don't mind slight inconvenience from a device that keeps me alive."

When the AID senses fibrillation, it can administer a shock within 18 seconds and up to three follow-up shocks within 2 minutes. After the fourth shock, the device must sense 35 minutes of regular heartbeat before it resets itself. "The shock is a heavy electrical jolt deep in the chest," says Jack. "It's like a blow that knocks me out for a second or two before my heart reverts to sinus rhythm. It's not pleasant, but it's not painful either, and the jolt has no aftereffect." The shocks require much less electricity than is used with standard external defibrillators. Placement of the electrodes in the heart rather than outside the chest prevents much of the power from being dissipated.

The AID can be turned on and off with an external magnet. Using telemetry, a physician can retrieve information about how many times it has shocked the patient and how long it takes for the batteries to charge the capacitors. The batteries last up to 3 years or for 100 shocks before the AID must be replaced.

Jack's AID lasted for 26 months before replacement was required. A former industrial arts teacher and physical plant administrator, Jack retired after suffering one heart attack and one episode of VT in February 1982. Twice that spring he was hospitalized locally for edema and then severe angina; ventricular tachyarrhythmias were recorded despite the use of antiarrhythmia drugs. His physicians referred Jack to the Stanford University Medical School for drug tests. The experimental drugs Jack tried at Stanford likewise did not correct his arrhythmia, so he underwent surgery involving electrophysiological mapping, resectioning of a dead portion of his heart, and three bypasses. When he

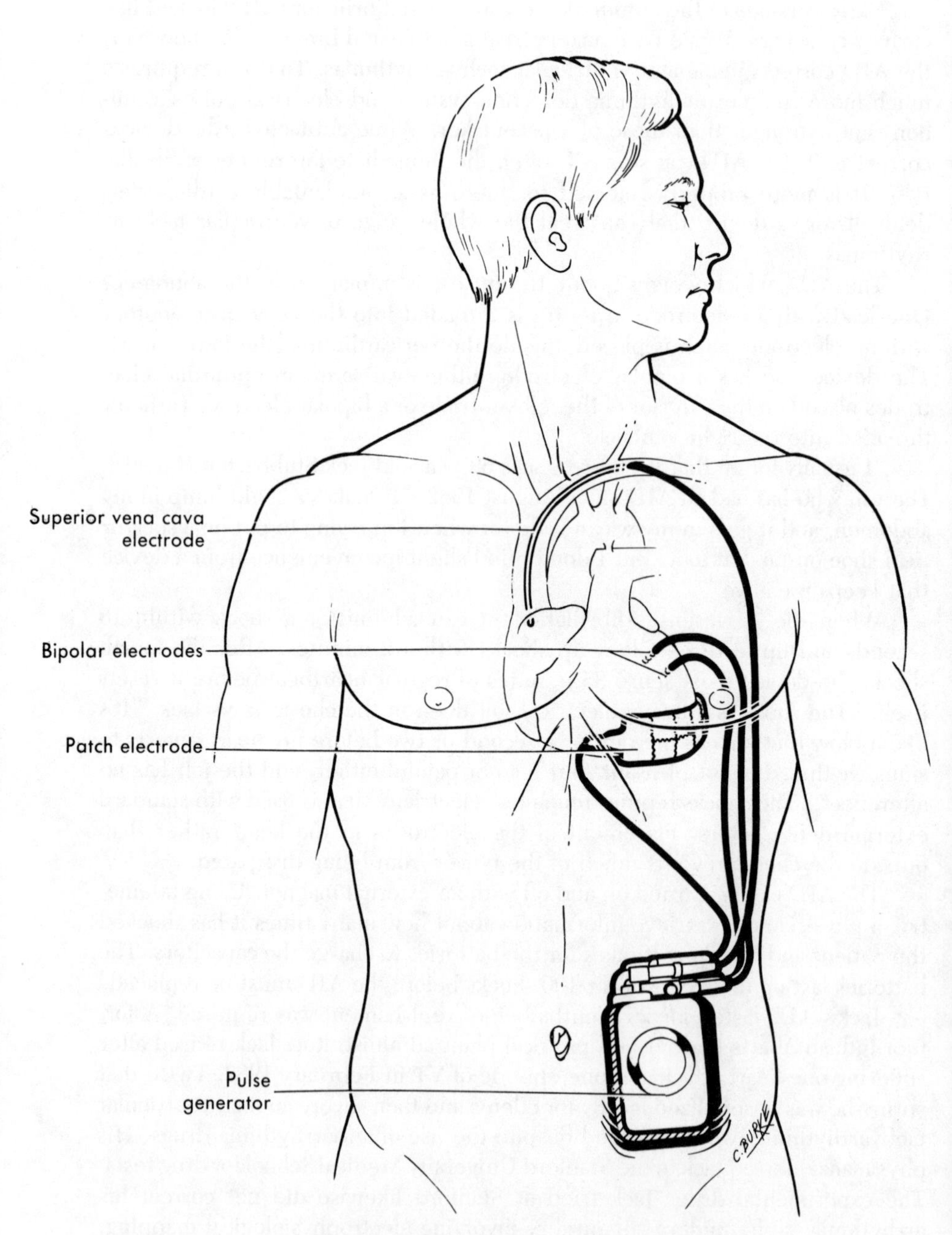

The automatic implantable defibrillator (AID).

recuperated, Jack still had arrhythmia, making him eligible to receive the AID.

"I had never heard of such a device. People gave me all kinds of advice about it, pro and con, so I was a little confused at first. But since I knew that the AID was a last resort to keep me alive, it didn't take me long to decide to try it," says Jack. "I was a little more concerned about the AID than Jack was," says his wife, Dorothy, a 62-year-old retired teacher and volunteer. "I was worried about the idea of his having a strange mechanical device that might break down. Since Jack is an industrial arts expert, he knew about and had more confidence in machinery than I did. I wanted him to live and to live happily." Jack became the eighth patient at Stanford to test the AID experimentally.

Since receiving the AID, Jack has had several VT attacks that resolved themselves before his heart reached the 176 beats per minute required to activate the device. Ten other VT episodes required a single jolt each from the AID to correct the arrhythmia. "I now have more lives than a cat," Jack says. The frequency of the episodes has risen from once every 6 to 9 months to once a month. With changes in diet and medication, Jack looks forward to having the intervals lengthen again.

In October 1984 the batteries in Jack's AID gave out, so he returned to Stanford for a replacement. "The surgery is very minor," he says. "They just open up the lower abdomen, remove the old implant, and put in a new one." Since he lives so far away, Jack spent 3 days in the hospital for the surgery; he believes that local patients would be sent home with instructions to return the next day for follow-up.

Jack and Dorothy find that the AID has lifted a tremendous burden from their minds. They had bought their property on the north shore of Oahu (20 minutes from the nearest hospital and an hour from a full cardiac treatment center) in 1952, hoping to retire there. When Jack's condition was diagnosed, they were told to move within 3 minutes of a hospital so that he could receive emergency treatment. "Now we can travel and live in the country, and I lead a normal life," says Jack. "I was always afraid that Jack would have an episode while I was out somewhere," says Dorothy. "Now I can go to a class or a luncheon or just walk the dog without having to worry."

"What really surprises me about the AID is that so few people have it," says Jack. "Thousands of people are dying because they can't use antiarrhythmia drugs. I want all of them to know that this device is a lifesaver and that they need have no fear about using it."

Newer models of the AID may include bradycardia pacing; meanwhile the present version provides an alternative treatment for tachycardia victims and some of the 400,000 Americans who experience VF each year. For those whose ventricles fail completely, other bionic treatments are available.

14

"Halfhearted" Lifesavers

As explained in Chapter 1, one of the purposes of this book is to locate the Jarvik-7 artificial heart in the context of bionic devices in general. The heart's more immediate context is that of the various types of blood pumps being developed at institutions across the United States. A brief history of this research should indicate why the attention paid to the Jarvik-7 artificial heart is truly unprecedented.

The earliest breakthrough in artificial heart research in the United States occurred in 1957: a polyvinyl chloride heart designed by Dr. Kolff and Dr. Tetsuzo Akutsu, now at the National Cardiovascular Research Institute in Osaka, Japan, kept a dog alive for 90 minutes. Various other researchers had been working on similar devices; in 1964 Congress appropriated money for an artificial heart program to be conducted by the National Heart, Lung and Blood Institute.

Four years later the Heart Institute contracted with three companies (as the Atomic Energy Commission did with two others in 1971) to develop nuclear-powered hearts. The goal was a fully implantable heart, with no external power source, that would last 10 years. The devices produced were not very successful, and the concept became increasingly unpopular. By the mid-1970s the commission was doing everything possible to discourage further work on nuclear hearts, largely by announcing that the devices had to meet impossible specifications. Anecdotally, the commission was worried that among other things, recipients of nuclear hearts would be kidnapped by terrorists.*

Meanwhile the artificial heart program was proceeding along separate but complementary paths of research. Total artificial hearts were being developed to replace entirely natural hearts irreparably damaged by end-stage cardiac disease. The studies were conducted mostly by individual research groups pursuing ideas of their own for which they had received grant support.

Top priority in national programs was given to external or implantable left ventricular assist devices (LVADs), the eventual goal being permanent pumps that could be implanted in the chest and abdomen. This research, which has

---

*The nuclear heart is noteworthy because it is the only type of device that can be *totally* implantable. References by the media and researchers to totally implantable artificial hearts actually mean electrical devices powered through percutaneous wire leads or transcutaneous coils.

received about 80% of the program funds, has been carried out mostly under contracts following government specifications.

In the mid-1970s Theodore Cooper, M.D., then Head of the National Heart Institute and later Assistant Secretary of Health, announced that researchers would not get more funding until they started using the technology they had already developed. Dr. Cooper helped redirect program efforts toward more urgent clinical goals. "Stone heart" (in which the heart locks into a tetany) and similar problems in weaning patients from heart-lung machines after cardiac surgery had fired interest in using assist devices on postoperative patients. "For these reasons, several groups began working on temporary left ventricular assist devices between 1974 and 1976, but only as a spin-off from the work on permanent systems," says Dr. Bernhard. "We thought that we'd compete with the intra-aortic balloon pump to save a life or two, not knowing at the time that there was reversible heart disease that would make these devices more widely applicable."

The artificial heart program, then, ought really to have been called the blood pump program. For one thing, there is actually no "artificial heart" as such; the Jarvik-7 and other models are ventricular pumps designed to be sewn to a recipient's natural atria. For another, the program has included—indeed emphasized—devices other than artificial hearts. Researchers have worked on or tested various blood pumps at the Children's Hospital Medical Center and Massachusetts General Hospital in Boston, Pennsylvania State University, the St. Louis University Medical Center, the Cleveland Clinic Foundation, the Texas Heart Institute, the University of Utah, Stanford University, the Pacific Presbyterian Medical Center in San Francisco, and other institutions. "The government has been pretty reasonable," says Dr. Bernhard. "The program has taken twice as long as anyone thought it would because the work is so difficult." Research has progressed particularly because of the development of improved polyurethane blood surfaces, brushless DC motors, and transcutaneous energy-conversion systems used in the devices.

Each type of blood pump—temporary left or right ventricular assist device (RVAD), permanent VAD, and "total" artificial heart—is an alternative mechanical treatment appropriate for some types of heart disease. Many researchers agree with the Heart Institute that VADs should have priority, at least at the present stage of investigation. Those working with the Jarvik and Utah hearts obviously feel differently, and their experience provides additional information useful to other researchers. Before discussing total artificial hearts, however, we should understand why VADs offer them such stiff competition.

Obviously the heart's right ventricle, which pumps blood only to the lungs, does far less work than the left ventricle, which must generate arterial pressures of at least 100 millimeters of mercury to push blood throughout the body. The left side of the heart does about 80% of its work and until quite recently was thought to cause most of its problems. This is why the Heart Institute has historically emphasized research on LVADs rather than RVADs. More work with these systems has shown that the right side of the heart could benefit from temporary assistance as often as does the left. Our discussion will similarly emphasize LVADs while acknowledging that temporary RVADs will be needed as well.

Temporary VADs

Although merely a spin-off of the major artificial heart program, the temporary VAD is the only device to date that has really proved itself to be lifesaving. Temporary LVADs currently being used on human patients are pneumatic devices designed to pump blood from the left atrium or ventricle to the aorta; an RVAD fills from the right atrium or ventricle and sends blood to the pulmonary artery. In most cases the pump is placed on the patient's chest and connected to these vessels by percutaneous tubes. It can conceivably remain in place for up to a month, depending on the patient's preexisting condition or the likelihood of infection. (The thick, pulsing air lines of a pneumatic device are more likely to cause infection than the thin wire leads that will power implantable electric blood pumps.) Recovery from reversible left ventricular failure takes up to 9 days (usually the right side of the heart recovers in 3 to 5 days), after which the patient returns to surgery to have the pump removed.

About half the patients treated with LVADs now survive; cardiologists do not yet know exactly how they recover. They think that improvement may result from decreased fluid retention in the myocardium (middle layer of the heart wall) and increased high-energy phosphate levels in the tissues.

Temporary LVADs may eventually be used to treat myocardial infarction (destruction of myocardial tissue resulting from decreased blood supply). In different uses they have helped patients suffering from cardiomyopathy (myocardial disease that destroys muscle cells, causing the heart to enlarge and eventually to fail) and cardiogenic shock (shock resulting from decreased cardiac output in heart disease). LVADs assist patients who are recovering from heart surgery and have made headlines when used as "bridges" to tide over patients awaiting heart transplants.

POSTOPERATIVE USE

Over 300 patients have been treated with LVADs postoperatively to wean them from heart-lung machines. We have already noted that the heart-lung machine oxygenates and pumps blood so that the natural heart will remain motionless and bloodless during surgery. After about 6 hours of use the machine begins to damage large numbers of red blood cells. If damaged clotting factors are not replenished with transfusions, the surgical wound will continue to bleed. Sometimes a patient's heart is not strong enough to take over pumping action after surgery, but leaving him or her on the heart-lung machine would destroy more blood components. In this case the patient can be treated with various cardiotonic vasoactive drugs (which increase cardiac output) or with an IAB.

The IAB was developed by Drs. Kolff, Moulopoulos, and Topaz in 1962. Thanks to both design improvements and initial clinical experiences, the pump was widely used on adults by the mid-1970s.* It was modified for application to infants in late 1984.

The IAB is a resuscitation device that increases and decreases the pressure in the aorta so as to ease the workload of the left ventricle. It consists of a balloon the size of a wiener with a thin gas tube running down its center. The balloon is attached to a catheter and controlled by a bedside console. It is inserted in the femoral artery of the leg and threaded to the aorta, where it controls pressure by inflating and deflating in synchronization with the heartbeat.

Although effective in treating some patients with cardiogenic shock associated with either postoperative complications or myocardial infarction, the IAB increases the efficiency and pumping dynamics of the heart by only 15%. It can be used only up to a month, and some patients cannot be weaned from it.

Patients dying of left ventricular failure who are past help from drugs or IABs may recover if 80% to 90% of their cardiac output is kept up for anywhere from several hours to a few weeks. In about half the cases an LVAD can reverse a myocardial depressant situation within 10 days. The surface area of the LVAD is much smaller and less likely to cause clots than that of the heart-lung machine, so it does not damage blood as much. The pump can start out doing all of the left ventricle's work, then be turned down slowly until the patient's heart can take over on its own.

Temporary LVADs are an excellent example of the practicality of bionics research. To begin with, work on these devices first made cardiovascular surgeons aware that some forms of heart disease were reversible. "The temporary LVAD is limited to use on reversible myocardial muscle damage, with swelling, edema, or ischemia [deficiency of blood supply]," says Dr. Hill. "It can't bring infarcted muscle back to life."

Similarly, it was only by using LVADs that researchers were able to discover that right ventricular failure frequently accompanies that occurring on the left side of the heart. Previously they had believed that most patients suffered left ventricular failure and that treating them with an LVAD and sufficient drugs would help the right side of the heart recover if similarly afflicted. The earliest assist devices were therefore called LVADs and were designed only to support the left ventricle. The few times that it was attempted on patients, biventricular assistance was not very successful. "The literature on biventricular pumping is so discouraging that many groups just gave up the

*The IAB was popularized chiefly by Adrian Kantrowitz, M.D., Chairman of Cardiovascular and Thoracic Surgery at Sinai Hospital in Detroit and Professor of Surgery at Wayne State University, and his associates.

notion altogether as unfeasible," says D. Glenn Pennington, M.D., Professor of Surgery at the St. Louis University Medical Center.

William S. Pierce, M.D., Professor of Surgery and Chief of the Division of Artificial Organs at the Milton S. Hershey Medical Center at Pennsylvania State University, had pointed out that right ventricular failure was a significant problem and had written a paper describing the difficulties that arose in treating it with assist devices. Looking at their own experience, Dr. Pennington's group found right ventricular failure to be common in patients with postsurgical complications. They have since succeeded in defining the role of biventricular assistance in postoperative heart patients. "We happened to have the most appropriate patients with whom to make the procedure successful," says Dr. Pennington.

The right side of the heart fails despite its comparatively light workload because of a combination of circumstances. One is that the pressure in the pulmonary artery of postoperative VAD patients increases during the first 12 hours after surgery; this resistance to pumping from the right ventricle requires it to work harder. The extra effort in turn increases the tendency for right ventricular failure.

The group found that the best predictor of biventricular failure is perioperative myocardial infarction (that which occurs a few days before, during, or immediately after surgery). This condition, too, can result from several circumstances. Assist device candidates who have received emergency surgery for acute myocardial infarction will have injured hearts even though they have been revascularized. "We learned that most of the time the perioperative infarcts were biventricular," says Dr. Pennington.

Using the Medtronic centrifugal pump and the Pierce-Donachy pump made by Thoratec Laboratories, Inc., the group found that patients with isolated left ventricular failure could be weaned from the heart-lung machine and survive after treatment with just an LVAD. Those with biventricular failure, however, required biventricular mechanical support: either an RVAD and an IAB, or two VADs. An LVAD and drugs did not work.

"We also found that with biventricular failure, the right ventricle recovered sooner, even though it might initially have been worse than the left," says Dr. Pennington. "That seems comparable to findings at other centers. The reason may be that resistance to blood flow from the right ventricle decreases; since normal pulmonary pressure is usually low, the right heart can recover faster."

Joe Hoggard cannot say whether or not *his* right ventricle recovered first because he was disoriented during the entire postoperative period he spent on biventricular assist. A 45-year-old boiler operator from Columbia, Illinois, Joe had suffered from progressive heart disease for several years before having a cardiac bypass. "It's hard to describe how I felt during that time," he says. "I just thought I was slowly getting old. I couldn't do certain activities without becoming completely exhausted, and I was very ill natured and short tempered."

In July 1983 Joe had a mild heart attack; cardiac catheterization revealed some minor blockages for which he was prescribed medication. His condition kept deteriorating until March 1984, when his physician sent him to the St. Louis University Medical Center for cardiac rehabilitation. Cardiologists there found some half a dozen severe blockages and quickly scheduled Joe for surgery on April 1. Although the operation went well, Joe could not be weaned from the heart-lung machine because of biventricular failure.

"I had never heard of such a thing as the VAD," Joe says. "They said nothing about it before the surgery, because they had no way of knowing that I'd need it. The surgeon came out of the operating room and told my wife that she had two choices: either let me die or start signing a whole stack of consent forms without wasting time reading them." Fortunately, Iva Hoggard is a registered nurse; her experience working with acute care procedures made her willing to provide informed consent. "I had told Iva that I didn't want to be put on machines and kept as a vegetable, but I'm sure glad that they put me on these pumps," says Joe.

Joe spent 4 days on biventricular assist; both his pumps were removed at the same time. "It didn't bother me to find out that I had been on the machines, because I was too sick to realize what it meant," he says. "As I regained strength, I came to recognize that the pumps saved my life."

Aside from being treated for postoperative arrhythmias, Joe has had no problems since the surgery. "I feel super great now," he says. "I'm better than I've been for the past 10 years, and I'm enjoying myself much more. I had a really serious operation, but I'm not done for by any means. I ride racehorses; I work full time and overtime and generally do whatever I please."

Biventricular failure does not occur just postoperatively; this is one reason that these findings about assistance are so important. A transplant candidate awaiting a donor heart might likewise suffer biventricular failure. "We once tried to use an LVAD on a young woman we were trying to maintain for transplantation; her right ventricle failed, and we had to add an RVAD as well," says Dr. Pennington. "If we had recognized the need for two VADs and put both in immediately, she might have survived."

These findings help justify the use of biventricular assistance. There are many disadvantages to treating patients with two devices: the cost is doubled (heart assist pumps presently cost anywhere from $10,000 to twice that amount); the power demands are greater; the surgery is more difficult; and the patient endures more trauma and may face more complications. If biventricular assistance is the only treatment known to work, however, there is no other choice.

This lifesaving knowledge could not have been obtained without heart assist devices. "I don't think that we understood right ventricular failure very well; obviously we didn't know how to treat it, and the fact that the right side of the heart recovers sooner than the left is extremely important," says Dr. Pennington. "Now we know that we might use an implantable permanent LVAD in a patient along with a temporary RVAD that could be removed if that side of the heart recovers."

All this means that the government should once again shift its emphasis, from LVADs to VADs that can be used for the right as well as the left sides of the heart. "In our summary report from our NIH contract, we emphasized the high incidence of biventricular failure in postoperative patients," says Dr. Pennington. "I think that the recent NIH experience would indicate that it's a real problem to be dealt with."

Although they have proven their value as research tools, temporary VADs have rather limited application. "Temporary VADs are useful and save a few lives, but the systems are terribly expensive, and there aren't many patients who need them," says Dr. Bernhard. "Only about 500 to 750 patients a year will have problems after cardiac surgery and not benefit from intra-aortic balloon pumps or drugs. There should be none. If we do a good job and continue to improve our knowledge, technology, and devices, there should be an infinitesimal mortality from cardiac surgery. That would eliminate 99% of the candidates."

The number of candidates will increase, however, if temporary VADs continue to be used on patients awaiting donor hearts for transplantation.

BRIDGE TO TRANSPLANT

Temporary blood pumps used as so-called "bridges to transplant" are associated in most people's minds with Denton Cooley, M.D., of the Texas Heart Institute. In 1969 Dr. Cooley had implanted an artificial heart made by Dr. Domingo Liotta in a patient named Haskell Camp until a donor heart became available for transplantation; he repeated the procedure in 1981 on a patient named Willebrordus Mueffels, using a heart made by Dr. Akutsu. In 1978 Dr. Cooley had used an LVAD manufactured by Thermedics, Inc., for the same purpose.

Two more recent uses of VADs as bridges to transplant occurred in the early fall of 1984 in California. Researchers at Stanford University implanted an LVAD made by Novacor Medical Corp. in a 51-year-old man suffering from chronic cardiomyopathy. The device, an electromechanical LVAD, had to be implanted rather than used percutaneously because of its design. It worked very well in that application, and the patient survived.

Simultaneously Dr. Hill used an LVAD manufactured by Thoratec to sustain 47-year-old Ronald Meehan of Sausalito, California, who had suffered a massive heart attack. The patient lived with the device for 2½ days before receiving a heart transplant and has recuperated well. "The use of bridges to transplant is one of the first leap-throughs to a new set of patients who previously had always died," says Dr. Hill. "These devices do the job, allow the people to live, and give us valuable clinical experience."

Thoratec manufactures more than one model of LVAD; the pump Dr. Hill uses was designed by Dr. Pierce. It can be used either externally or as an implant. "There are a variety of medical reasons that we might want to apply it one way or the other," says Dr. Hill.

The present pneumatic model was actually developed for chronic use as an electrohydraulic device with an implantable drive system. Its pulsatile flow causes few blood clots, so the patient using it need not take anticoagulant drugs. "The pump is quite versatile, since it can be used inside or outside the body or as a pneumatic or electrohydraulic device," says Dr. Hill. "This feature gives it certain long-term advantages as a temporary device or bridge to transplant."

The spring of 1985 brought additional application of blood pumps in heart transplant candidates. A bridge to transplant performed without FDA approval under unusual circumstances took place in March in Tucson, Arizona. Jack G. Copeland III, M.D., Chief of Cardiothoracic Surgery at the University of Arizona had performed a transplant on Thomas Creighton, a 33-year-old auto mechanic whose heart had been weakened by two attacks and chronic cardiomyopathy. When the transplant failed, Dr. Copeland put the patient on a heart-lung machine for nearly 8 hours while trying to obtain an artificial heart. Since the team was unable to obtain a Jarvik-7 artificial heart quickly enough, Cecil Vaughn, M.D., of St. Luke's Hospital in Phoenix, instead implanted an artificial heart designed by Kevin Cheng, D.D.S., a dentist affiliated with St. Luke's. Thomas Creighton underwent a second transplant 11 hours later but died of "shock lung" and subsequent heart failure, a complication of his time on the heart-lung machine, the following morning.

In April 1985, 16-year-old Michael C. Jones received a donor heart at Jewish Hospital in Louisville, Kentucky, after his natural heart was damaged by a virus. He was kept alive for 5 days on biventricular assistance until a donor heart could be implanted by Laman A. Gray, M.D. Although given only a 20% chance to live with the new heart because his kidneys were not functioning, Michael has survived his ordeal.

Another survivor whose condition was equally serious is 54-year-old Donald Croskrey, a ceramic tile setter from Branson, Missouri. Don was suffering from ischemic heart disease and began developing symptoms about 4 years ago. "I was getting short of breath and weak while going up a flight of steps, and I couldn't carry heavy objects that would have been easy for me when I was younger," he says.

On April 3, 1985 Don had a heart attack and was taken to his local hospital. On April 11 he had a second attack and was flown by helicopter to St. John's Hospital in Springfield, Missouri. "There I progressed just so far—I'd have good and bad days," he says. "When I felt progressively weaker 4 days in a row, the doctors told me that a transplant was necessary to save my life." On May 13 Don was flown 200 miles to the St. Louis University Medical Center in a helicopter that just managed to sneak between two severe storm fronts. The following day he had a third, nearly fatal heart attack and went into shock. "Without a VAD, I would have died at that point," Don says. "I remember hearing the nurse twice tell the doctor that I had no blood pressure and no pulse."

Like Joe, Don had explained to his wife, Dolores, that he did not want to prolong his and the family's agony by being kept alive on machines. "When the surgeon told me about the pump, I was initially scared to death," Dolores says. "In an emergency they can't explain things very well, and I couldn't listen well at that point either. I knew that we had to do something really quickly to save Don, but I wasn't sure that this was what he would really want." Dolores went in to see her husband, who was able vaguely to understand her questions and nodded to indicate that she should give approval for use of an LVAD. "I got just enough of a response out of him to make me feel that signing that paper was the right thing to do," says Dolores. "I wondered afterwards and hoped that my feeling was correct—and it was, because the pump was what saved his life." Don was kept in surgery until the physicians were satisfied that he would not need an RVAD as well. He remained on the LVAD for 36 hours until a donor heart arrived from Texas.

Don was semiconscious for much of the period he spent on the LVAD but remembers very little. "I know that the family came in to see me; they asked me questions, and I could respond by squeezing their hands. I don't remember the roomful of machinery, the loud noise they said it made, and the way it made my bed shake," he says. "It was scary to see the pump—not as bad as I thought it would be, but bad enough," says Dolores. "But it was wonderful to go in after Don's surgery for the LVAD, ask him a question, and have him respond to me under sedation. Once he was on the pump, I was sure that he had made it through the worst of it, and by the time he had the transplant, I was positive that everything would be all right." Apparently the pump's action helped Don's condition improve, thereby better preparing him for the transplantation.

About 28 hours after receiving his new heart, Don began regaining consciousness; he has steadily improved since. He was released from the hospital on June 1 to an auxiliary house and expects to be sent home at the end of the month. "We've been looking at a photograph of me and my daughter taken 2 weeks before Easter," he says. "I looked all puffed up and very sick, but my appearance had changed so gradually that no one in the family had noticed it. Now I've lost 40 pounds, my color is back, and my wife says I get better looking every day." A month earlier, Don had been too weak to write and recognize his own name. Now he has daily physical therapy at the hospital and is regaining strength.

"I wouldn't hesitate a second either to recommend this pump to someone who might need it or to go through it all again," Don says. "It's a marvelous, marvelous invention that pulled off a miracle for me."

Both Don and Dolores feel that VADs should be better publicized to spare others from additional worry and bewilderment in cardiac emergencies. "The unknowns are what scare people. If people were prepared for VADs or had seen pictures of them, they wouldn't be so alarmed and frightened at the idea of using one," says Dolores. "I think this treatment should be brought out more; maybe it will inspire people to donate organs," says Don. "And it will cause less confusion if people know about it ahead of time. Their first impression is that it's a piece of life support machinery, when actually it's a means to an ultimate cure."

As bridges to transplant, temporary VADs could be used for 7 to 10 days on up to 300 or 400 patients per year. The effect of their use on transplant survival rates is not proven. "I think that the primary factor that would decrease a bridge patient's chances for survival would be an infection," says Dr. Pennington. Implantable systems like Stanford's would make infection less likely, although none of the VAD patients who had transplants has become infected because of the holes in the chest. VADs therefore look promising for transplant patients in particular as well as heart disease victims in general. "I think that these intermediate systems for bridge to transplant will become steppingstones to chronic pumps," says Dr. Hill.

LIFE AFTER ASSISTANCE

The patients presented here represent others who have survived heart failure and ventricular assistance. Last year the NIH contractors did a follow-up study on 15 patients from the St. Louis University Medical Center, the Cleveland Clinic Foundation, and three Boston hospitals: Children's Hospital, the West Roxbury VA Medical Center, and the Boston University Medical Center. About half of these patients were actively employed full time; most of the rest were retired and active; and only a few were cardiac disabled. "The point is that a patient placed on an assist device who recovers and leaves the hospital has a good chance of living well rather than as a cardiac cripple," says Dr. Pennington. "We operated on one woman when she was 68 years old and used a VAD on her. Four years later her biggest complaint is that her back hurts when she mows her lawn. People think that these pumps are horrible machines that will have them in hospitals for the rest of their lives, and that's just not true."

"Most of the cardiac cripples I've run into in therapy are psychologically handicapped," says Joe Hoggard. "They won't do anything because they don't feel safe unless they're hooked up to monitors. I can't live like that—I have too much of my life left and too many important things to do."

Misconceptions about VADs arise in part from the difficulty of trying to predict which patients will need ventricular assistance after surgery. The patients and their families do not have the opportunities in advance to learn about VADs, discuss that option, and give informed consent. Obvious candidates for VADs are usually too sick to sign consent forms; at other times, surgeons rush out of operating rooms to inform families that if they do not sign the forms immediately, the patient will die. "In almost every instance, we got permission from the family rather than the patient," says Dr. Pennington. "Most people

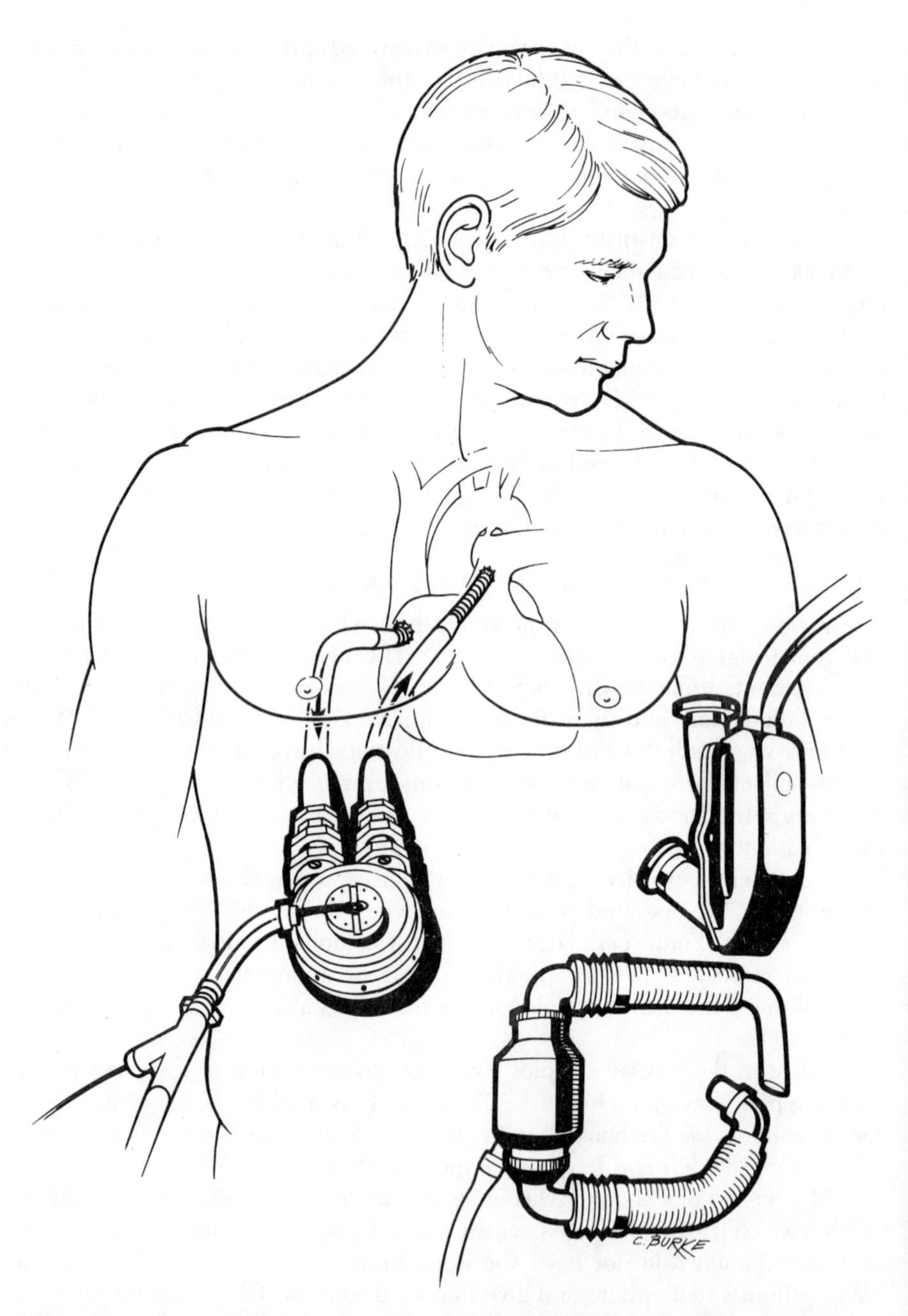

A temporary pneumatic right ventricular assist device (RVAD) can pump blood externally from either the atrium (as illustrated here) or the ventricle. The models shown have been developed at Pennsylvania State University, Thermedics, Inc., and the Cleveland Clinic Foundation.

say that they'll take a chance on a VAD rather than face certain death. But people greatly fear being kept alive in hospitals beyond hope of recovery; they don't want to be dehumanized in that way."

One of Dr. Pennington's patients was a 28-year-old woman who had to be placed on biventricular assistance after cardiac surgery. She woke up in intensive care to find two pumps sitting on her chest; as soon as she was able to, she motioned for pencil and paper. "Each time I came in to see her, she would write a note saying, 'Please take these pumps away, and let me die,'" says Dr. Pennington. "I kept her on the assist; she recovered, and now she's one of our most grateful patients."

It would be tragic if ventricular assistance were refused because the patient or the family thought that a prolonged vegetative state would result. Although an external pump attached to the heart is not a pleasant form of treatment, it is important to remember that a temporary VAD is finite therapy, used when recovery seems likely or transplantation is anticipated. "At first we feared that patients might survive only on VADs, so that eventually we'd have to switch off the pumps, but that hasn't happened," says Dr. Pennington. "Those patients who don't survive almost always die of multiple organ failure."

Those who do survive have benefitted from temporary VADs in incredible ways that they themselves probably do not understand. "All of them would otherwise be dead, and yet now they have good hearts and are living active lives," says Dr. Pennington. "They are positive reinforcement for everyone in this business. We all went for a long time without having any patients survive, so when the first few live and return to normal, it's very exciting."

Permanent LVADs

Research on permanent LVADs addresses problems completely different from those involved with temporary VADs. "There are so many candidates for these devices that people don't like to talk about them, and the transplant enthusiasts become overwhelmed," says Dr. Bernhard. "The estimate of 50,000 a year or so that is tossed around is a conservative one that the experts can defend."* Thousands of people of all ages have irreversible heart muscle diseases like end-stage coronary artery disease or cardiomyopathy. Far fewer donor hearts are available for transplant; some of these patients are not candidates for the procedure, while others who are reasonably young and healthy are dying because cardiologists do not know how else to help them. This group is the focus of the Heart Institute's efforts. To date, nobody has established for sure that they *can* be helped.

Even if nuclear-powered blood pumps were not controversial, they would

*Dr. Bernhard also points out, however, that the Heart Institute's most recent estimate, published in May 1985, is that 17,000 to 35,000 heart disease patients under age 70 would be candidates for permanent blood pumps.

be too bulky, heavy, and technically unreliable for permanent use. Researchers have therefore spent several years trying to perfect electrically driven permanent LVADs. "These devices pose a major problem in engineering, blood contacting surfaces, biomaterials, electrical systems, and electronics," says Dr. Bernhard. "It's an order of complexity greater than anything we've ever done, in comparison to which the temporary pump is a toy."

Various components of permanent LVADs have been bench tested for as long as 5 years according to protocols that make the NIH look like Big Brother. Each LVAD is assigned a registered number and placed on a mock circulatory system that functions constantly for 2 years. These prototypes are all connected to the NIH mainframe computer; any malfunction is immediately recorded in Bethesda, Maryland. Unless adequately explained to the Heart Institute, "multiple events" of this sort eliminate the device from further development. "Over this same time period, we're doing stringent animal studies," says Dr. Bernhard. "At the end of all this, 3 years from now, some systems should qualify with the NIH, the Heart Institute, and the FDA."

Many different models of permanent LVADs are presently being developed or tested (permanent RVADs are less necessary than temporary ones). A "generic" device of this sort is designed to be implanted in the chest or abdomen, usually under the diaphragm, with conduits running through that muscle to the heart. It has a blood pump—either a diaphragm or a sac—that can function on different modes, the basic goal being to synchronize it with the heart. A brushless DC motor activates the pump to send blood from the left atrium or ventricle to the aorta.

Whether to connect a VAD to the atrium or ventricle has always been a controversial issue, since each technique has certain advantages. "If a patient has an acute myocardial infarction and we can't get him or her off the heart-lung machine or out of shock, we wouldn't want to remove a large piece of ventricle in order to thread in a cannula," says Dr. Pennington. "In this instance, atrial pumping would be much preferable. On the other hand, if the patient had a big scarred heart that was not worth saving, and we planned a bridge to transplant, we would use ventricular assist and leave the atrium for transplant."

The motors presently being developed for VADs are generally either electromechanical or electrohydraulic. An electromechanical LVAD has a DC motor-driven cam or roller screw that converts reversing rotary motion into linear pusher-plate strokes. (Using solenoid to energize two opposite pusher plates pumps the blood with less strain on the sac.) In an electrohydraulic device (which will be described more fully in Chapter 15), a motor-driven turbine pressurizes silicone fluid that compresses the blood sac, or a gear pump does the same with a pusher plate.

In the distant future, fuel cells might be used to power LVADs: blood glucose would be passed through a system that transforms chemical into mechanical energy.

The permanent LVAD also has a bag that injects gas into the motor to preclude a vacuum when the pusher plate device moves the plate in order to pump. The gas moves out as the plate moves down. This design allows large changes in volume to occur with minimal changes in pressure in order to control filling and consumption.

A separate system in the LVAD controls motor speed and reversing rate and synchronizes the device with the natural heart, which is left intact.

All this technology has an external DC power source that runs on batteries or can be plugged into a wall outlet. The first generation of permanently implantable devices will be percutaneous, with a wire connecting the LVAD to the power source. In the future each system will have a secondary coil that will be implanted under the skin and a primary coil that will be taped over it. An oscillator worn on the patient's belt will convert the DC current to AC, the only type that can be transferred transcutaneously.

Some permanent VADs being developed are variations on these themes. A permanent LVAD produced as a collaborative effort by the Cleveland Clinic Foundation, the University of Washington Joint Center for Graduate Study in Richland, and Whalen Biomedical, Inc. is a thermohydraulic system. "Thermal systems had been originally developed to run VADs that could last 10 to 15 years without batteries or refueling," says Raymond J. Kiraly, M.S., Director of Engineering in the Department of Artificial Organs at the Cleveland Clinic Foundation. "Instead of continuing this approach, we're now using these systems with thermal batteries." Each prototype contains a mass of molten salt that is periodically melted when the device is connected to an electric heater; the thermal battery will provide the patient with complete mobility for 8 to 10 hours. The design is much more complicated than that involving an electric motor.

The Cleveland group is also working on improving a centrifugal pump that was developed at the University of Minnesota, Minneapolis, then marketed and later discontinued by Medtronic. The pump has a steady, nonpulsatile flow; the impeller runs at about 10,000 rpm, converting velocity into pressure to pump the blood without valves. "This type of flow may damage the blood a small amount, but no more than any other of the blood pumps we've been using," says Kiraly.

The centrifugal device is simple and compact: it has one moving part and a simple electric motor the size of a flashlight battery. The group plans to integrate this motor once they have eliminated clotting and leakage that has been occurring at a shaft seal. The device also needs a sensor to adjust the flow and pressure characteristics in response to changes in demand. Once these technical problems are solved, the pump will be totally implantable. "It could be one of two pumps used as a total heart, or one pump used as an assist," says Kiraly. "We've tested it on about a dozen patients recovering from cardiac surgery."

LVADs designed to be permanent have been tested on humans only temporarily. Most researchers believe in starting with temporary implantations, continuing tests on animals, and working up to permanent implantation once

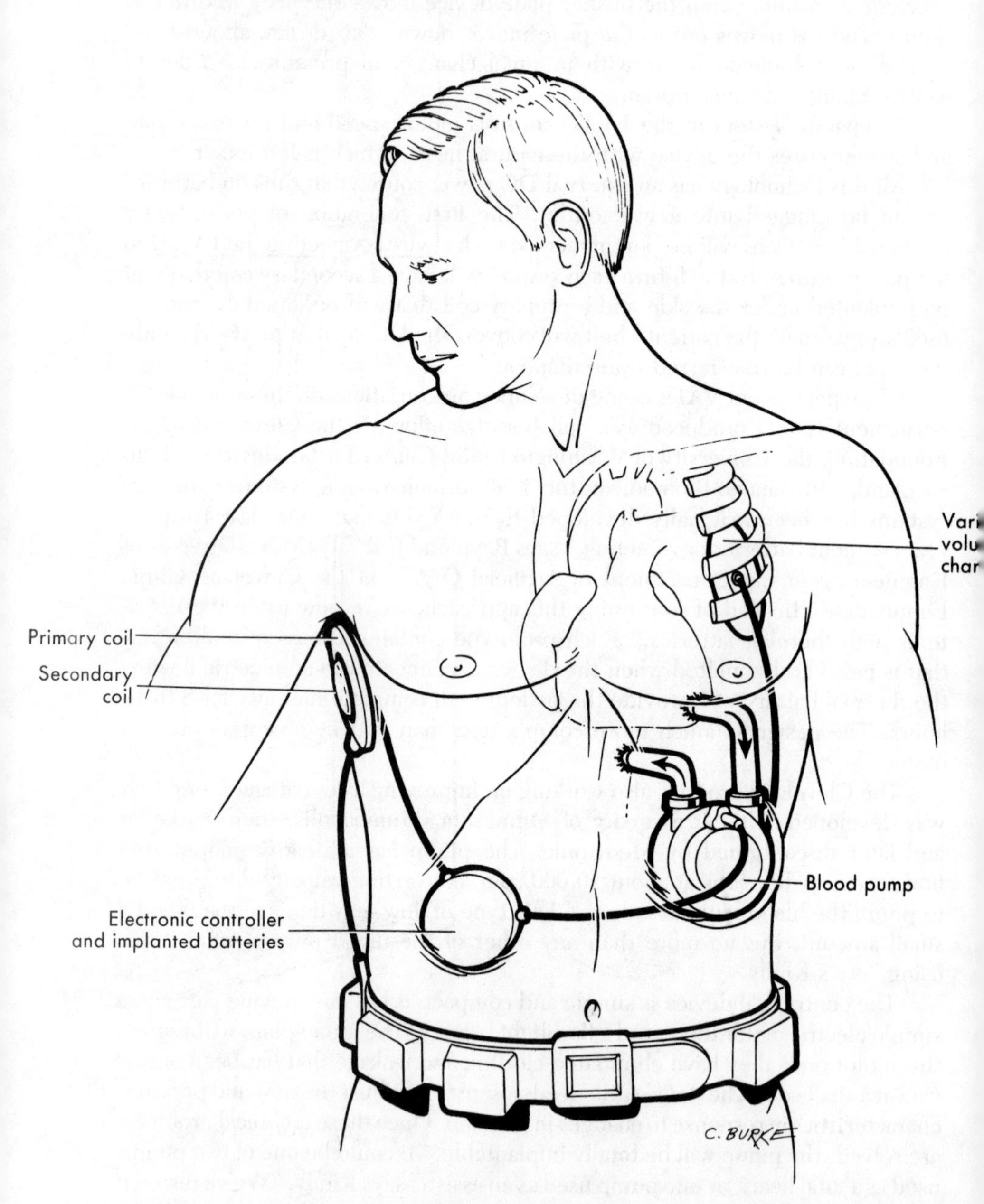

A permanent transcutaneous electric left ventricular assist device (LVAD) developed by researchers at the Children's Hospital Medical Center in Boston and Thermedics, Inc. This LVAD is powered by current running from the doughnut-shaped primary transformer coil taped on the skin to the implanted secondary coil.

the devices can offer significant improvement in their recipients' quality of life. This means that most blood pump programs involve various models of total artificial hearts and VADs at different stages of development, all aimed at providing alternative treatments for the largest number of patients. The more options available to cardiologists, the more their patients will benefit. Nevertheless, controversy still exists over the relative advantages of permanent VADs and total artificial hearts. "LVADs are highly experimental," says Dr. DeVries. "They are by no means any more therapeutic than artificial hearts are at this point."

We have noted that a total artificial heart is essentially a single-unit biventricular assist device; the main difference is that it replaces the natural ventricles rather than assists them. "Certain patients may be well served by having their hearts removed, because that simplifies implantation of a permanent device," says Dr. Pennington. "There is a definite need for acute VADs, but I see no need for a permanently implanted biventricular assist system instead of a total artificial heart," says Dr. Olsen. "Permanent VADs imply that the ventricular myocardium is not likely to recover or will continue to deteriorate. The natural heart is not at all static; it constantly wears out and regenerates."

The major advantage of VADs, however, is that they leave open far more options. A VAD can always be implanted temporarily until a patient's prognosis becomes clearer. Tests performed on brain-dead patients by Jack Kolff, M.D., of Temple University, have shown that it is feasible to implant two pumps, then remove one of them if one side of the patient's heart recovers. "With acute myocardial infarction, we can't tell how much of the heart will recover," says Dr. Pennington. "It makes sense in these cases to preserve the natural heart to see what happens."

The cardiac surgeon has the choice of removing the VAD also if more surgery is required or if a donor heart becomes available; otherwise, the device can be left in permanently. The patient can receive one VAD as a heart assist or two on either side of the natural organ as a total heart. As total hearts, VADs can be used to run important tests that are impossible with natural hearts or single-unit artificial ones. The natural heart remains as a safety factor: if the VAD fails, the heart might keep the patient alive until he or she can get to a hospital. Leaving the natural heart intact also avoids all kinds of potential ethical problems that have been troubling the FDA and institutional review boards (IRBs).

Although the technology of a VAD is similar to that of a total artificial heart, the assist devices are easier to use, less expensive, and applicable earlier in the patient's treatment. The possibilities of doing clinical work with the VAD are much greater than with the total artificial heart. Researchers learn much more from clinical than they do from animal tests, so work on the VAD enables them to progress faster.

Temporary VADs have already saved the lives of hundreds of patients, and their use as bridges to transplant has opened up new possibilities for transplantation. Some 80,000 people per year might be candidates for temporary VADs before transplantation if the number of donor hearts could be increased from the current total of about 1000 annually. Their use of the devices might stimulate the public or government to improve donor collection processes and continue with this type of research.

The point of such comparisons is simply that although it has received enormous publicity, the Jarvik-7 artificial heart is just a preview of what is to come. The most appropriate pump for any heart patient will be the simplest one that provides the circulatory support required. "I would like to see a broad-based effort involving all aspects of the circulatory system that develops a whole arsenal of devices ranging from the simplest temporary pump to the most complicated totally implantable one," says Dr. Pennington. "For a whole population, that is what we would need to select from. We should have heart research centers located strategically around the country that offer patients every type of treatment known. Entering one of these hospitals would lower a patient's chances of dying of heart failure to practically nothing."

"Right now we're at the very, very beginning of developing a whole family of devices: acute and chronic VADs, acute and chronic hearts, and transplants," says Dr. DeVries. "We must determine *which of the modalities work,* what patients have the best chances with what types of devices, what complications to expect, and how to go about choosing a device. I think that in the future we'll have many alternatives, perhaps with a great deal of controversy over which ones work the best, but that's healthy and good, and I'll welcome it. There's no single answer; they may all be good, and as we learn more, our decisions will probably change."

Although this is the prevailing view, VADs still seem to be the current favorites of most researchers. "The next 5 years will be very exciting, because we have all these options, and researchers taking different approaches will be pooling their information," says Dr. Hill. "By 1995 we'll have many heart patients with LVADs; perhaps 30% to 40% of severely ill patients will receive two VADs or total hearts. We should definitely develop both types of devices, but if given priority the VADs will make our research much faster and more efficient."

15

Total Artificial Hearts

Blood pump research cannot proceed too quickly or efficiently. Heart disease has indisputable status as the top killer in the United States. Increased prevention and improved treatments have reduced deaths by 25% over the past 10 years, but heart disease still kills about 1 million Americans each year. Clinical experiments now in progress, however, may dramatically change the prognoses of patients with irreversible end-stage heart disease who are not transplant candidates. These patients may someday join those now at the Humana Heart Institute in Louisville, Kentucky, who are living with artificial hearts implanted by Dr. DeVries, formerly of the University of Utah. Theirs may be the world-renowned Jarvik-7 artificial heart, the first to be permanently implanted in humans, or one of several other models being developed by Dr. Jarvik and other researchers.

Artificial hearts recently increased their prestige in the national blood pump program thanks to a strong endorsement by an independent 14-member panel convened a few years ago by Claude Lenfant, M.D., Director of the National Heart, Lung and Blood Institute. The endorsement is one result of the series of studies ordered by the Institute since the artificial heart program began. This study is the first to take place since the Jarvik-7 artificial heart was clinically tested; the conclusions are the most positive to date. The panel has urged the federal government greatly to expand its research efforts to develop fully implantable, permanent artificial hearts, noting that thermal engine devices rather than the proposed electric ones would last longest and be most compact.

Astounded by the strong endorsement, Dr. Lenfant intends to take seriously the directive to continue the research and implement new programs to develop artificial hearts.

The major artificial heart programs in this country are at the University of Utah, Pennsylvania State University, Hershey, and the Cleveland Clinic Foundation. Taking different approaches, researchers at these and other institutions are developing similar means to a common goal. "Interestingly," write Dr. Pierce and his colleagues in a 1982 article, "the hearts developed by the half-dozen groups worldwide that have achieved reasonable results following animal implantation demonstrate many similarities in design." In most cases these designs are based on previous work with VADs of the kind described earlier.

Naturally, there is competition between researchers (and, of course, manufacturers) to create the best device to treat the tens of thousands of Americans who could benefit from artificial hearts each year. Who is ahead depends on the point of reference. "Neither Hershey nor Cleveland has our record of survival with pneumatic systems," says Dr. Olsen. "But in terms of eventual clinical implantation of electric artificial hearts, Dr. Pierce is clearly ahead. He has kept a calf alive for 222 days on an implantable electric system. That clearly puts him in the forefront on that heart."

The prestige and profits that await the creators of the best artificial hearts mean little compared to the benefits these devices will provide for heart disease victims. The race to develop artificial hearts is one in which almost everybody wins.

Cleveland Clinic Hearts

At the Cleveland Clinic Foundation, researchers are working on implantable biolized blood pumps of different types. "We take a diversified approach that includes devices for various applications," says Kiraly. "They can be temporary or permanent, replacements or assists, and they work with different drive systems." The VAD program, on which their efforts are concentrated, is oriented toward engineering development, while the artificial heart program is oriented more toward research. The permanent VADs now under development are collaborative efforts in which the Cleveland group makes the pumps and their colleagues elsewhere make the drive systems.

These pumps include a pulsatile device that is being used with two Nimbus, Inc. drive systems, one electrohydraulic, the other thermal pneumatic. The titanium device uses a pusher plate and diaphragm to pump blood in synchronization with the heart. "The pump has a complex geometry because it was designed as a permanent system," says Kiraly. "We've been attaching tubes to use it as a temporary percutaneous assist with pneumatic actuation."

Two of these VAD pumps are combined to make a total artificial heart. Pumps used for this purpose are flatter and have different blood ports to allow for arterial and atrial connections. The pneumatic actuators are essentially the same.

Because each pump is self-regulating, one of the factors to be considered in combining them is whether to let each ventricle run independently. The present configuration allows the pumps to run synchronously or in alternate beating modes at a rate established by either one. "For a total system, it might be better in some ways if both pumps were synchronized so as to run at the same rate," says Kiraly. "There are many ramifications to building a whole system.

We need to look at what configuration provides the lowest atrial pressure and whether we should use one motor or two."

The total heart has a magnetic position sensor and a new electronic control system enabling each pump to operate at a constant stroke volume and variable rate. Tests so far suggest that it is superior to other designs in approximating normal blood circulation. The record for animal implantation with this pneumatic heart is 206 days.

The device will be made implantable by the addition of an energy conversion system. "The big difference between the Utah-Humana approach and ours is that we'd like to have our system completely implantable before we test it on humans," says Hiroaki Harasaki, M.D., Ph.D., Associate Staff, Program Director in the Department of Artificial Organs at the clinic. "A major design difference is that our electrically driven pump is shaped differently from one that is pneumatically driven and has a free diaphragm. Although we want to try a different system, we are very encouraged by the trials with the Jarvik-7 heart."

The total heart prototypes are about 6 years behind the VADs, primarily because of work to be done on the electrical control systems. "We know the hardware but must work out the bugs, optimize the devices, and demonstrate their performance," says Kiraly. One of the prospective systems has an implanted coil that receives transcutaneous electrical energy through the intact skin.

Dr. Nosé is extremely confident that despite its comparatively low priority in the blood pump program, the development of a total artificial heart will proceed on schedule. He trained with Dr. Kantrowitz, who believes that ventricular assist is the better approach, and Dr. Kolff, who believes the same of the total artificial heart. As their successor, Dr. Nosé resembles most researchers in striving for diversity. "Our mission is to help patients of various kinds, from those requiring a few days' assistance to those in need of total hearts," he says. "Any approach that helps these patients is fine, but I consider our eventual goal to be an implantable total artificial heart."

"As far as I'm concerned, the research phase of our heart program is over; we are in the development stage of the program, and the device will be ready for use by 1991," he adds. "All the programs we have started since 1977 have met their objectives, and since this one is planned out, I know that we'll complete it. We're starting a preclinical study of an implantable LVAD with an NIH contract. In 3 years, when that study is completed, we'll start another 3-year study to make the total artificial heart applicable for patients. We'll have a few models, one of the earliest of which will be electrohydraulic. Our thermal engine heart, however, may take an additional 3 years."

As noted earlier, researchers at the Cleveland Clinic Foundation tend to work slowly and carefully rather than jumping on investigative bandwagons. It is possible, then, that 1991 might well bring heart disease victims a very benefical product.

Hershey Hearts

"Next to our group in Utah, the Hershey group has been most active and interested in work on the total artificial heart," says Dr. Jarvik. "Dr. Pierce was also the first to apply heart assist early enough to save many patients' lives."

Biventricular postoperative heart assistance is, in fact, one of the two stated objectives of the artificial heart program at Hershey. The second objective is the use of artificial hearts as bridges to transplantation. Pneumatic hearts currently available have bulky drive systems, percutaneous air supply hoses that can increase the risk of infection, and huge price tags; for these reasons Dr. Pierce does not wish to use air-driven artificial hearts as permanent devices. His group's pneumatic device, called the Penn State heart, recently received FDA approval for use in six patients as a bridge to transplantation.

The Penn State heart is one of four projects currently underway at Hershey. Like other groups, the researchers are also testing temporary pneumatic LVADs, permanent electric VADs, and electric artificial hearts. As noted, the temporary LVAD is quite popular; currently it is being used in 15 other research centers in the United States.

THE PENN STATE HEART

The Penn State heart is basically similar in design to the Jarvik-7 artificial heart, to be described in more detail later. There are also significant differences between the two models. The blood sacs in the Penn State heart have no seams, to reduce the risk of clots; their design creates a good washout each time to avoid blood stagnation. Although the heart uses standard cardiac disk valves, the disks themselves are made of plastic rather than carbon to make them more durable. A unique feature of the Penn State heart is that it is electronically controlled to increase or decrease blood flow according to the needs of the patient.

The automatic control system allows the pump to fill completely with each beat, while a variable beat rate maintains normal aortic and left atrial pressure. The system uses a negative-feedback loop that controls the output of each ventricle. The aortic pressure control for the left pump and a left atrial pressure control for the right pump are both derived from the left pneumatic power unit hose pressure so that implanted transducers are not necessary.

The calf implantation record with the Penn State heart is 270 days.

THE ELECTRIC HEART

The Hershey group's implantable prototype of an electric artificial heart is an extension of their work with their LVADs. Electric assist pumps and electric artificial hearts can use similar energy converters; each type of device requires compactness and reliability. The electric heart, however, is more complex by virtue of its second pump and more sophisticated control system. Each blood pump is activated by pusher plates; these are powered by an oscillating, low-speed, brushless DC motor that has a cam or ball screw motion translator. The motor revolves clockwise until one pump fills and the other empties; then it reverses direction to repeat the cycle. The advantage of this configuration is that it uses standard engineering techniques that are likely to provide long functional life; the disadvantage is that design constraints make the system rather large and heavy.

The outputs of the natural right and left ventricles differ slightly because of the bronchial circulation; they vary between individuals and change as people increase their circulatory needs with exercise. The outputs of artificial ventricles also differ because the discrepancy in pressure between the aorta and pulmonary artery causes different amounts of blood to back wash through the incompetent mechanical valves. The Hershey heart therefore uses a reversing electric motor drive to permit variation in the beat rate; limiting motor oscillation varies the stroke volume of each ventricle by as much as 10% in mock circulation.

Because an electrically powered artificial heart requires 10 to 20 watts of power, similar to the amount used by an AID, the group is testing lightweight, compact batteries to use with the device. Meanwhile the heart itself has kept a calf alive for 222 days. It will not be ready for use in humans, however, for another 5 to 10 years.

Utah Hearts

THE JARVIK-7 ARTIFICIAL HEART

As products of Dr. Kolff's artificial heart laboratory, the Jarvik hearts are, by Dr. Nosé's reasoning, Dr. Kolff's invention. In one of his articles Dr. Kolff points out, however, that the efforts of 247 people went into the development of what finally became the Jarvik-7 artificial heart. Around 1970 Clifford Kwan-Gett, M.D., Research Associate Professor of Surgery at the University of Utah, created a design based on several earlier devices that was to evolve into several later ones. Dr. Jarvik was the principal investigator for the Jarvik-3 (1973), the Jarvik-5 (1975), and the Jarvik-7 (1977) artificial hearts. Several other staff members also contributed to the ultimate success of the first clinically implanted artificial heart. Its quick connect system, for example, which allows for easier surgical implantation, was designed by Thomas Kessler, former Head of the Fabrication Department at the University of Utah, and Jerrold Foote, then a department engineer. Other staff members contributed to the development of biomaterials, techniques of fabrication and surgical implantation, design of the heart driver, and procedures for postoperative care. Dr. Jarvik therefore was understandably amused in 1983 when he alone was named Inventor of the Year for the Jarvik-7 artificial heart by Intellectual Property Owners, a trade association of patent owners.

The misconception that Dr. Jarvik was the sole inventor of the artificial heart was reinforced by Dr. Kolff's policy of naming the various hearts he has sponsored after their major contributors: Kwan-Gett, Nosé, Olsen, and so forth. "I warned him not to do this at the same time I warned him not to give pet names to the laboratory animals," says Dr. Nosé. "He would just say, 'But Yuki, I like this calf,' and he wouldn't listen to me." (The Utah staff still prefer names because a wide variety of animals live for quite a while; names make it easier for the staff to remember certain experiments and discuss their results.)

University of Utah staff members like to reminisce about the memorandum that Dr. Kolff sent to all contributors to the artificial heart program back in 1980. The memo alerted them that when clinical implantations of the Jarvik-7 artificial heart began, the media would concentrate on only two or three people involved with the device, and the surgeon would get most of the attention. "I also pointed out," says Dr. Kolff, "that if anything went wrong, the surgeon would be the only person who ended up in jail." Dr. DeVries has not served time yet, but he was certainly upset after the first implantation in 1982 when his car was vandalized and some of his colleagues at other institutions criticized his work.

Fabrication. The Fabrication Department at the University of Utah makes various types of artificial hearts that come in several sizes: the Jarvik-5 is quite large, the Jarvik-7 somewhat smaller, the new Utah 100 smallest of all. Each heart is handmade over a period of about 3 weeks.

The process requires highly polished stainless steel molds. The Jarvik-7 artificial heart is made of a polyurethane called Biomer, which is soluble in an extremely strong solvent called dimethylacetamide. A technician makes the housing by pouring the liquid over the mold and pulling off the solvent using a warm-air oven. The layers of the housing are built up one by one, and the inside is reinforced with Dacron mesh. The technician then places the completed housing over a concave stainless steel mold and solution casts the material again on the inside to form the blood-contacting surface. This creates an extremely smooth inner layer.

Each of the four layers of the diaphragm is extremely thin, individually made, and separated by graphite, a dry lubricant. The four-layer diaphragm is soft and flexible, yet strong enough to withstand the driving pressure.

Each ventricle also has a polyurethane base and a Velcro patch that holds it to the other ventricle inside the chest. (Once the heart is implanted, scar tissue will grow in to secure them together.) The completed device includes four tilting disk mechanical valves and four quick connects (which are sewn to the recipient's tissues and snapped onto the artificial ventricles). Two lines from an external power unit deliver compressed air to drive the pumps; skin buttons at their exit point from the recipient's abdomen help relieve stress and decrease the chances of infection.

The relatively simple fabrication process is subject to rigid quality control. The quick connect system and the process for pouring the blood membrane surface have been copied worldwide. The Utah group is now using vacuum forming to build ventricles more quickly and relatively inexpensively. They are also considering the use of injection molding for easier fabrication.

The Jarvik-7 permanent artificial heart, powered by air from a large mobile console or from a portable driver now being tested.

Design. Although shaped differently from their natural counterparts, the pumping chambers of the Jarvik-7 artificial heart correspond to the left and right ventricles. Each chamber has ports through which blood flows in and out, a pumping diaphragm, and a rigid base. During systole, an air port in each base delivers compressed air into the chamber to expand the diaphragm. The expansion increases pressure on the blood reservoir in the ventricle, causing the inflow valve to close and the outflow valve to open.

During diastole, the pressure in each blood chamber drops so that the outflow valve closes and the inflow valve opens. Blood stored in the atrium flows into the ventricle so that the cycle can start over again.

The Jarvik-7 artificial heart pumps steadily and automatically, opens and closes the valves and adjusts stroke volume according to the changing needs of the recipient. The control module permits selection of the heart rate (beats per minute), the pressure of compressed air, and the percentage of each beat spent in systole. The heart can beat for over 200 million cycles (roughly 4 to 5 years) before failure and handle pressures of 200 millimeters of mercury while driving.

The console currently being used with the device weighs 323 pounds (originally it weighed 375 pounds). The Heimes Heart Driver, a portable console that includes all the backup systems, weighs only 12 pounds, can be carried in a camera case, and is currently being tested on implant recipients participating in clinical trials.

THE JARVIK-8 ARTIFICIAL HEART

Symbion, Inc., the Salt Lake City firm presided over by Dr. Jarvik, makes its own hearts and contracts with the University of Utah to perform animal experiments with them. One of its current projects is the Jarvik-8 artificial heart, which will have several specific fundamental improvements over its immediate predecessor.

As noted, the Jarvik-7 artificial heart is a handmade device designed to last 4 or 5 years. The Jarvik-8 artificial heart is intended to last twice as long and to be an extremely reliable, commercially reproducible device. Design modifications will improve fit and flow conditions. More of its fabrication methods will be standardized, and technicians will be better able to inspect it to preclude flaws. "There will probably be designs that mix technologies as we go from the Jarvik-7 to the Jarvik-8," says Dr. Jarvik. "Initially we'll build the Jarvik-8 heart by hand, study its basic functions, and then transfer to mass production." The Jarvik-8 artificial heart has not yet been used in animal tests.

THE UTAH 100 ARTIFICIAL HEART

Back at the University of Utah, researchers are developing and testing the Utah 100 artificial heart. Developed in 1983, the Utah 100 has no connection with the Symbion work but shares some characteristics in common with the Jarvik-7 artificial heart. Both hearts are pneumatically driven polyurethane devices that have the same functions, pumping capacity (100 milliliters per stroke, hence the name "Utah 100"), and longevity to date in animal studies (about 9 months). Either device can be used with the Heimes Heart Driver as well as with the larger console.

The basic difference between the two hearts is that the Utah 100 is longer, narrower, and shorter than the Jarvik-7 artificial heart, making it appear elliptical. It better uses the space in the body and therefore fits into smaller chests. The fit advantages of the elliptical shape had previously been demonstrated to a certain extent in the Jarvik-3 heart. In the Jarvik-5 heart, Dr. Jarvik had concentrated on the precision of various parts; a round shape allowed the technicians to machine components to mold and fit very well, so that shape was retained with the Jarvik-7 artificial heart. "Since it returns to an elliptical shape, the Utah 100 is more difficult to machine and perfect in the mold fit, but that will be worked out," says Dr. Jarvik, who was not involved in its design. "I'm not sure that it's the best shape, but it's a good one."

In late 1984 the Utah 100 went through its first major modification, resulting in even better fit and slightly higher stroke volume. Changes in the blood diaphragm have increased its durability, while those in the quick connects have decreased the chances of clot formation in the heart.

The outer housing, inner pumping diaphragm, and base of the Utah 100 must all be perfectly aligned to preclude failures; the shape requires particular dimensions to create proper folds in the diaphragms and avoid creases or deformities. In these respects the Utah 100 more closely resembles the more recently developed Jarvik-8 artificial heart than it does the Jarvik-7 artificial heart.

Because its housing is smaller than that of the Jarvik-7 artificial heart, the Utah 100 actually has greater pumping capacity for its size. It is adjustable to meet increased cardiac output and blood flow needs as a laboratory animal grows; the second calf implanted with the device lived for 268 days and tripled its weight. "We're proud of the Utah 100," says Pamela Dew, Head of the Fabrication Department.

A MAGNETIC PUMP

Dr. Olsen, who has contributed to the design of the inflows and outflows on the Jarvik hearts, is building a blood pump that he considers revolutionary. It consists of a rotor suspended in magnetic bearings, commutated like a brushless DC motor, with a configuration that pumps fluid in a constant, nonpulsatile flow. The device requires no valves. "Our prototype pumps 6 liters of water per minute at 90 millimeters of mercury pressure," says Dr. Olsen. "This will be a totally implantable, highly efficient, no-wear, long-running device that has the potential to be a very valuable blood pump."

THE ELECTROHYDRAULIC ARTIFICIAL HEART

"The pneumatic system is not ideal for chronic implantation," says Dr. Olsen. "It will work for some degree of chronicity, but the concept leaves something to be desired." For this reason, the University of Utah researchers, like others, are working on a percutaneous electrohydraulic system. It will resemble the pneumatic hearts in that some of its components will remain outside the body. "I think that even a percutaneous design is safer than total implantation," says Dr. Jarvik. "Many complex components inside the body create problems with reliability. Positioning, fixation, infection, and individual component failure problems can occur. Having as many of the components and as much of the redundancy as possible in a compact external system enables us to service or replace them much more easily."

The electrohydraulic system being developed has an energy converter with only one moving part. The impeller of an axial flow pump is attached to the rotor of a brushless DC motor; both are supported by one set of bearings. The pump has reversing rotation that sends the hydraulic fluid from one ventricle to the other (in a VAD, this configuration would move the fluid from a reservoir sac into and out of a blood pump). Like the compressed air in a pneumatic device, silicone fluid activates the diaphragm to pump blood. The advantages of this design are the compact size and the few moving parts. An apparent disadvantage is the need to accelerate, reverse direction, and reaccelerate the pump in milliseconds, but this has been accomplished with reversal from 10,000 rpm clockwise to 10,000 rpm counterclockwise in less than 25 milliseconds. Mechanical tests have shown that the pump can run 5 years of reversals with no bearing failures.

The energy converter takes only as much space as a medium-sized flashlight battery. An electric wire runs through the chest to connect it to batteries weighing 2 to 5 pounds that can be worn on a vest or a belt and changed once or twice a day.

"The electrohydraulic system is not defined by the mating of that energy converter to any particular design," says Dr. Jarvik. The system has been used with both the Jarvik-7 and the Utah 100 designs. As part of the Utah 100, the device can fit into smaller animals and pump more blood per unit by weight.

Dr. Jarvik was principal director of the electrohydraulic heart until February 1985; he now works full-time as president of Symbion. The company has a licensing agreement with the University of Utah, which owns the electrohydraulic heart project. Because of this agreement, Symbion may exercise its right to produce and market the electrohydraulic heart when the device becomes clinically feasible. The company expects development of the electrohydraulic heart to take several more years.

Meanwhile Symbion is working on making the transition to the Heimes Heart Driver for the pneumatic heart. "The electrohydraulic systems are far enough away from human research that we're concentrating for now on the medical care of patients with compressed air systems," says Dr. Jarvik. "The issues of totally implantable systems are secondary right now." The same is true of VADs. Symbion owns the rights to what little work the Utah researchers are doing on temporary VADs and is modifying the Heimes Heart Driver to pump them.

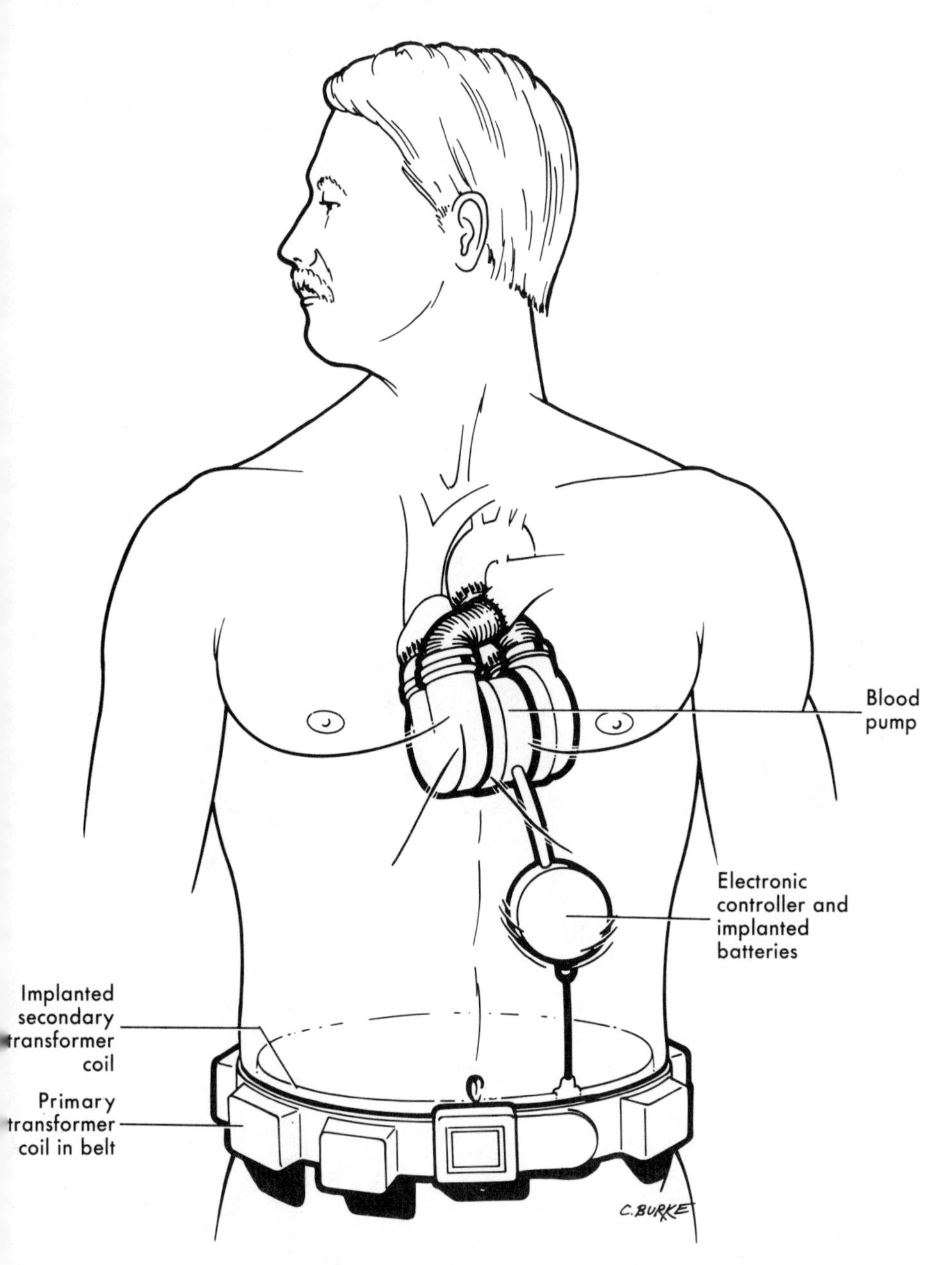

Artist's conception of a permanent transcutaneous electric artificial heart developed at Pennsylvania State University combined with a power system developed by Novacor Medical Corp.

Company policy is to take a progressive approach, from the blood pump powered by a large external driver, to the portable Heimes Heart Driver, and on to the improved Jarvik-8 artificial heart, which ultimately will be powered by the miniature electrohydraulic energy converter. "We all have to learn to walk before we can run," says Dr. Jarvik. "Most technologies have developed one step at a time." This and other research in artificial hearts will be speeded by the medical knowledge gained from patients who have already received them.

16

The People Who Live with Jarvik-7 Artificial Hearts

The largest number of human implantations of the Jarvik-7 and other Utah hearts have been performed not by Dr. DeVries but rather by Dr. Jack Kolff at Temple University. Dr. Jack Kolff has long been testing the Jarvik-7 artificial heart in brain-dead patients and has made some important discoveries in the process. It was he, for instance, who found that placing the heart slightly more to the left inside the chest helps it fit better. "The experimental implantations have been extremely useful and very beneficial for the patients in clinical trials," says Dr. Willem Kolff. The tests also suggest that an artificial heart or VAD could be used to keep organs alive in brain-dead patients who have donated them for transplant.

Far better known, of course, are the implantations of the Jarvik-7 artificial heart in living patients that have occurred at the University of Utah and the Humana Heart Institute.

Recipients and Families

The first clinical recipient of the Jarvik-7 artificial heart was Barney B. Clark, D.D.S., a 61-year-old retired dentist from Des Moines, Washington, who was suffering from cardiomyopathy. Barney received the heart on December 2, 1982 at the University of Utah Medical Center. He lived with it for 112 days before dying of multiple organ failure and drug-induced complications at age 62 on March 23, 1983.

Although his life was not long extended by the artificial heart, Barney and his family felt that the experiment was worthwhile. "I'm sure that given the chance, Barney would do the same thing over again," says Una Loy Clark, a 64-year-old homemaker who has recently been very active working for the American Heart Association and other organizations. "We knew that he was extremely ill and likely to die, and we knew also that the surgery was highly experimental. Barney hoped to live, of course, but he did not expect to receive much personal benefit from the artificial heart. I think that what happened was pretty much what he expected. He had told me that if he survived the surgery, he did not expect to live very long. Although Barney was by no means a martyr, his main purpose was to contribute to medical science by helping develop a therapy that was too late for him."

Una Loy fully agreed with her husband's decision to participate in the experiment. "I'm a very religious person," she says. "From the beginning, my calmness about the idea and the lack of repugnance I had made me feel as if the experiment was just meant to be. Barney was always very decisive, and once he made up his mind to volunteer for the surgery, I felt very secure because a great load had been lifted from my mind. I felt in my heart that we were doing the right thing, and I'll always feel that way."

Barney and Una Loy knew about the Jarvik-7 artificial heart because they were from Utah and kept up with the goings-on in Salt Lake City. When Barney became ill, he mentioned that he thought it unlikely that the device would be available in time to help him. At first he mistakenly supposed that with an artificial heart he would spend the rest of his life bedridden in a hospital in close proximity to several other patients, all of them tethered to huge power consoles. When Barney met with Dr. DeVries and Dr. Lyle Joyce, visited the animal barn, and got accurate information, however, his attitude changed. "He was very sick and had very little stamina by then, but he was so interested in the experimental animals that I could hardly drag him out of there before he wore himself out," says Una Loy.

While living with the artificial heart, Barney experienced the successes and setbacks so typical of medical experiments, at some times enjoying renewed health and strength, at others suffering physical and psychological anguish. His family has vivid memories of both good and bad experiences. Una Loy remembers two happy times in particular. "My knees buckled a little bit when I first saw Barney after the surgery with all the machinery and tubes, but instead of being grayish, his skin was pink, and he looked wonderful. He began making all these motions, trying to tell us that he loved us. That was a very joyful time." Another memorable day was March 6, 1983, the Clarks' thirty-ninth wedding anniversary, when the hospital cooks baked an enormous cake, pieces of which were given to all the patients.

"The bad times were when Barney had the seizures, when the heart valve broke, and when the nosebleeds went on for a month, and we wondered if they would ever stop," says Una Loy. Some nights she would leave her hospital room in the middle of the night to walk down to intensive care and check on Barney. Mornings were especially difficult. "Every morning was a new challenge," Una Loy says. "I'd walk down the hall to intensive care, watching the faces of the people walking past me to see if anything was wrong." Una Loy found herself living on an emotional roller coaster. "I never knew what to expect from one moment to the next, let alone one day to the next," she says.

The Clark family handled the stress very well. One of the sons, Steven Clark, M.D., had a particularly difficult time because of his medical knowledge and his inability to be with his father as often as he would have liked. "He told me that he could hardly bear it," says Una Loy. "He would see his father at his worst times, then he would have to return home to conduct his practice in a state of distraction."

Since the public was well aware of the difficulties experienced by Barney and his family, the media reported negative as well as positive reactions to this first clinical trial. Dr. Jarvik, however, was greatly encouraged by the response. "It's a real distortion of the facts to say, as some people have done, that there was a lot of ambivalence about Barney Clark's implantation," he says. "It was an extraordinarily widely publicized event. Over 90% of the response was positive. There are serious, severe critics of our work, but the overwhelming public opinion is that research should continue, and the medical profession has supported us. I'm as prejudiced as can be, but what I hear is that the public wants the artificial heart; they want it to work well at a reasonable cost, and since they don't know if it will be possible, they want us to find out."

The public was extremely interested in Barney himself and in whether the artificial heart could keep him alive and reasonably happy. Social changes that have occurred since then have created a different atmosphere in which the clinical tests are continuing at the Humana Heart Institute. The public, the patients, the implantation team, and the other staff are better educated. The use of a baboon heart to sustain Baby Fae stimulated great interest in heart implantation and experimentation. People began thinking about what a heart should be and whether they themselves would rather have a mechanical than an animal heart. No longer considered shocking futuristic experiments, heart implantations are now looked on more favorably.

Now that the clinical tests of the Jarvik-7 artificial heart have moved from an academic to a private hospital, the public is asking different questions about them. "People are interested in whether the experiments are worthwhile, whether they should be conducted in a for-profit hospital, and what their implications are for human ethics, choices, cost allocation, and medical economics," says Dr. DeVries. "The interest has broadened beyond single patients to include the tests and their impact on society as a whole." Surviving recipients of the Jarvik-7 artificial heart are starting to provide some answers to these and other questions.

On November 25, 1984 the Jarvik-7 artificial heart was implanted in William J. Schroeder of Jasper, Indiana, now 53, a retired quality assurance specialist at an Army ammunition depot. Bill had long been interested in the latest technology and experiments. "If he discovered something new, he'd tell a whole bunch of people that they should find out about it and see if it worked," says Margaret Schroeder, a 53-year-old retired department store clerk. "He was interested in helping everybody with new things." Bill had read extensively about Barney, not suspecting at the time that his own ischemic cardiomyopathy and coronary artery disease would make him a candidate for an artificial heart. When his physician suggested that an implantation was his only chance to stay alive, Bill immediately investigated the possibility. "He felt that the heart really would work and still believes that," says Margaret.

As a heart recipient, Bill achieved some technological firsts of his own. He was the first patient to test the Heimes Heart Driver (for up to 3 hours per day). Although he suffered a stroke 18 days after his implantation, Bill was also the first artificial heart patient to be released from the hospital. On April 16, 1985, after 132 days with the artificial heart, Bill moved with Margaret to a specially equipped apartment owned by Humana across the street from the Heart Institute. Margaret still spends nights at the apartment in order to rest better. Since May 6, however, Bill has been back in the hospital recovering from a second, massive stroke.

By mid-June, Bill was speaking a few words at a time. "I came in to see him after spending a few days at home in Indiana," says Margaret. "I asked him if he missed me, and he said, 'I sure did,' which is why I don't leave him too often. I think he understands everything, but he's just starting to shake his head and talk to communicate back." Bill is able to sit up and is starting to move his afflicted right arm and leg—with closed eyes and gritted teeth. "He was so down when he went back into the hospital, and he's come back so far," says Margaret. "He must be a very special person to hang on like that. We're all so proud of him!"

The third artificial heart recipient, Murray P. Haydon, is a 58-year-old retired auto worker from Louisville who was dying of cardiomyopathy. Murray received the heart on February 17, 1985. His major problem since the surgery has been pulmonary insufficiency that has kept him on a respirator.

The fourth implantation of the Jarvik-7 artificial heart took place at the Karolinska Thoracic Clinic in Stockholm, Sweden. On April 7, 1985, Leif Stenberg, a 53-year-old businessman, received the heart during surgery performed by Bjarne K.H. Semb, M.D., a Norwegian specialist. The patient seems to have suffered fewer complications from the implantation than have his American counterparts.

The oldest artificial heart recipient to date did not fare well with his device. Jack C. Burcham, a 62-year-old railway engineer from LeRoy, Illinois, had his implantation on April 14, 1985 because of heart disease and atherosclerosis. His surgery was difficult because his chest was smaller than the physicians had anticipated. Ten days after the operation, Jack died of cardiac tamponade: undetected blood clots had stopped the flow of blood through his artificial heart.

The Jarvik-7 Artificial Heart as a Research Tool

Although he lived only briefly with the artificial heart, Jack made a significant contribution to medical science. For years cardiologists had known that closed chest massage works. No one was sure whether the procedure affected the heart or the thorax—until Jack died and his artificial heart was turned off. "I did closed chest massage on him, and it didn't elevate his blood pressure at all," says Dr. DeVries. "That proves very conclusively that closed chest cardiac massage acts by constricting the heart, not the chest."

As previously noted, in addition to performing the functions for which they are intended, bionic parts also serve as unique research tools providing medical information that would not otherwise become available. Given the objections that have been raised about the clinical tests of the Jarvik-7 artificial heart and the quality of life it offers, it is especially important to emphasize that the device is also being used for this purpose. Drug experiments performed with Bill and complications suffered by all the artificial heart recipients have taught the researchers a great deal.

Drugs given to a patient with an artificial heart, for instance, will not affect that device but may cause vasoconstriction of the peripheral circulation that raises the blood pressure, or they may affect the pulmonary vascular bed more than the systemic arterial bed. A good example is the drug isoproterenol (Isuprel), which increases cardiac output. When infused into a heart disease patient, isoproterenol causes the heart rate to rise; until recently no one knew why the blood pressure rises as well. "If we give isoproterenol to an artificial heart patient, the heart is not affected, and the pressure goes down secondary to lowering his peripheral vascular resistance," says Dr. DeVries. "Basically this means that the main effect of isoproterenol is to increase the blood pressure by action on the heart. Now that's an *important* thing to know, and it's a fact we could not have learned by testing patients with natural hearts."

Some of the more disturbing results of the artificial heart implantations are the complications that have afflicted each of the patients. All four American patients were returned to surgery after their implantations for hemostatic operations: Barney for nosebleeds (and earlier for a broken heart valve), the other three for chest bleeding. All the patients had postoperative acute tubular necrosis (kidney failure) for about a week (as did Leif). They were immunosuppressed for about 10 days and had to be kept in isolation. Each developed some type of infection: pneumonia, urinary and skin wound infections, empyema (accumulation of pus) in the chest, or sepsis (blood poisoning). They all developed hemolytic anemia requiring transfusions because the Jarvik-7 artificial heart damages blood cells.

Three of the four patients suffered neurological complications ranging from Barney's slight seizures to Murray's transient ischemic attack to Bill's massive stroke (Leif has had a neurological episode that has not been specified). Only Jack, who lived only 10 days, showed no complications of this sort. Three of the patients also had pulmonary insufficiency complications. Murray lived with the heart for almost 112 days without complications before developing mild pulmonary insufficiency that has kept him on a respirator. The condition resulted from the blood transfusions he needed during his second surgery. "In reaction to Bill Schroeder's stroke, we had overanticoagulated Murray Haydon," says Dr. DeVries.

Because any one of these conditions is major and complicated in and of itself, it is amazing and disheartening to see the artificial heart recipients having to battle one such setback after another. Dr. DeVries, however, supervises his patients from a different viewpoint. "By the time we implant artificial hearts in these patients, they are dying," he says. "Their kidneys and lungs and brains just don't work very well, and we have three strikes against us even before we start. Considering the type of patients we're working with, I'm really amazed that we haven't had more complications than we've seen."

From working with these patients, Dr. DeVries and his team have learned a tremendous amount about the complications themselves and how to help prevent them. Murray's case in particular illustrates the fine adjustments that must be made in dosages of anticoagulants to avoid bleeding, on the one hand, and blood clots leading to strokes, on the other. The team has learned that if bleeding complications occur, they must return the patient to surgery immediately rather than waiting for the bleeding to stop on its own. They have likewise learned that to avoid infections they must keep patients isolated longer than they had originally thought, promptly remove the Foley catheter from the urinary tract, and use antibiotics only when necessary rather than prophylactically. Finally, they have found that the patients must be fed immediately at the time of the surgery because they are all malnourished.

In reading press accounts of the succession of setbacks that have befallen the artificial heart recipients, the average person might recall the cliché, "The operation was a success, but the patient died." As the nature of medical experimentation becomes clearer, however, it will be easier to understand why the Humana team is likely to term this series of tests a success.

Informed Consent

Of course, all these experiments and complications make it especially important to educate artificial heart candidates and their families to ensure informed consent. Interestingly enough, however, all four American patients grew somewhat bored with such red tape. Having already faced the prospect of death and decided that they wanted to live and help others, the candidates were not very interested in studying a long, detailed consent form.

They seemed least interested in hearing about possible complications of the heart implantation. "We went through all the possibilities ad infinitum, but they remembered only about half of them," says Dr. DeVries. "The impression they give is that they've already made up their minds, and reminders of what can go wrong only depress and discourage them." Margaret corroborates this impression. "I don't think it would have mattered what the doctors told Bill about the artificial heart," she says. "He was really sick and knew he was dying. As he saw it, the heart gave him a chance he would not have had otherwise. I think that he felt that he would either die or get better. He was such a strong person that I don't think he stopped to think whether he'd be bedridden for the rest of his life or anything like that."

The entire Schroeder family has taken their cue from Bill's attitude. "When Bill had the second stroke, the idea of shutting off the heart was mentioned to me a few times, but I just didn't even listen or think about it," says Margaret. "I just couldn't do that unless they proved to me that everything in him was dead. If he can fight like this in his condition, we surely can stick beside him all the way. When he gives up, we'll have lots of time to think about things like that."

Publicity

Bill's attitude toward his heart implantation became something of an issue for the family when Margaret's position on the subject was misrepresented in the press. "It really bothered us when the press misunderstood something I said and blew it all out of proportion," she says. "They made it sound as if Bill would not have participated in the experiment if he had it to do over again and knew that he would have complications that might cause hardship for the family. He could not have known what would happen, and we backed him 100% on whatever he decided. He *would* do it over again, and so would we."

Whether or not the discussion of this issue in the press was accurate, it has supplemented the informed consent procedure by making future candidates for the Jarvik-7 artificial heart better aware of what they themselves must consider. This beneficial effect of publicity must be balanced against the wishes of the patients and their families regarding privacy. "Our patients have asked many questions about what kind of publicity to expect," says Dr. DeVries. "Murray Haydon in particular didn't want a camera in his face every time he turned around. People don't want to feel like zoo animals—they have serious concerns about it, as I do."

The only other negative experience the Schroeders had with the press occurred one day when Margaret wanted to take Bill out of the hospital for some fresh air. The hospital grounds were covered with reporters and cameras. "I got really upset, because I didn't want to think of Bill as being imprisoned for any reason," says Margaret. "I asked the press to please let us come out, and they agreed to that. Actually they've been really nice to us."

In general, the Schroeder family has benefitted from all their attention. Bill was excited about being on television; he enjoyed his hats and other gifts and has had fun talking to the press. Meanwhile Margaret has been kept busy answering all her mail. "We've had thousands of letters from all over the country and everywhere overseas, even Russia and Japan," she says. "I didn't expect all this attention. We're just a country family, and I'd rather be busy at home than out in the limelight, but if Bill hadn't had those strokes, he'd be out making speeches to the heart foundations. All the letters and prayers we've received have been very supportive for all of us."

Una Loy enjoyed a similar positive experience from the publicity surrounding her husband's implantation. "The publicity greatly surprised me but didn't bother me at all, possibly because I was so protected in the hospital," she says. "So many people have heart disease and take personal interest in the experiments that I felt as if it were a joint affair, something that should be shared with the world. I'll never get over the feeling I had of people's love and caring as we went through this. Their prayers, letters, gifts, and phone calls made me feel as if we were all in this together and gave me much strength and moral support."

As the biggest fan of the implant recipients, Una Loy is disappointed by the physicians' decision to cut back on the amount of information being released about the artificial heart experiments. "I hung on the television broadcasts when the implants were being done," she says. "I know that people are tired of hearing about the series, and the publicity is very difficult for Dr. DeVries, but I'm very concerned about Mr. Schroeder and Mr. Haydon and want to know how they're doing. I was much happier when we had more information."

The decision about how to handle publicity is one that must be made between the physician and the individual patient and family. Obviously Bill finds publicity a boost; just as clearly, Murray wants his privacy respected. What is best for the patient must be determined in advance.

The Future

As of this writing, Dr. DeVries has not yet selected the fifth candidate for the Jarvik-7 artificial heart (he is authorized by the FDA to perform a total of seven implantations). Potential candidates include patients who cannot be weaned from the heart-lung machine after surgery, those who develop irreversible cardiac shock, or those with end-stage heart failure. The patients must be ineligible for transplants.

"I understand that Humana is assessing the ethical, legal, social, and medical issues involved in the three implantations that have taken place there," says Una Loy. "I agree with this responsible action on their part. It indicates that proper attention and concern are being given to technology, that recipients and their families are being afforded every possible consideration and protection, and that proper guidelines will be established for future experimentation. I continue to support the artificial heart and believe it to be a medical breakthrough of tremendous significance to the world."

Meanwhile, Dr. DeVries can suggest the prognoses of the surviving American patients. "Murray Haydon is coming off the respirator, will be released from the hospital, and will enjoy a very good life," he says. "Bill Schroeder will continue to recover, and although ultimately he'll be severely impaired, he will be able to appreciate being alive."

Bill's family likewise appreciates still having him with them. "If Bill hadn't had those strokes, we would have been just fine," says Margaret. "All the children and I are right behind him, taking it day by day. We have to get over these setbacks and get ahead. Bill's my husband, and I love him a lot and I'm really prejudiced, but I do wish that people knew how great a man he really is."

The implications of the Jarvik-7 artificial heart, particularly the first experiment, have engendered considerable criticism as well as praise. The objections seem to be aimed not at the concept of a total artificial heart, but rather at the timing of its clinical tests. The Jarvik-7 artificial heart is an experimental device that is far from perfect. Researchers argue that problems should be worked out in an external version of the artificial heart before it is implanted. The external power source in particular seems to detract considerably from the quality of life experienced by the recipients. Although the Heimes Heart Driver will eventually enable its users to be far more mobile, they remain at risk for infections along the percutaneous drive lines.

Dr. Kolff would like to see the development of a truly implantable heart (an atomically driven device), but he is still very much in favor of air-driven pumps. "Thousands of people are dying of heart failure," he says, "and it's very little solace to them to know that 5 years from now there will be an implantable artificial heart. By using a simple device like the air-driven heart, we can help people now and learn a great deal in the process. This heart model also requires far fewer constraints to assure durability."

Dr. Olsen meanwhile feels that the likelihood of infection is not a deterrent to permanent implantation of the Jarvik-7 artificial heart. "I think that the incidence of infection will not be linear," he says. "I would hesitate to say that the longer a recipient keeps the artificial heart, the greater the chances of infection." Many researchers disagree, but the development of electric hearts with percutaneous wires should make this less of an issue.

If the human implantations of the Jarvik-7 artificial heart were premature, so too would be criticism of these first few clinical tests. As noted, they have already been successful from a research standpoint. All great scientific and medical discoveries have taken time; development of an artificial heart is no exception. "The artificial heart implantations are not premature—they are justifiable," says Dr. Nosé. "We must apply whatever is best for the patient. Heart patients on drugs are often bedridden and worried that their hearts will stop any minute. Implant recipients can at least walk around without having this concern. The Jarvik-7 is crude, but the quality of life it offers is 10 times better than that obtained with drugs. If it works, we should use it."

Right now the Jarvik-7 artificial heart implantations are still in a state of uncertainty. Despite the complications suffered by each recipient, the series has not yet reached the point at which the physicians or the FDA would feel it appropriate to discontinue the work. The patients and their families simply wait to see what the future brings. "I guess the worst part of it is not knowing what's going to happen tomorrow," says Margaret. "The situation can change at any time because Bill can get an infection or a temperature—it's really strange. It's like going up and down hills and around curves without anybody knowing what we're going to meet next."

Future candidates must understand that the Jarvik-7 artificial heart does not yet prolong life indefinitely; laboratory animals have lived with it for only 10 months. Recipients must remember that the Jarvik-7 artificial heart is a long-term life-support system. They should be prepared to decide whether to shut it off in case complications become unbearable. As heart disease victims grapple with these unprecedented issues, the world will watch with interest.

"Nothing worthwhile ever comes easily," says Una Loy. "I know that the artificial heart recipients have not received the quality of life they would have liked. But it will take many people to accomplish the goals, and it will be wonderful if in the end the heart does become a therapeutic treatment. Many people will suffer a lot to attain this, but if we succeed, it will have been worthwhile, and if we don't, at least we'll have shown the courage to fail. I know that my husband derived great satisfaction from his role in these experiments. I'm certain in my heart that he's very proud to have contributed to making his life really count."

"Basically I believe that we're at the very beginnings of this research," says Dr. DeVries. "C. Walton Lillehei, who invented one of the first heart-lung bypasses, knew this when he sent me a telegram at the start of the experiments. It said, 'Congratulations on the end of the beginning.'"

SECTION

V

RENEWED SENSES

17

A "Feeling Eye" and Other Vision Systems

In noting how " . . . in the night, imagining some fear, / How easy is a bush supposed a bear!" Shakespeare referred to the intricate relationship between what we see and what we think we see. While recalling recent dreams or memories, we actually "see" images that are not before us regardless of whether our eyes are open or closed. The eyes move to follow the actions in dreams. Optical illusions occur because the brain misinterprets certain images; thoughts, feelings, and expectations all affect the visual processes.

Our simian ancestors used sight more than their other senses and bequeathed us eyes of an appropriate size and complexity, along with brains involved in processing and interpreting visual stimuli. This eye-brain connection is important to researchers trying to construct artificial vision systems. Vision starts at the surface of the eyeball and proceeds to the back of the brain, which responds with suitable feedback. The bioengineer must select a point along this continuum at which to implant a vision device.

The Visual Continuum

Very simply, vision begins when light strikes the cornea, the surface of the eye that does about three fourths of its focusing. It passes through the pupil, whose size is controlled by the colored iris, to the crystalline lens, which further refines the focus.

Between them the cornea, pupil, and lens angle the light rays into a cone that converges at a nodal point within the eye. The rays continue past this point, spreading out to form a mirroring cone. Now the rays are upside-down and reversed as they pass through the vitreous body, a gel-like substance that forms the inner eye, and finally focus on the macular area of the retina, the innermost of three tissues that form the back of the eyeball.

The retina forms from the brain embryologically and is actually a specialized part of it. With the complex processing of retinal images, the visual function ceases to resemble the mechanical action of a camera and instead becomes electrochemical. The retina consists of light-responsive nerve endings: 130 million rods, which detect low levels of light, and 7 million cones, which provide detailed vision and color discrimination. Complicated signals from the retina about shades and colors are altered, prepared, and finally sent through the optic nerves to the visual cortex, two small areas in the lower back of the brain. The visual cortex does the detail work of processing and subjectively interpreting retinal images. Damage to the visual cortex results in blindness even if the eyes and visual nerves are working.

The entire process of seeing a particular image takes about 1/100 of a second.

This simplified description should explain why bionic sight requires more than just a television camera plugged into an eye socket. The eye can perform more functions and react more sensitively than the most sophisticated and delicate physical instrument. It detects space, depth, time, and movement both straight ahead and peripherally. It modulates light, focuses images, and presents them in extremely subtle colors. Cooperating with the brain, it makes the most of poor input and interprets what it sees. It can sign off with an afterimage and do much more. These are a few reasons that most people dread the thought of losing their vision and many scientists busily invent devices to maintain it.

Permanent bionic parts help contribute to eye health. Plastic "explants" can be sewn around an eye to buckle it after a retinal detachment. The thin floor of the eye socket can be fractured by a blow, limiting movement or causing double vision; the implantation of a curved section of silicone rubber, Teflon, or tantalum mesh corrects this condition by supporting the eyeball. Several kinds of implants can successfully channel fluid from the eyes of glaucoma victims. People whose tear ducts are damaged from scarring or trauma can have them replaced by tiny silicone tubes. Certain eye diseases or operations may decrease the amount of vitreous fluid, which may someday be supplemented by polyvinyl alcohol hydrogel or another suitable material.

Treating the eye bionically is one thing; reproducing vision is another thing altogether. Researchers developing artificial vision must decide what aspect of seeing best lends itself to imitation. Some start at the beginning: they improve or restore vision with lenses or implantable prostheses that perform the tasks of the cornea and crystalline lens. Some have bypassed the eye entirely, attempting to train other senses to do its work. Finally, others have approached vision through the back door, as it were, starting with the brain's end of the continuum and trying to restore sight through the visual cortex.

Lenses

Eyeglasses and contact lenses rival fillings and dentures for the status of best-known spare parts of the human body. An estimated one out of five schoolchildren, two out of five college students, and two out of three adults need corrective lenses; few people older than 45 years can read without them.

Compared to eyeglasses, contact lenses give patients a psychological lift while often providing better vision. Soft contacts are used for daily or extended wear, depending on their gas permeability. Hard lenses are making a comeback now that new gas-permeable rigid materials are being introduced. "I think that these gas-permeable rigid lenses will replace most soft lenses within the next 3 to 4 years," says Sanford E. Kaps, F.A.A.O., M.R.S.H., an optometrist in private practice in Hackensack, New Jersey. "They provide sharper, more stable vision for most patients, especially those who have astigmatism or who need high corrections."

In addition to correcting typical vision problems, contact lenses can restore vision to many patients who are functionally blind as a result of genetic, neurological, or pathological problems. They are useful for albino patients and those with irregular astigmatism, keratoconus (corneal pathology), and nystagmus (rapid, extreme side-to-side movement of the eyes). "For many patients, contact lenses are the only way to achieve functional vision," says Dr. Kaps. "New materials will permit more and more patients to benefit from them." Innovative contacts are even replacing bifocal lenses or reshaping the cornea for sharper focus.

Intraocular lenses for cataract victims are highly successful and frequently used. By age 65, about half the population begins to develop cataracts; they have started to form in everyone over 70. In 1983, 650,000 Americans had cataracts removed; 85% of them received intraocular lenses with a 95% success rate.

The crystalline lens of the eye normally consists of clear protein enclosed in a capsule. Cataracts are changes in the molecular structure of the lens protein that gradually turn it opaque. Their removal causes vision to become hopelessly blurred. To restore their vision after cataract surgery, patients can choose between cataract glasses, which cause side effects; contact lenses, which may be difficult to handle; refractive surgery, which is considered experimental; and intraocular lenses.

Intraocular lens implants almost match the focusing abilities of the natural lens and provide normal depth perception and peripheral vision. Available in a variety of designs and prescriptions, the lenses can be fashioned to correct other vision problems, including nearsightedness and farsightedness. The tiny plastic disks are attached by different kinds of tabs, hooks, or loops to the eye's anterior chamber (the section between the cornea and iris), iris, or lens capsule. They are implanted permanently, often in an outpatient procedure, and begin to improve vision within a few days.

Artificial Corneas

Each year a few thousand people may lose their eyesight because of damage to their corneas caused by disease or chemical burns. Corneas for transplantation are no easier to find than donor organs, and some people's eyes are so heavily vascularized that they reject foreign corneas (usually such transplants work because the cornea does not contact the blood, which carries the immune agents). People in this predicament are often told by local specialists that nothing can be done for them, either because the ophthalmologist is ignorant or because he or she does not wish to provide a referral. "Even if the cornea is severely diseased, an artificial cornea can be placed with a high degree of success as long as the back of the eye is functioning," says Howard Schneider, M.D., F.A.C.S., F.I.C.S., Attending Ophthalmologist at Mount Sinai Hospital in New York City.

The idea of implanting artificial corneas in diseased or damaged eyes is 200 years old but did not become practical until synthetic materials made it possible to construct prostheses. In the early 1960s, Hernando Cardona, M.D., who is now in Colombia, began work on keratoprostheses, or artificial corneas. Dr. Cardona's "nut-and-bolt" prosthesis consisted of a cosmetic contact lens whose transparent middle section was attached to a cylinder lens. The cylinder lens screwed into a perforated plastic flange that was inserted behind the cornea and stitched into it. The problem with the nut-and-bolt keratoprosthesis was that the cornea could not hold it in place; it became soft, and the device would extrude.

For the past 10 years, Dr. Schneider has developed and used a penetrating keratoprosthesis that is much more successful. The implant is a cylinder with a Teflon skirt placed through a 3½-millimeter opening in the cornea. Dr. Schneider covers it with a portion of periosteum (a specialized connective tissue that covers bones and is a source of collagen) from the tibia, then with a layer of conjunctiva (the membrane that lines the eyelid). The device holds very well. If the patient has an extremely dry eye, the eyelid is stitched shut around the prosthesis.

The success rate for penetrating keratoprostheses is between 90% and 95%. "If an ophthalmologist tells a patient that he or she has advanced corneal disease and that there is no hope for restoration of eyesight, by and large that's a perfect indication for an artificial cornea," says Dr. Schneider. Unfortunately he knows of only three other ophthalmologists who use the procedure: one in New York, another in Connecticut, and the third in Houston.

Mobility Aids

People who are past help from surface or implanted lenses can use any of hundreds of external devices to obtain information about their surroundings. These range from mobility or reading aids to bacon crispers, insulin needle guides, and sock tuckers. Such devices are not truly bionic; rather they are external tools designed to help blind persons lead more normal lives. But the principle behind them, namely, using hearing or touch to substitute for the lost vision, has been used to create genuine prostheses that imitate the experience of seeing.

The Tactile Vision Substitution System

Every child has played the game of reading a message that a playmate has "written" on his or her back with an index finger. The skin can "read" messages because it is embryologically derived from tissue resembling that of the retina and is the organ functionally most similar to it. Although the skin cannot discriminate more than five absolute points of stimulation, it can, like the retina, pick up many thousands of relative stimulation points. This ability has enabled researchers to experiment with a fascinating visual prosthesis, the Tactile Vision Substitution System (TVSS).

The TVSS was developed by Dr. Bach-y-Rita and Collins while both were at the Smith-Kettlewell Institute of Visual Sciences. "We wanted to show the plasticity of the central nervous system," says Dr. Bach-y-Rita, "the brain's ability to handle information from one sensory system as if it were coming from another." After 8 years of research, he and Collins developed prototypes of the TVSS and began testing it in the late 1960s. "We started working with the early [that is, congenitally] blind because their condition was a clean experimental model: they had no internal representation of vision," says Dr. Bach-y-Rita. "We wondered if we could make tactile information seem three-dimensional to them, to teach them what vision was like. Then we got into more practical applications of our system."

The first models of the TVSS had three parts: a television camera, a commutator, and a grid of mechanical stimulators mounted on the back of a dentist's chair or wheelchair. The camera recorded visual images; using dots, it reproduced them and sent their signals to the commutator, which sent corresponding vibrations to the grid against which the subject sat. (Later the grid was moved to the stomach with better results.) The vibrations from the 400-point grid "tattoed" two-dimensional image outlines on the subject's skin. Location cues provided depth perception (for example, the closer of two objects appeared lower on the grid). In different versions of the system, the camera was either mounted on spectacle frames or hand-held on a boom (the boom camera had a zoom lens).

The researchers also built a portable TVSS that tapped the image on the skin with weak electric currents. An objective lens was mounted on an eyeglass frame; it connected to a camera and commutator worn behind the subject's right arm on a shoulder strap. The camera, commutator, and battery-pack belt altogether weighed about 5 pounds and could be worn comfortably all day. They connected to a 1024-point electrode matrix worn on the subject's abdomen.

Subjects with both early and acquired blindness were taught first to control and direct the cameras. They learned to discriminate lines, then shapes, then solid geometric forms. Common objects (a cup, a chair, a telephone) were shown to them in different positions at various distances and finally in groups. After extensive training (in some cases from 100 to 300 hours), the subjects had surprising results with the systems. Those using the 400-point mechanical TVSS were tested on their recognition of objects, letters, grills, checkerboards, and people. Many learned to identify individuals by their height, hair length, posture, and outstanding features; locate them in a room; and describe their movements. They could also identify human heads from photographs.

The subjects testing the electric system did equally well; after training, one could locate, walk to, and pick up an object within 10 seconds. "I found this the most intriguing system," says Jim Gammon, now a 36-year-old Visual Services Coordinator for the Disabled Students Program at the University of California, Berkeley. "It seemed to have more contact and capability, and although it wasn't really useful outside the laboratory, I always hoped that they would let me try it outdoors."

Jim and his fellow subjects learned to understand perspective, parallax, and size constancy. Highly trained early blind TVSS subjects could shake outstretched hands and bat balls that were rolled toward them. They could picture the flickering of a candle flame and see red blood cells or fly wings under microscopes. Within a few hours of trying the TVSS, they found that vibration on the skin ceased to be an important stimulus in itself; instead, they visualized the objects they were examining in three-dimensional space. One subject even ducked an "approaching" object when the zoom of the camera was accidentally jolted by someone else.

Bill Gerrey, now a 37-year-old electrical engineer at Smith-Kettlewell, even learned to assemble and inspect miniature diodes on an assembly line in an electronics manufacturing firm. The television camera he used was placed in the ocular of a dissecting microscope, and the tactile array was fixed to a workbench against which he could press his stomach. His tasks included placing pieces of solder and tiny diode wafers into small glass cylinders that passed under the microscope. The wafers were gold on one side, silver on the other; Bill had to make sure that the correct color faced up. His supervisor was satisfied with his work in both assembly and inspection during his 3 months on the job but felt it impractical to hire him permanently.

In his 1972 monograph, Dr. Bach-y-Rita describes certain experiments with the TVSS that he and Collins learned to discontinue. On one occasion, for example, two early blind college men were using the system to describe photographs. The researchers tried to pep up the proceedings for their subjects by using *Playboy* centerfolds. Although the students could describe the photographs in considerable detail, they were dismayed to find that unlike their sighted friends, they were not turned on by the "visual" stimuli. "It's like swearing in a foreign language, or acquiring vision late in life and having no warm or cold associations with colors," says Dr. Bach-y-Rita. "These things mean nothing without an emotional context, which must be learned very early in life or over a long period of time."

Similarly, the researchers found that it was not always a good idea to let subjects see people who were close to them. One fellow eagerly escorted his girlfriend into the laboratory and proceeded to spend 5 minutes scrutinizing her with the TVSS. To everyone's embarrassment, he remained silent during his exploration and seemed rather glum.

Dr. Bach-y-Rita himself was somewhat disappointed the day Jim demonstrated the TVSS for one of the many groups of VIPs who visited the program. He set up a display that Jim cooperatively identified as a board with a light bulb on top of it. "No," said Dr. Bach-y-Rita, "it's a picture of me."

Most of the subjects recruited to work with the TVSS were highly intelligent college students studying engineering, psychology, or other relevant fields. Besides being fun, the experience was academically useful for them. "Much of what I was studying became more concrete because I could actually check it out and see what it probably would be like visually," says Mike Cole, 38, now Director of the Living Skills Center in San Pablo, California. "The experiment added another dimension to my reading, and we had a great intellectual environment within which to discuss perception."

Bill found that having studied optics at a university, he could use the TVSS to help him picture mirror images, lens theory, and optical illusions. "I was able to understand concepts that I could appreciate at school only after a lot of hard thinking. I saw things that I could never have guessed ahead of time and developed a truly experiential appreciation of them," he says. Lawrence A. Scadden, Ph.D., the graduate student who conducted the experiments, is now Director of Rehabilitation Engineering at the Electronics Industries Foundation in Washington, D.C., and feels that working with the TVSS helped launch his career. Blind since age 4, Scadden also used the system as a subject. "I'm a perceptual psychologist," he says, "so the experience helped me academically and in understanding other people's writing." Another student, Gerard Guarniero, wrote a philosophy dissertation on the relationship between visual and tactile space based in part on his use of the system.

Each of the subjects had a favorite experience with the TVSS. "Using the camera taught me a lot about zoom, focus, and aperture. When my wife bought a 35mm camera, I was actually able to teach her about it. That was a piece of color, if you will, in my intellectual experience that I'm really grateful for," says Mike. "To me the most fascinating exercise was to hold a cube in my hand while I trained the camera on it," says Scadden. "That gave me proprioceptive information about its shape as I rotated it before the camera."

Scadden found that the amount of information the subjects derived from the TVSS was correlated more closely with their spatial ability (to get about on their own) than with either intellectual or verbal ability. For the first time, the subjects were able to understand spatial concepts like the fact that approaching an object makes it appear larger and higher in the field of vision. They could recognize that a desk does not change shape while appreciating the different perspectives from which it might be seen. "To me an oblong table is always oblong. The idea that walking around it causes it to change shape is a very different experience for an early blind person," says Bill. "Most people don't recognize that spatial information is available to senses other than vision—hearing, for example," says Scadden. "We used tactile information to present monocular information about distance and depth."

This last statement is a loaded one. Dr. Bach-y-Rita had hoped to prove that the TVSS was a true replacement for vision, a "feeling eye" rather than a device that made use of another sense. Reports of some of the subjects do not make clear whether he succeeded. "The TVSS was not a visual mode being translated into tactile, it was a tactile mode of conveying information," says Scadden. "I never got beyond a certain plateau with the TVSS no matter how much I practiced, so I'd get discouraged and think that I was doing tactile things that had no relation to vision and were a pain in the ass. I myself don't think of it as a vision system, but if vision means checking something out and getting a good idea of what it's like without touching it, then that is in fact what we did," says Mike. "It could have used ultrasound, infrared, or some other means of communicating the information," says Jim, "but it was really a translation of the visual mode into the tactile."

These men feel, however, that whether or how they were able to "see" with the TVSS was less important than the educational experience it provided them. They were paid for their time and entered the project with a healthy skepticism fed by unnecessarily grandiose claims made to vision experts, the CIA, and other big shots about what the TVSS could do. To them it was an experimental prototype, a way to explore concepts and methods rather than a practical reading aid or mobility device. They were committed to the TVSS not because they believed that they could see, but because it involved a well-intentioned group of intellectuals who provided a thrilling professional and social atmosphere for them.

Dr. Bach-y-Rita and Collins eventually developed a 1024-point TVSS designed as a vest, with the television camera mounted in eyeglass frames. "This was a 32 by 32 grid," says Collins. "It gave us four times better resolution than an 8 by 8 grid—not a factor of 64 to 1024, but one of 8 to 32." The researchers planned other refinements of the system as well. "We had presented the information in a visual mode and translated it to tactile receptors so that it was still presented in a visual mode," says Dr. Bach-y-Rita. "That's not the way skin handles information; it processes more information sequentially than it does simultaneously." (In other words, a letter *A* pressed on the skin is harder to read than a letter *A* drawn on the skin.) He and Collins investigated systems that could trace rather than imprint pictures on the skin.

These improvements did not come to pass; instead, Dr. Bach-y-Rita and Collins set aside the TVSS for 8 years. Their funding had run out, and although they had demonstrated that the TVSS would work, it was nowhere near being

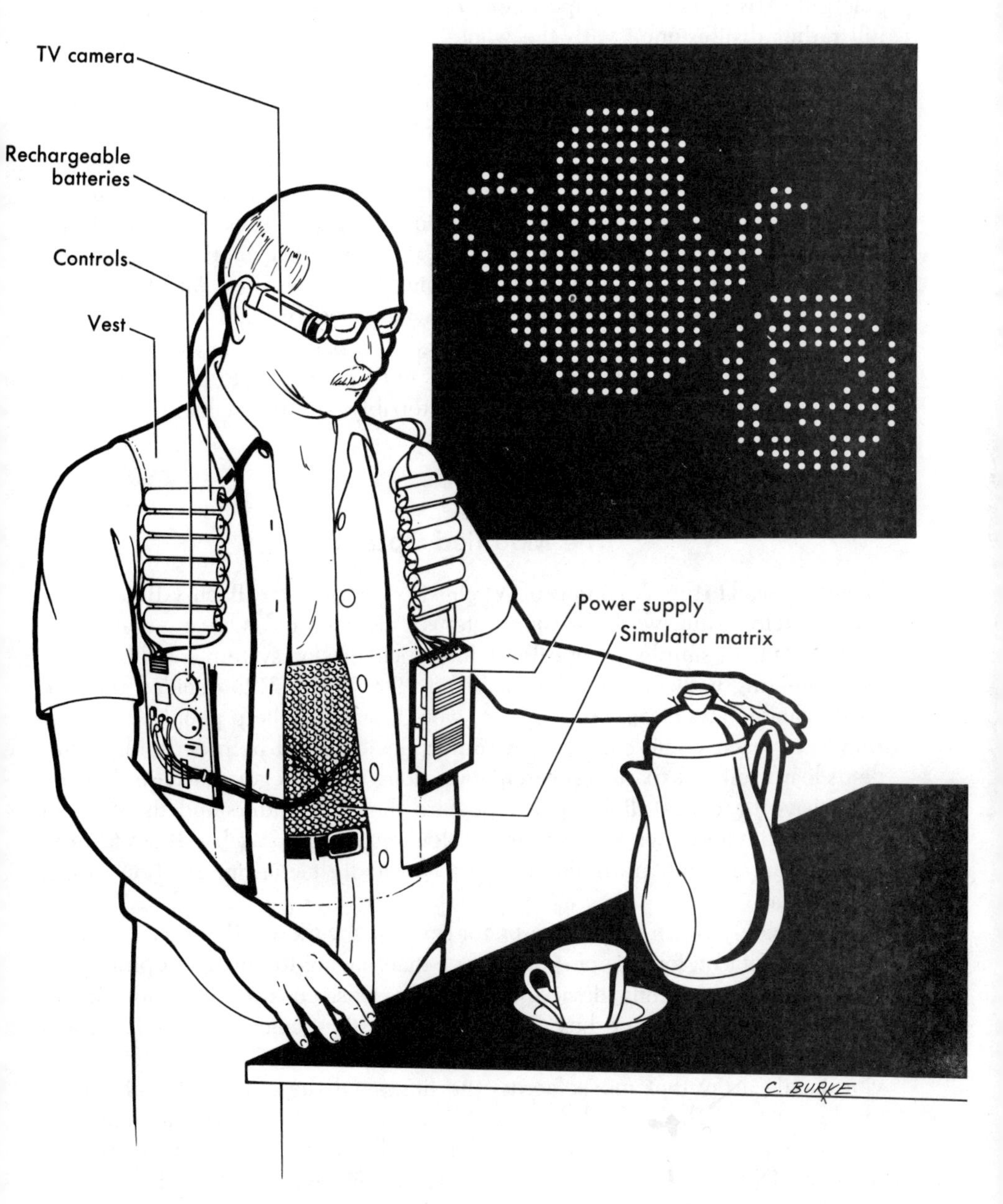

The 1024-point electric Tactile Vision Substitution System (TVSS). The inset shows how objects recorded by the television camera could be imaged on the abdominal skin by pulses from the electrode arrays in the undergarment.

practical. His "practical" experiment on the assembly line had in fact left Bill rather disillusioned with the whole TVSS experience. "It was an interesting reseach project and an instrument for study that has applications even today, but it was an awful thing to do," he says. "I had to devise all kinds of aids and work for 3 months to learn what it took a sighted teenager 15 minutes to learn in order to work on that line." "We definitely demonstrated that certain kinds of visual concepts could be presented tactually, but certainly not quickly enough to ever make it a feasible substitution for vision," says Scadden. "To be practical, we'd need a computer providing the spatial information quickly, but then the nervous system would not be doing the processing. And I sincerely doubt that even learning spatial concepts will ever be valuable to the blind in their everyday orientation and mobility tasks." Dr. Bach-y-Rita meanwhile faced the tremendous handicaps that a scientist must overcome to get a prototype from a laboratory into application. "I'm terribly impressed by the delay of getting anything into practicality," he says.

The Modified Optacon

Having moved to the University of Wisconsin, Madison, Dr. Bach-y-Rita recently decided to resume work on vision systems for two reasons. One was that he had devised a clear, simple, practical task for which a vision system could be used. After meeting with former TVSS subjects, Dr. Bach-y-Rita decided that the most valuable aspect of such a system would be not to help people "see" but simply to teach visual concepts. A similar device could help early blind students learn self and spatial representation along with categories (these children often have to feel 50 different balls before they can understand the concept "ball"). Ultimately such a system might demonstrate Dr. Bach-y-Rita's favorite theory, namely that the brain can learn to handle tactile information as if it were visual.

Dr. Bach-y-Rita's second reason for resuming work was that he could use a modified Optacon for this purpose rather than having to develop a prototype. The Optacon, a reading device for blind persons, consists of a tiny camera attached to an electronic box. The user holds the camera in one hand and moves it slowly across a line of print. The device transforms the letters into a vibrotactile array that moves across the fingers of the other hand as it rests inside the box. The procedure is useful for papers like bank statements and computer printouts that do not come in Braille, but it is very slow. Several thousand Optacons have been sold, but far fewer are actually used.

The devices are, however, readily available, as are people trained to teach others their use. Dr. Bach-y-Rita's group modified an Optacon camera by adding a clip-on lens that permits spatial information to be delivered to the user's fingertips in a 6 by 24 vibrating array. The camera and lens attachments are mounted on a lightweight helmet or headband. This permits scanning and image stabilization to be controlled by head and neck muscles, freeing the hand to manipulate objects. These modifications simply make the two-dimensional Optacon into a three-dimensional representation of the environment. "The users feel three-dimensional objects the same way the retina does; there are a number of theories how," says Dr. Bach-y-Rita. "It's comparable to poor-grade monocular vision."

Subjects using the Optacon can get much the same information as that provided by the TVSS in a readily available, cheaper form. Dr. Bach-y-Rita has done some preliminary designs for the system and sees other potential uses for it. He has federal grants to develop a teaching package for visual concepts and to demonstrate how to get more information through a somatosensory system. "It's really fun coming back to this," he says, "but I learned a lot from being away."

The "Seeing-Eye Computer"

Dr. Bach-y-Rita, who detests prototypes, has always been interested in the theoretical aspects of visual prostheses. Collins, who assembles a few prototypes each week, has always preferred to work on devices that provide mobility and other functions for the blind. He had found that the portable TVSS was impossible to use outdoors. Unlike the laboratory, the real world offers an overload of information to which the blind person must react in real time. A mobility system would need to preprocess this information down to essentials that could be used quickly. To perform this task, Collins and his graduate student, Mike Deering, developed a "seeing-eye computer."

The computer does the hard part of identifying obstacles, while the skin is given feedback about their locations. An early system used ultrasound, which bounces back from a few points perpendicular to the source, providing almost enough information to guide the user. But more information was needed for the artificial intelligence program, and the latest system uses television. Sixteen stimulators are placed on the user's forehead along an eyeglass frame that has a miniature television camera and a little earpiece. The electronics package will be carried in a shoulder bag or on a belt. When the user approaches an object, he or she is tapped on the forehead in its precise direction. The number of taps informs the user of its distance; they increase as he or she comes closer. Meanwhile, the speaking miniature computer identifies the object in a single monosyllable: "pole," "curb," "car," and so forth. The synthetic speech is binaural, so that it seems to originate from the object. (The system can also be designed to offer only verbal cues like "pole, 2 o'clock, 9 feet.") If the path is totally blocked, the voice advises the user to be extremely careful or to stop and search.

Collins hopes to improve this microcomputer system with higher speed and resolution, more powerful processors to analyze the visual scene more thoroughly, and larger dynamic range. Other planned modifications will show the texture and slope of the ground and give the subject control of his or her field of view.

Like the TVSS and the modified Optacon, this system would need to be accepted by blind individuals themselves. Blind people agree that none of the mobility aids presently available tell them anything they could not learn by using canes and other senses. "The mobility problems are the wrong ones to address," says Bill. "The real problem blind people have is figuring out where in hell they are. Did they cross seven streets, or eight? None of the mobility aids, including the ultrasound or TV systems, provides signs or solves other major problems."

Other researchers, however, had ambitious plans to create a portable vision system that could teach spatial concepts *and* aid mobility without having to train users or involve another sense. Its simplicity for users was based on its connection with the brain itself.

Electronic Vision

Being hit on the head or rubbing the eyes causes a person to see tiny points of light, or "stars." These bright spots, called phosphenes, actually originate within the brain. Electrical stimulation of the visual cortex produces phosphenes, but for decades scientists showed little interest in this phenomenon. For one thing, phosphenes are not crucial to understanding vision; for another, electrical stimulation of the visual cortex is possible only during brain surgery on nearby tumors, which are rare. (Experiments under any other circumstances fall into the aptly named category of "skull taboo.")

The scientific community was once dead set against the idea that phosphenes could be the basis for a new type of visual prosthesis. Researchers at MIT were able to organize a conference on the subject in 1966 only after plotting like conspirators and assuring participants that the proceedings would be confidential. While these scientists debated the feasibility and safety of experiments with phosphenes, two of their British colleagues performed an experiment at Cambridge University that was destined to squelch their reservations.

According to their study published in 1968, Giles Brindley and Walpole Lewin had implanted an array of 81 electrodes against the right visual cortex of a 52-year-old blind nurse. They recorded what their subject saw when they stimulated the electrodes or when she moved her eyes. Although the results of their experiment were predictable, they had made a great contribution by ignoring rampant skepticism and boldly demonstrating that a brain implant was possible.

Using phosphenes rather than skin taps to outline the visual world would provide a crudely spatial display, with one important advantage over the TVSS. Rather than involving another sense, a phosphene display would appear to the brain itself—the portion of the nervous system that specializes in interpreting spatial perceptions. If blind persons could be stimulated to see points of light, and if those points could be arranged in patterns corresponding to camera images of the visual world, blind individuals could achieve unprecedented freedom and mobility. This possibility intrigued William H. Dobelle, Ph.D., who was then at the University of Utah. In 1969 Dobelle organized a research team to begin work in this area; he hoped to build a system that would permit blind people to "see" with their visual cortexes.*

The first step was to develop the proper instruments and techniques for performing visual cortex stimulation. Dobelle wanted to map the behavior of phosphenes in persons with normal vision without exposing them to unnecessary surgical risks. The Utah team therefore requested surgeons throughout the United States and Canada to notify them of upcoming brain operations near the visual cortex. When the patients agreed to the experiments, the team would be called, often at the last minute, to fly with their bulky equipment to various hospitals.

Since the brain does not experience pain, the patients would be awake under local anesthesia during their surgery. While a patient's skull was open, Dobelle would place electrodes against the visual cortex, stimulate them, and have the patient describe how many dots he or she saw, where they appeared, where they moved, and so on. In his limited time, Dobelle would map the area to determine what size and pattern of electrodes would produce what results.

While performing 37 of these experiments from 1969 to 1973, the team made some fascinating discoveries about phosphene behavior. Stimulation of one electrode usually produces one phosphene, whose brightness or dimness varies with the electrical current's strength. Phosphenes move with the eyes but maintain their position relative to each other. Using electrode arrays in blind volunteers, the team would later discover that larger numbers of electrodes cause some phosphenes to disappear and others of various brightnesses to form groups. Although the electrodes were mounted in straight lines, the blind subjects described phosphenes appearing in random patterns sprinkled over their fields of vision.

During the preliminary experiments, Michael G. Mladejovsky, Ph.D., and the team at Utah were working on the instrumentation for the visual prosthesis. They developed a 64-channel computer-controlled stimulation system, a computer graphics program, and other equipment designed to map the phosphenes. The size and location of the visual (and auditory) cortex vary between individuals; since an electrode array would be placed slightly differently in each person's head, the phosphene map would have to be custom made. Fortunately, the map remains accurate because the phosphenes do not shift position.

The team also made 64-electrode arrays of Teflon and platinum that were

---

*This discussion is based on published sources. Dobelle did not respond to several interview requests, and his associates declined to review this account for accuracy.

compatible with brain tissue. The arrays were embedded 8 by 8 in thin Teflon strips with Teflon-insulated leads.

By 1973 the team was ready to perform temporary implants on blind volunteers, starting with adults with acquired blindness who could remember what vision was like. Two blind men had participated in the Utah Artificial Eye Project for over 4 years. They went through rigorous screenings and briefings to ensure that they qualified for long-range testing and that they and their families fully understood the experiments and risks. Dave was a 43-year-old electronics technician and piano tuner who had congenital cataracts and had gone blind at age 15. Doug, a 28-year-old graduate student in social work, had lost both his eyes at age 21 in a Vietnam land-mine explosion.

In the fall of 1973 the volunteers and team flew to Ontario. Operations on both men were performed simultaneously by neurosurgeons John P. Girvin, M.D., of the University of Western Ontario and Theodore S. Roberts, M.D., of the University of Utah. Each surgeon made a small incision in the patient's skull, inserted the Teflon ribbons against the right visual cortex, and stimulated the electrodes to help ensure proper placement. Dave and Doug were under local anesthesia and could respond to questions. The surgeons then closed the incisions so that only the wires protruded through the skin; the ribbons had been designed for easy removal without additional surgery. The wires were connected to the computer, which would select phosphenes to form patterns.

Dave and Doug participated in 3 days of experiments; each described phosphenes in his left field of vision because the electrodes stimulated his right visual cortex. Doug's experiments were successful. Four of the electrode wires broke, and his array shifted, but once his phosphene pattern was mapped, he saw lights for the first time in 7 years. He described phosphenes ranging in size, color, and tendency to flicker. The scientists constructed phosphene patterns forming geometric shapes and simple letters that he identified correctly.

Dave's array was poorly connected to his brain and slipped too quickly to permit mapping or pattern representation. He did, however, see 43 flickering phosphenes—a crucial finding. The researchers were not sure whether the visual cortex changes after long disuse; since Dave had not seen for 28 years, his experiments suggested that any blind person could use the device Dobelle had in mind.

By 1975 Dobelle and his team had implanted arrays in two more blind volunteers. One subject's array was placed too far forward in his brain. The other subject was 33-year-old Craig, another participant in the artificial eye project, who had been blinded by a gunshot 10 years earlier. Because his was a long-term implant, Drs. Girvin and Roberts performed more complex surgery to install it. Craig's Teflon ribbon had 64 electrodes; the connecting wires ran out of a hole at his skull base, then under his scalp to a round, dime-sized graphite socket above his right ear. A computer wired to a television camera was plugged into the socket.

When the camera was aimed at a pattern, Craig saw phosphenes that reproduced that pattern: a white line on a dark background, geometric figures, even individual letters and simple sentences written in a special visual Braille alphabet. Craig could read this Braille alphabet five times faster than he could read tactile Braille. Although like Doug's his array had four broken wires, he did not suffer any ill effects from it, but the visual system was useless to him when he was not in Dobelle's New York laboratory being tested with the external equipment.

Craig was one of four blind volunteers who received long-term implants; the two more recent operations were performed in December 1978. Eventually Dobelle wanted to implant two arrays of 256 electrodes each (one for each visual cortex), with more finely tuned brightness and contrast. He imagined an electronic vision system consisting of a tiny camera mounted inside an artificial eye and a microcomputer and battery pack attached to an eyeglass frame. The computer would process the camera images into accurate phosphene pictures similar to the animation on an electronic scoreboard. Its users would be able to identify shapes, recognize faces, and read more quickly and easily than before. Rumors among Dobelle's friends and colleagues, however, are that he has discontinued experimentation with visual prostheses; he is instead working on auditory prostheses, hybrid pancreata, or catamarans, depending on who is asked.*

Returning to our skeptical TVSS alumni, we find little enthusiasm for Dobelle's concept. "Presenting Braille characters through phosphenes sounds to me like teaching dolphins to talk—it's just something to do rather than an experiment that will have practical benefits," says Bill, himself a talented engineer. "If that system is ever made to work," says Mike, "it won't be produced to benefit the blind—it will be used to sell something. Often we benefit from commercial products simply by accident."

These and other subjects had extensive training on the TVSS, and some obviously did extremely well. It is hard to tell whether the experiments succeeded because of the ingenuity of the system, the intelligence and motivation of the subjects, or both, since the same cannot be said of people trying to use vision aids in general. Dobelle has estimated that fewer than 20% of blind persons can read Braille, fewer than 10% can use a cane to travel, and fewer than 1% can use guide dogs successfully. While Dobelle's proposed system would have done everything for its user, it posed other problems: enormous technical difficulties, high expense, risky surgery, and possibly low acceptance rates.

Visual prosthesis researchers must answer not just to funding agencies who want well-researched proposals with wide application, but also to an audience that demands practicality. "I just wish that an engineer would come to blind

*Dobelle is now in New York; Dr. Kolff meanwhile hopes to reactivate the neuroprosthesis program, including electronic vision, at the University of Utah. He believes that it is definitely possible to implant very fine electrodes into the brain to stimulate it with currents that are one thousandth the strength of those used previously.

folks and ask, 'What do you want?' They never do that," says Mike. "Many devices designed for the blind are just engineering projects that come into being because a transducer or some other device exists and the engineer just wants somethings to do," says Bill. "An actual assessment of needs is very important in a research and development setting." It must be quite demoralizing to blind people to know that research potential is being wasted—even more so for the "heartbreakers," those who travel from one research center to another hoping to find a miracle to help them see. (Ironically, people who do acquire vision late in life often cope very badly, and a few even commit suicide.)

Well-intentioned sighted people may in some ways be more "shortsighted" than blind individuals themselves. "Sighted people would love to give vision to the blind, who many times are actually at peace with their handicap and should be allowed to stay that way," says Mike. "It may mean a lot to sighted researchers that they can help blind children visualize things under microscopes, but to the kids that may be just another tedious task." The product does not always translate to the consumer in the way the sighted person would like, although the blind person may enjoy it for different reasons. It is possible, too, that the blind may not recognize their genuine accomplishments, like Bill's work on the assembly line, because they involve hard work and have no long-term practical benefits.

These issues apply to the field of bionics in general. "A similar analogy might be applied to cochlear implants for the deaf," says Jim. "There is sometimes too much emphasis on making handicapped people bionic not to open real opportunities for them, but just to make them like everyone else. We want to be accepted as we are."

If the true goal of bionics researchers is to improve the quality of life, perhaps they need to define more clearly just how that goal can be reached. Certainly provision must be made for basic research, as long as it does not acquire a track record of failures. We have noted that funding agencies are too quick to dismiss innovations as impossible, and researchers must avoid being dragged into that narrow-minded trap. On the other hand, they must identify and aim to fulfill genuine needs. *Then* they can begin bridging the gap between the feasible and the practical. The area Jim just referred to is making progress toward that goal.

18

High-Tech Hearing

Despite the ubiquity of Muzak, noise pollution, electronic beepers, and other distractions, most of us do not recognize the importance of sound in our daily lives. Deafness is an affliction even worse than blindness, primarily because of the isolation it imposes on its victim.

Sound plays a crucial role in our understanding of ourselves and our world, starting right from prenatal development of hearing ability. It is the primary means by which we learn about events in the environment and relate to our surroundings. A child who is born deaf or loses hearing while very young does not learn that objects produce sounds, each of which represents a meaningful occurrence. He or she must be taught that sound is omnidirectional and can reach us even from objects that are out of sight (a major reason that almost all warning signals in our society are auditory). Deaf children do not develop good oral-aural speech or language skills; often those in their midteens read only at third-grade level. "Children who grow up without communicating with their parents experience a huge loss. At age 18 or 19, when they should be ready to go out on their own, they function at the level of five- or six-year-olds. To stay deaf is to sacrifice a heck of a lot," says David Columpus, a vocational rehabilitation counselor for the deaf in the California State Department of Rehabilitation. "I'm just learning how very much people lose because of their deafness, most especially when deafened from birth," says Consuelo Wild, a clinical assistant at the House Ear Institute in Los Angeles. "I feel that young children born deaf or deafened at an early age deserve everything that can be done for them. They have great potential, but without early intervention the gap between them and their hearing peers widens to a point where it can't be bridged."

Of course deafness decreases the ability not just to receive signals but also to send them. People usually learn to speak by focusing on the desired outcome—a word—and imitating others' pronunciation until it sounds correct. Deprived of such feedback, deaf children must be taught how to position their mouths and tongues and control the action of their vocal cords and breathing. This is like learning to throw darts by studying the correct hand and arm positions rather than by aiming at a target.

Estimates on the size of the profoundly deaf population in the United States range from 50,000 to ten times that number. Like the blind, the deaf have various aids available to them: hearing aids and dogs, gadgets to notify them when the telephone, doorbell, or alarm clock is ringing, pagers that vibrate, and so on. A new computerized device that can be attached to standard telephones allows deaf persons to punch in messages on a keyboard and read replies off a liquid crystal screen. None of these, however, compares with an actual restoration of the sense of hearing.

We have seen how electrical stimulation of the visual cortex can theoretically be used to create a practical form of sight. Using a similar principle, researchers have tried for years to develop devices that will restore enough sound to the deaf to aid them in lipreading and eventually enable them to understand speech without it. The result has been a worldwide population explosion of cochlear implants—devices that electrically stimulate the auditory nerve. Although still crude compared to the human ear, cochlear implants have aroused much enthusiasm and considerable controversy. Some professionals question their value, and even their most enthusiastic proponents cautiously avoid raising the hopes of deaf persons about their capabilities. It seems, however, that cochlear implants offer undeniable benefits as a means of delivering sound, especially speech sounds, to the deaf and as research tools uniquely qualified to help scientists better understand the hearing process. These benefits are emphasized as much by the fervor and biases of the scientists who develop cochlear implants as they are by the descriptions of deaf persons who use them.

The Hearing Process

We have seen that the muscles transform electrical energy into mechanical energy; the ear too is a transducer, one that transforms the mechanics of sound into electrical impulses. The ear is divided into the external, middle, and inner ear. The auricle and ear canal forming the external ear gather sound and direct it toward the tympanic membrane of the eardrum.

The middle ear consists of the eardrum and the bones (ossicles): the hammer (malleus), anvil (incus), and stirrup (stapes). Acting as levers, these tiny bones increase and transmit sound vibrations from the eardrum to an opening in the inner ear called the oval window. This is actually a process of converting sound vibrations into fluid waves.

The inner ear chamber, which is filled with fluid, contains vestibular (balance) and auditory mechanisms. The chamber, or cochlea, is a pea-sized structure shaped like snail shell and divided lengthwise. Its two compartments are separated by a very thin shelf of bone and the basilar membrane, on which sits the organ of Corti covered with hair cells.

Vibration of the hammer, anvil, and stirrup sends fluid waves into the cochlea, where they stimulate certain of the 20,000 delicate hair cells. As the waves flex the basilar membrane, the hair cells generate electrical signals that are picked up by the fibers that form the auditory nerve. The current passes through several complex interconnections in the brain stem to the auditory cortex, the portion of the brain that recognizes it as sound.

High-pitched sounds stimulate nerve fibers at the end of the cochlea closest to the oval window; low-pitched sounds stimulate those at the far end. Complex sounds stimulate fibers along the length of the cochlea. Various combinations of stimuli are interpreted by the brain as the pitch, loudness, quality, and direction of sound. No one knows exactly how the brain makes these interpretations.

Types of Deafness

Deafness can result from congenital defects, infections, illnesses like meningitis, certain drugs, head injuries, loud sounds, or aging processes. It is categorized according to what part of the ear is affected.

CONDUCTIVE DEAFNESS

Conductive deafness refers to the inability of the outer or middle ear to transmit sound. It can be treated medically (as with a hearing aid) or surgically.

The form of conductive deafness believed to have afflicted Beethoven is caused by otosclerosis, in which spongy bone forms in the middle ear and fixes the small bones. This condition can be corrected by replacement of the stirrup with a prosthesis. The large variety of implants available for this procedure, called stapedectomy, are made of stainless steel, Teflon, platinum, tantalum, or Plastipore. As noted earlier, a new middle ear prosthesis made of Bioglass has recently received FDA approval for marketing; this device can be trimmed for use in stapedectomy as well. Prostheses are also available to substitute for all three bones.

Inflammation of the middle ear that is not draining properly can be relieved by one of several types of middle ear ventilation tubes.

SENSORINEURAL DEAFNESS

Sensorineural deafness (often misnamed "nerve deafness") results from damage to the inner ear. There are two types of sensorineural deafness. Sensory deafness involves only the hair cells in the cochlea. Hearing aids are almost useless, but a cochlear implant can bypass damaged hair cells and activate the auditory nerve. Its success in this application may depend on the number and location of surviving nerve fibers: large numbers of fibers close to the implanted electrodes might give better results than small numbers of fibers located far away.

Neural deafness, caused by damage to the nerve fibers themselves, cannot be treated by an implant.

CENTRAL DEAFNESS

A third, less common affliction is central deafness, which results from damage to the brain stem or auditory cortex. Like neural deafness, this type cannot be treated by an implant, which depends on intact nerves.

How Cochlear Implants Work

The various types of cochlear implants have the same basic design. Sound is picked up by a tiny microphone worn on an ear loop or tie tack. The microphone sends the signal to a pocket-sized speech processor, which translates the input into electrical signals. A detachable cable connects the processor to an external coil magnetically held against a receiver implanted under the skin behind and above the ear. Using radio waves, metal induction, or some other signal transfer, the coil relays the signals to the internal electronics.

The receiver is connected to a thin wire lead of electrodes surgically inserted through the mastoid bone into the cochlea to transmit the electrical signals. The electrodes are placed according to what regions of the cochlea are to be stimulated. Alternative methods are to place an electrode near the round window outside the inner ear or to use an electrode array that pierces the cochlear nerve itself.

All these devices have filters to screen out background noise, compressors that automatically dampen sudden loud noises, and controls that permit wearers to raise and lower the volume. Some allow the wearers to adjust individual channels. Those used in experimental studies are usually designed for customization to the subjects' needs as determined by laboratory tests.

Cochlear devices are implanted in the more badly damaged ears of patients who derive no benefit from hearing aids. To work properly they must overcome several technical difficulties: the ear's processing of sound is incredibly sophisticated; the cochlear fluid, in which the electrodes must operate for years, is like seawater; the signals must be timed to produce appropriate frequencies and must be those that the brain will accept as speech.

Cochlear implants are commonly distinguished as being either single-channel or multichannel and intracochlear or extracochlear.

SINGLE-CHANNEL AND MULTICHANNEL IMPLANTS

Whether a cochlear device is single-channel or multichannel depends on how the device is designed and what is put through it once it is implanted.

An implant can have one or more electrodes: the number of wires or separate points for signal distribution. The number of channels refers to how the information is processed to be sent to the electrodes. Single-channel processing can be used with either a single-electrode implant or a multielectrode one (in the latter configuration, identical signals are sent to all the electrodes, one pair at a time). Obviously, no such thing as a multichannel device with a single electrode exists; a multichannel implant sends different signals to various electrodes.

By sending discrete bits of information to several electrodes, a multichannel implant has the ability to provide more information than does a single-channel one. Multichannel implants presently being tested are built for a fair amount of flexibility in their coding schemes and use several different formats. The most common configuration takes advantage of the way in which frequen-

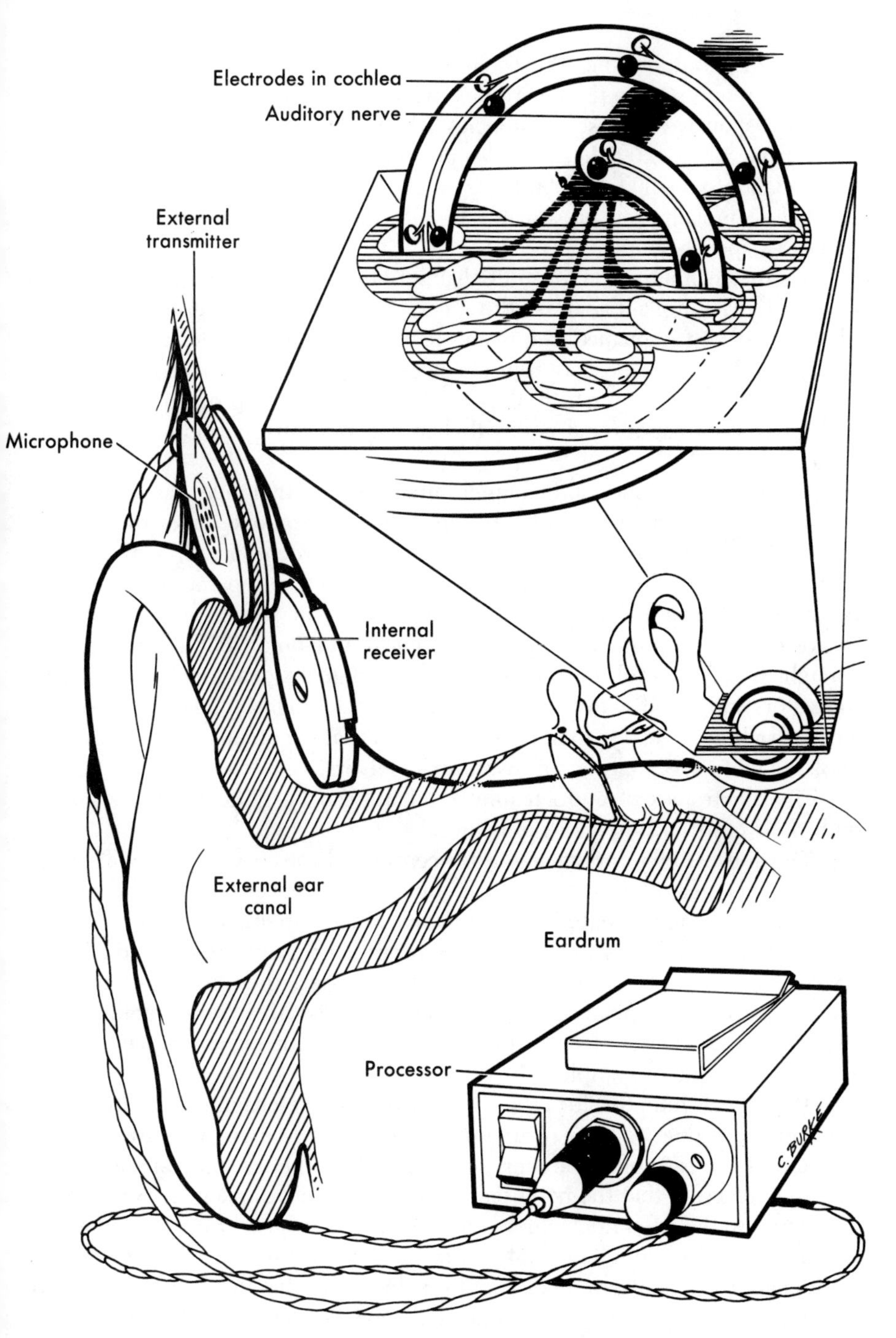

A multichannel cochlear implant with detail of the electrode wire wound inside the cochlea.

cies are distributed along the cochlea. The incoming sound is broken into frequencies and sent to the electrodes, which stimulate restricted areas on the cochlea responsive to those frequencies. The frequency range along the cochlea varies from 50 to 20,000 hertz (cycles per second); most important elements of speech fall into the 300- to 3000-hertz range. By stimulating selected regions on this band, implants can presumably imitate the way in which the natural ear processes speech sounds.

Another configuration for multichannel implants treats all incoming sound as speech. The implant analyzes the speech and divides out the second formant (the front cavity resonance frequency that is very important for intelligibility) and sends that to the most appropriate electrode. Then it takes the fundamental frequency (the rate at which the vocal cords vibrate, producing inflection and other information) and codes that as the rate of repetition at that electrode.

As we will see, one multichannel implant currently in use has a multiplex transmitter receiver. A multiplex system is like a telephone line with one wire carrying many channels on it. Each channel is synchronized; a bit of it is interspersed with the other channels, then decoded to separate them all out. Because each channel is stimulated individually in sequence, multiplexing is slower than simultaneous multichannel stimulation. If it occurs quickly enough, however, multiplexing can modify a digital signal (one that involves transistors and on-off switches) to look like an analog signal (one that is continuously variable). One researcher has suggested that eventually all multichannel implants will have multiplex transmitter-receivers.

The electronics of multichannel implants can be quite complex, raising both their costs and their risk of failure.

The success of a multichannel device depends on its wearer having intact nerve fibers near the areas being stimulated by the electrodes and on the information being presented appropriately. Some patients may not be able to benefit from all channels; no way exists to predict this outcome ahead of time. As researchers learn more about hearing processes, they might someday be able to prescribe implants as they do hearing aids. Meanwhile, opinions are divided as to whether it is better to start a patient on an implant that is extremely simple or complex.

Individuals tend to affirm the superiority of single-channel or multichannel implants according to their own vested interests, whether these are developmental (in the case of researchers) or financial (in the case of manufacturers). "Everyone agrees that the potential for single-channel implants is very limited compared to that of multichannel," says Dr. Jarvik, whose firm Symbion manufactures a multichannel implant. "There is still controversy about whether multichannel implants are better, but results are starting to show that they are far superior to single-channel implants," says Michael S. Hirshorn, M.D., President of the Nucleus-Cochlear Corp., which manufactures the most sophisticated multichannel implant being clinically tested.

"There's a common misconception right now that a basis exists for saying that multichannel implants are better than single-channel," says Karen I. Berliner, Ph.D., Director of Hearing Services Research at the House Ear Institute, whose single-channel device is the only cochlear implant that has FDA approval for marketing. "I don't think that there's clear evidence for that, nor do I think that single-channel versus multichannel is ultimately the issue—it has more to do with the combination of signal processing techniques, electrode design and placement, and patient physiology. For the past 5 years many researchers have been disappointed in multichannel work because of the general attitude that more complex implants would be so much better."

One investigator cited by others as a model of scientific objectivity is Bruce Gantz, M.D., Associate Professor of Otolaryngology and Head and Neck Surgery at the University of Iowa Hospitals. As we will see, Dr. Gantz and his colleagues are conducting tests of the various cochlear implants now available. "We've tried to be very cautious in evaluating the results," he says. "We feel that we're seeing a trend because of the consistency with which patients with multichannel implants understand open set speech [that which is not lip-read]. I have not seen any published data demonstrating that the 3M-House single-channel implant can perform at the levels we're seeing with the two multichannel systems that we have tested. The Hochmairs in Vienna, however, have reported that their single-channel coding scheme results in open set speech in some patients. We can't really state right now that multichannel implants are superior to single-channel implants." So far lipreading performance appears to be similar between the two types of implants.

INTRACOCHLEAR AND EXTRACOCHLEAR IMPLANTS

Implants can also be classified according to whether they are intracochlear or extracochlear. Most implants are intracochlear: they enter the cochlea through the round window. An alternative is to place the electrode outside the cochlea; most of these implants are single-channel, but theoretically a multichannel extracochlear device is possible.

Extracochlear implants are not yet being marketed in the United States. Preliminary tests in other countries suggest that their performance is comparable to that of the single-channel 3M-House device. 3M Co., which manufactures the House implant, is working on an extracochlear device developed by the researchers in Vienna that is now undergoing clinical trials.

The placement of extracochlear implants potentially would cause less damage to delicate ear structures than does that of intracochlear implants; this advantage may prove especially significant when the recipients are young children. "Almost all congenitally deaf children have some low-tone island of hearing," says Dr. Gantz. "We can detect that a 1-year-old child is profoundly deaf, but we're not sophisticated enough at this point to measure how much residual hearing he or she might have. Placement of an intracochlear implant probably destroys any residual hearing. If we had an extracochlear device, we could implant it in very young children and perhaps provide electrophonic hearing by electrically rather than acoustically stimulating the hair cells available."

A patient deafened by meningitis or another such condition might have bone growth in the cochlea that would preclude his or her using an intracochlear implant; an extracochlear device might be more appropriate.

Potential disadvantages of extracochlear implants include their larger power requirements to reach the auditory nerve, which mean shorter battery life and other inconveniences.

The Challenge of Cochlear Implants

Cochlear implant research in the United States began in the early 1960s with the separate investigations of Drs. William F. House, Blair Simmons of Stanford University, and Robin P. Michelson of UCSF. The attitudes they encountered resembled those faced by the first researchers to investigate visual prostheses. For years other scientists refused to have anything to do with electrical stimulation of the ear. Like Brindley and Lewin, Dr. House in particular barged into a controversial area, ignoring the protests of other researchers that such work was practically impossible and downright immoral. To this day researchers face the same confusion as did their counterparts working with visual prostheses about whether they are creating sensory aids or sensory substitution devices. Donald K. Eddington, Ph.D., Director of the Cochlear Implant Research Laboratory at the Massachusetts Eye and Ear Infirmary, believes the latter: "The 'fallout' from the work done so far is that the patients can use the devices as a sensory aid," he said at a 1982 meeting, "but since that was not the goal previously established it's not the direction of our maximum effort."

This confusion results in part from the difficulty researchers face in evaluating cochlear implants—their own as well as those being developed at other centers. "We find that the variables are greater than the number of patients, so we're trying to focus on, test, and base our hypotheses on as few variables as possible," says Dr. Gantz. The biggest problem is that no one really knows how the brain hears and understands speech, making it difficult to program an implant to perform this function. It will require much experimentation to find the locations along the cochlea that determine the volume, pitch, texture, and quality of sound. Even if implants are improved in these respects, patients must then be taught to hear the sounds. Unlike scientists working with visual prostheses, those trying to stimulate hearing cannot look to the appropriate brain cortex for help. Not only is the auditory cortex less accessible than the visual cortex, it leaves much processing of sound to the hair cells and other parts of the cochlear nucleus. Four or five mysterious levels of processing act on sound before it reaches the auditory cortex.

"We think that the major advances over the next few years will be not technical ones but rather in speech processing strategies in learning how to put sound through a missing mechanism that is not well understood," says Stephen Rebscher, M.A., Coordinator of the UCSF–Storz Cochlear Implant Project of the Coleman Laboratory at UCSF. "The implanted components of these devices may not have far to go in terms of technology." Researchers disagree, however, about the numbers of patients who should be implanted at this time. Some believe that few patients should be studied until the devices are more sophisticated or speech processing is better understood, while others believe that the present implants offer sufficient benefit to justify clinical application.

Even understanding speech processing might be misleading, since cochlear implants are being tested on extremely abnormal ears. All cochlear implant work tries to reproduce sound as it was perceived before the onset of deafness. Another possibility is that there might be a totally different way to code the signal to provide more information for distinguishing sounds, one that is nothing like what sound is now. A young child might use such a system, or even existing systems, better than an adult. "We are trying to interface high technology with badly damaged auditory systems, using what we understand now about the natural ear," says Berliner. "But we mustn't get locked into these processes in case they don't apply." (Similarly, little research has been done on speech production; feedback to improve a patient's speech might be different from signals most effective for understanding the speech of others.)

Meanwhile researchers face other problems: animal models of auditory processes are not really comparable to human ones; humans cannot be examined as thoroughly as animals; and tests on humans have highly variable results. Each patient is unique, so even those using the same implant will respond differently. Tests are strongly affected by how they are given. A tremendous difference in response occurs if lipreading is not used or if patients have practiced or are given the same test three or four times. Some examiners repeat each word twice before recording the response, a practice that can double the patient's score. Dr. Gantz and his colleagues are trying to standardize their tests: in open set testing, for example, they believe that a recorded, unfamiliar voice should be speaking new words and sentences to the patient. The only way to compare implants meaningfully is to give the same test (like the MAC or the Iowa test batteries) to all their recipients.

Extremely conservative scientists and audiologists are not even sure that cochlear implants are worth all this investigational bother. As we are about to see, however, many implant recipients would strongly disagree with them.

19

Cochlear Implants and a Listening Belt

All the cochlear implants described here are currently undergoing some phase of human trials for FDA premarket approval. In 1984 the 3M-House implant was approved for marketing to adults. Children are still undergoing clinical trials with this device, as are adults with the Nucleus implant (which is expected to receive approval within a few months). The Bioear, Ineraid, and UCSF implants are approved for investigational testing on experimental patients. Dozens of medical centers across the country are testing cochlear implants. "Before these trials are completed, there is no basis for any of us to make claims about what our device will or won't do," says Berliner.

The common denominators of published studies and anecdotal reports indicate, however, that cochlear implants have much to offer profoundly deaf persons. Although they may hear buzzing or static that sounds like that of a poorly tuned radio, implant recipients can perceive environmental sounds from which they can distinguish speech sounds. Many can identify men's and women's voices, certain accents, and changes in volume and inflection. Recipients can lip-read more accurately; a few can understand many individual words and connected sentences spoken behind their backs. This ability holds true for videotaped as well as live conversation when the recipient can lip-read. "I can pick out speech on television when I sit close and concentrate hard," says Earl Trottier, who has a UCSF multichannel implant. "I haven't tried films yet," says David Columpus, an Ineraid recipient, "but I do pretty well with television, especially when women are speaking. I enjoyed watching the Democratic convention because both Jesse Jackson and Gary Hart have relatively high-pitched voices, and by sitting close I was able to follow their speeches almost perfectly."

Hearing-impaired persons and their counselors agree that the deaf feel extremely deprived because they cannot hear music. "The music a recipient can hear with an implant has no comparison to what he or she remembers if postlingually deaf," says Consuelo Wild, who has a 3M-House implant. "But several adult patients I have counseled have told me that they can recognize hymns sung in church. I love to sing, and if I'm singing in the car with the radio on, my pitch sounds very normal to me. I don't recognize music from another source unless I know it already. I can pick up the beat well enough to dance to it easily, but it lacks the pure tone of music."

Implant recipients find that single voices or instruments like a piano or guitar are the easiest to hear; a band or orchestra is just so much noise. "I've heard an ensemble of brass instruments quite clearly," says David. "I don't know whether strings, for instance, would sound as good, but it seems that instruments in the same family come across well."

For those who can manage it, an extremely exciting postimplantation experience is that of talking on the telephone. While lip-reading, an implant recipient cannot be sure how much comprehension results from hearing and how much from sight; telephone conversations, on the other hand, require auditory speech discrimination. David can understand over half the conversation when speaking on the phone with his audiologist and can carry on extended conversations with his wife. If the teletype in his office is down, he switches to the telephone with some of his business associates. When he received a new speech processor, the first thing he did was call his two daughters and his mother. "It was the first time I had talked to my mother over the phone without help in 20 years, and we were both thrilled," he says. "Much of my understanding has to do with the context of the conversation, and I'm still somewhat insecure about speaking on the phone. I missed it terribly, but I didn't expect the technology to make it possible so quickly." Other deaf persons should not expect it either; David's capabilities with his implant are quite remarkable.

Consuelo and other implant recipients with less discrimination use codes to speak over the telephone. "I can hear the other person's voice, but all one-syllable words sound the same," she says. "We're taught to use the phone by asking the other person questions to which he or she responds, 'Yes, yes' or 'No.' That limits conversation, but I think it's great to know that if there is an emergency, I can communicate well enough over the phone to get help."

Of course, the other side of listening over the telephone is speaking distinctly oneself. Earl, David, and Consuelo are all postlingually deafened, and each has excellent speech. "People tell me that my speech has improved with my implant. It's better modulated and has more expression and tone. Part of that is getting used to how my voice should sound to me to communicate best with other people. I've learned, for instance, that if I speak as loudly and distinctly as others sound to me, I'll be too loud," says David. "With his implant Earl speaks more softly and slowly and seems to be getting back his natural voice pattern," says his wife Emily. Consuelo sounds almost *too* natural. "My having good speech causes people to forget that I'm deaf," she says, "and sometimes they think that I'm ignoring them because I haven't heard what they've said. I think the hardest part of being postlingually deaf is that it's an invisible handicap; unless you share your problem, you may be labeled stuck-up, weird, or unsociable."

The experiences of these implant recipients are just examples of how cochlear implants are helping deaf persons. We have noted that it is practically impossible at this point to compare the performances of different devices; what we *can* do, however, is mention some of the interesting features of those being clinically tested in this country.

Single-Channel Implants

THE 3M-HOUSE IMPLANT

Dr. House has been testing both single-channel and multichannel cochlear implants since 1961, and he and his colleagues have implanted the devices in 377 American adults and 152 children to date. His extensive laboratory tests with his earliest patients committed him to using single-channel implants for several reasons. The patients had subjectively decided that the signal processing scheme in that type of implant worked best for them, and it was comparatively easy for an engineer to make it into a wearable, completely implantable device that required no percutaneous access. The high-frequency 16-kilohertz carrier signal in the implant needs relatively little power to be transferred through the skin. Simplicity also decreases the chances that the device will fail.

The 3M-House implant has one electrode that goes into the cochlea and another that is placed in the temporal muscle or an adjacent area as a ground. "We've experimented with different devices since then, but clinically this single-channel implant has been very effective," says Berliner. "There's some misconception that we're married to single-channel implants. We strongly favor this implant's long history of clinical benefit. Since it takes a long time to improve a device and apply it clinically, we consider it appropriate to offer this implant to patients now until eventually we develop a better one."

The 3M-House device is the only one available for implantation in children; currently it is being tested in clinical trials at carefully selected centers. Its use in children makes this comparatively simple gadget extremely controversial. The AMA endorses implants only for postlingually deaf adults, who comprise the majority of recipients. Many audiologists think that prelingually deaf children cannot benefit from cochlear implants. "Those who object most vocally and strenuously to implantation in children are responding to a large extent on an emotional level and are often not willing to find out the facts," says Berliner. The facts mustered by the House group in response to objections are straightforward and cited also by deaf adults who work in the field.

A major objection to implantation in children is that insertion of a cochlear implant can damage the ear; if children have residual hearing, such injuries might preclude their using better aids as they grow older. This risk must be weighed against both the fact that only one ear is implanted and the potential benefits of sound to the child. Animal tests show that levels of current necessary for stimulation do no discernible harm, and while insertion damage may occur, no one knows whether it would cause a decrease in perception. Charles Graser, an early recipient of the 3M-House implant, has used his device daily for 13 years, and 20 other patients have used theirs daily for 5 years or more with no perceptible damage. The small amount of bone growth that may be stimulated by the presence of the implant seems insignificant. "We took out a 3M-House device that extended 17 millimeters into the cochlea and put in a Nucleus device without any problems," says Dr. Gantz. "Although we do have to be cautious, I'm not terribly concerned about bone overgrowth and other factors that had caused so much consternation in animal studies."

Critics also point out that children are more prone than adults to middle ear infections that could theoretically spread along an electrode lead into the inner ear. Patients with the 3M-House device, however, have had such infections treated by their local physicians with no complications. A very good barrier is maintained between the middle and inner ear, and tissue encapsulation of the electrode isolates it from middle ear spaces. The likelihood of infection spread is therefore small.

Children wearing cochlear implants might outgrow their devices, some researchers suggest. Dr. House uses one size for implanting adults and children, partly because his work with adults has given him a common denominator for calibrating the implants. In children the extra length of electrode wire is left lying in free space in the middle ear, well anchored at both ends; as the skull grows, the electrode should uncoil and stretch out without pulling loose. Young children have had the implant for over 3 years with no problem. The most rapid time of skull growth that would affect the cochlear implants occurs before age 2, and the House group does not implant children younger than that because it is hard to be certain how deaf they are.

Some critics believe that it would be better to wait until improved devices are available before implanting children. At this point, however, it is not clear what would constitute "improvement." Most other implants will not be approved for children for several years, and none would restore normal hearing. A more complex device might cause even more damage and be harder to remove for replacement. This issue could be resolved only by a perfect implant; meanwhile the 3M-House implant could prepare children for improved devices by helping maintain growth and maturity in what is left of their auditory systems.

Researchers have also asked why Dr. House has implanted many children instead of intensively studying a few. This is partly because the kinds of measurements required by the FDA occur over long periods of time even in normal children. Within the age group of 2 to 18, there are many types of deafness and several other variables that make comparisons difficult. "It may seem that we've implanted many children," says Berliner, "but for any age group that is appropriate for comparing language or cognitive skills, speech production, age of onset, communication, or other crucial variables, we really have very small numbers. And among these children we've seen almost no risks and very positive benefits."

Children who have received the 3M-House cochlear implant can tell the difference between the volume, continuity, length, and speed of sounds. They can count the number of sounds they hear and the numbers of syllables in words and sentences. Sounds like a door knock, a siren, and their own names are readily distinguishable.

The children's voices become quieter and more modulated, relaxed, and natural, and they progress faster in speech exercises. Parents report that their children treat objects more gently, tend to watch faces as people speak, and seem curious about speech.

"In contrast, we know what happens to children who do not receive sound. The auditory system will degenerate over time. Every child who waits to have an implant loses time in which he or she could be helped by having some hearing, however useful it may be," says Berliner. "I'm sure that almost every parent whose child has a hearing problem would think it a real boon to provide the child with a means of communicating fluently with the family," says David.

It is partly for the sake of her deaf children that Consuelo was implanted with the 3M-House device. Consuelo has a hereditary, progressive hearing loss that has also afflicted her father, her four brothers, and her two sons. Now 38, Consuelo began using hearing aids when she was 9; by age 23 she was profoundly deaf. She continued using an aid on her right ear but became concerned about her life-style change when she became a single parent. "I went back to school for another degree and became worried about finishing my studies and going to work while my hearing was growing progressively worse," she says. "If the aid on my right ear wasn't working properly, I would panic and be afraid to drive or even go anywhere. I was terrified of losing all my hearing." Since Dr. House had treated all three generations of Consuelo's family, she went to the Institute for testing.

To receive an implant, a deaf person must not be able to benefit from a hearing aid. Consuelo's case was borderline. She can no longer use a hearing aid on her left ear, so 4 years ago she opted for implantation on that side and continued use of an aid on her right ear. "I decided on the implant for several reasons," she says. "The sounds I receive from the implant and aid are quite different, and I wanted to teach my implanted ear what those sounds were before I lost all my hearing in the right ear. It would have been more traumatic for me to have adjusted to total silence, then the implant. Because my sons are hearing impaired, I wanted to gain experience with the implant and also help the research in this field. Knowing that better implants will be available by the time my sons need them makes my life and work much more rewarding."

To her surprise, Consuelo has found that when her hearing aid breaks down she can hear almost as well with just the implant. "I probably manage better now than I did before the implantation even though my hearing has continued to deteriorate," she says. "For the first time in years I can hear on both sides." Consuelo has thick, shoulder-length hair; her speech processor is small enough to be tucked inside her bra, and if she wears a high-necked blouse, neither the hearing aid nor the implant shows. Her speech and discrimination abilities are so good that the average person conversing with her would have no idea that she is totally deaf. "I'm much more comfortable in groups of people now, and much less tense with strangers," she says. "With two devices instead of one, I feel much more relaxed and confident. The thought of not using my hearing aid doesn't scare me nearly as much."

THE BIOEAR IMPLANT

The Bioear implant, developed and manufactured by Biostim, Inc., in cooperation with Stanford University, is a multimode, single-electrode, single-channel system. Its configuration and performance are quite different from those of the 3M-House device.

The Biostim unit is a complex microelectronic receiver-stimulator, hermetically sealed in a titanium capsule. This package provides the patient with regulated outputs, distortion-free signals, analog or pulsatile stimulation codes, several unique conditioning techniques, and the flexibility of being able to use any stimulation code devised in the future. The 16-kilohertz carrier wave in the 3M-House transmitter accompanies the sound signal into the cochlea while the implant is used; by decoding this carrier wave the Biostim device eliminates its possible long-term detrimental effects on the central nervous system.

Researchers at Biostim are also working on a ten-electrode monopolar multichannel implant and an eight-electrode multiplex multichannel system. The company's philosophy is that multichannel systems must await the clinical results of the Bioear for future direction.

Multichannel Implants

THE UCSF IMPLANT

Researchers at the Coleman Laboratory have developed cochlear implants over the past 15 years, including a single-electrode single-channel system; a multielectrode single-channel system; and a multielectrode multichannel system presently being produced by the Storz Instruments Co. and applied clinically at UCSF and other centers.

The multichannel implant has four completely independent channels, each of which has a bipolar (positive and negative) pair of electrodes. Bipolar stimulation restricts each channel field so that they do not interact, making the sound volume suddenly very loud or soft. Because the implant has 16 electrodes, it can eventually be upgraded to six or eight channels without replacement. The researchers believe that if speech processing strategies improve, six or eight channels may be sufficient for high levels of speech understanding in a large number of patients. In the present device four separate antennae are arranged around the ear to transmit and receive the four channels of information; a six- or eight-channel device would be a much smaller implanted system. Besides working on devices with more channels, the UCSF group is developing implanted electrodes that promise greater long-term reliability and smaller processors that will produce less distortion with greater speech information.

The UCSF researchers believe that the connector in their device is the best design available with respect to ease and reliability of replacement. Failed electronics can be replaced without disturbing the electrode array. The researchers have intensively tested several experimental patients with the multichannel system during the past 4 years and have signed an agreement with the Storz Co. to begin larger clinical trials.

Before receiving the implanted receiver module and four channels, the patients participating in the experimental research series undergo 3 months of full-time testing in the laboratory. The implanted electronics are initially attached to a percutaneous cable that is run up over the head and out behind the other ear. This system enables the researchers to determine which four channels will give the best results for the individual patient (the remaining four are disconnected). It also allows them to assess many different speech processing strategies and to measure the implanted electrodes directly. All of these tests lead to the development of the implant system for application in the clinical group of patients. The research subjects with intact auditory nerves have all found that more channels give better speech understanding and more pleasant sounds.

In a joint project with Duke University the UCSF group is also using laboratory tests to try to stimulate different strategies for speech processing. One aspect of this testing is that of simulating the different implants currently being used. To factor out all the variables arising from different types of electrodes and unique patients, they imitate the various implants on a single channel with 16 individual contacts. The researchers have found distinct, logical differences between the available implants. "The testing we're doing is the most sophisticated I know of," says Rebscher.

Not everyone agrees. "Because of the central reprocessing that occurs, it may take a patient as long as a year to learn to hear with a cochlear implant," says Dr. Gantz. "It's not worthwhile to compare coding strategies for different implants when each is tested for only an hour, or even a day, rather than a long enough period of time. I object to statements that one implant is better than another on the basis of such brief studies."

At least one UCSF patient, however, is pleased with the test results. It is difficult to tell what Earl enjoys most: hearing with his cochlear implant, demonstrating how the speech processor works, or showing photographs of his operation. A 67-year-old retired automobile mechanic from Novato, California, Earl has had the implant since April 1984. For the 16 years preceding his operation, he had been totally deaf from a hereditary condition after wearing hearing aids for 25 years.

Like other experimental patients with the UCSF implant, Earl at first wore the percutaneous cable and was tested in the laboratory. Through the cable he was tested with a simulation of the Ineraid implant, but the signal it provided was fluctuating and unstable. With the UCSF device, he gets good reception using four channels driving electrode pairs 1, 4, 5, and 7 with bipolar stimulation. To activate his customized speech processor, Earl turns on one channel at a time, adjusts the volume on each, then turns on all four at once to obtain an overall comfort level. He does hear some intermittent buzzing from two of the channels.

Earl came to the UCSF Cochlear Implant Project as a referral from his hospital plan. "I was told not to expect to understand speech—just to hear sounds and have help with lipreading," he says. "I wasn't at all worried by the prospect of surgery—if I had been, they might not have accepted me into the program."

Earl's hearing with the implant has exceeded expectations. Since he is a poor lip-reader and would not learn to sign, Earl used to pretend to understand subjects in conversations. Patiently he would sit for hours while visiting with friends, watching them talk but unable to join them. "That was very frustrating for me, and I'm much happier now that I can hear speech," he says. "I really enjoy being able to communicate with my family again."

"The implant is a miracle that has made an enormous improvement in our lives," says Emily Trottier, a retired secretary. "With his implant turned on, Earl can hold conversations with me and with our friends. This morning I tried to tell him something simple before the device was turned on, but it took tremendous effort. It makes a world of difference not to have to beat my head against the wall trying to get a simple idea across to him."

Earl had one of the UCSF technicians make him a few different battery chargers so that he would not have to worry about his device losing its power. "I want to use it all the time," he says.

The researchers are still trying to make Earl's speech processor more compact and improve its quality. "The buzzing makes it hard to concentrate," Earl says. "Sometimes I can hear the phone or doorbell without being able to identify them. But I can still hear complete words; I can hear myself speaking, and I can differentiate people's voices. I've recommended the implant to a woman with a single-channel device who can hear sounds but no speech, and I'd definitely encourage others to try it."

THE INERAID IMPLANT

The Ineraid implant was developed by Eddington and his colleagues while he was at the University of Utah and is now manufactured by Symbion. It has eight electrodes—six contact points in the cochlea and two grounds. The device has a relatively large sound processor, a 13-ounce pack that is worn on a belt. "We feel that decreasing the size of the processor is less important right now than improving the implant's performance and reliability," says Dr. Jarvik.

Instead of being transmitted through the skin, sound information is sent to the Ineraid implant through a dime-sized percutaneous button in the scalp.

Until the researchers develop a transcutaneous system, the plug offers certain advantages in that laboratory equipment can be connected to it directly and it is more secure than a magnetic coil, which can slip out of alignment. "In general, the plug is not uncomfortable," says David. "I sometimes feel discomfort if I sleep on my left side and the skin pulls, but I don't do that very often. I shower; I jog; I swim the side stroke; I'm exposed to salt water and spray, but I've had no problem with its being wet. In the 7 years I've worn the plug, the area became infected two or three times when hair grew into it. I just trimmed the hair and used some ointment to clear it up."

A 53-year-old resident of San Diego, David had normal hearing until the age of 31, when he developed Ménière's syndrome and began using hearing aids in both ears. For the past 12 years he has been totally deaf. While working as a librarian in the Deafness Center at the University of California, Northridge, he heard about Dr. House's research and went to the Ear Institute for an interview. There he was told that the 3M-House implant could provide him with sound perception but that he might not be able to understand speech. "That didn't interest me," says David. "I was more attracted to research that was being done in multichannel implants." Four years later he was contacted about participating in experimental tests at Utah, and in 1977 he had his implantation.

For the next 7 years David and another subject traveled to Salt Lake City every few months to participate in most of the research being done with recipients of the Ineraid implant. Tests showed that David's speech discrimination was good with four channels. The speech processing system was not yet portable, so the only times he could use the implant were when he was plugged in to the laboratory equipment. "Hearing with the device for the first time didn't affect me as much as having it turned off—readjusting to silence meant overcoming a real sense of loss," says David.

For the past 18 months David has had a portable speech processor. The processor keeps shrinking as the electronics improve and the batteries last longer. David does less monkeying with the processor now that he can no longer adjust the channels himself.

"It's been exciting to work with Don Eddington and make various discoveries over the years," says David. "I'm not good at lipreading, and it's wonderful to be able to talk to everyone. When we went out to dinner with friends, my wife used to have to sign the conversation to me while her meal got cold—now she can eat, and I can carry on conversations by myself." David attributes his speech discrimination to steady improvements in the implant's electronics rather than practice with the device. "All I've really learned is to adjust to loud background noises and to use contexts the way lip-readers do," he says.

Like Consuelo and Earl, David depends on his implant. "I've had times when the electronics needed fixing or adjustment and I'd be unable to use the implant for a few weeks," he says. "That throws my life all out of kilter, and I get very depressed. Hearing is once again an extremely important part of my life that has definitely been worth all the time I've put into the research."

The Ineraid implant has changed David's work habits and may affect the development of his field. "I used to need a full-time interpreter in my office," he says. "Now I use one about 8 hours a week, mostly on the phone. It's exciting to know that many of my clients could benefit from this device. Right now it's in the position that heart pacemakers were in 20 years ago, but I can see how it will affect the whole field of deafness in the future."

"I certainly believe that the Ineraid gives patients a much broader range of sound than would a single-channel implant, although we haven't proved that yet," says Dr. Jarvik. "I'm absolutely sure that our system will give patients something very valuable both therapeutically and emotionally. We find that it is important as an aid to lipreading; I'll be reserved about nonvisual speech discrimination because we must be careful not to promise too much, but I think that ultimately the Ineraid will do very well there also. So far some of our patients have done extremely well."

THE NUCLEUS IMPLANT

Although nobody knows if it is the best, the Nucleus device, manufactured by the Nucleus-Cochlear Corp., is certainly the most complex and sophisticated cochlear implant now in clinical trials. It uses a connector similar to that of the UCSF implant and has basically the same configuration, although it has only one platinum receiving coil where the UCSF implant needs four (this means that the transmitting coil is smaller and simpler). The Nucleus implant has 22 electrodes stimulated in pairs to give up to 22 individual channel locations. It may be referred to as a nonsimultaneous multichannel device because the stimulation is multiphased in a manner similar to that used in fiberoptic communication.

To avoid the potential problem of 22 channels interacting in the cramped cochlear space, the researchers designed the Nucleus implant as a multiplex system. Instead of using all 22 channels at once, the device stimulates one pair of electrodes at a time. The signal is transferred from one channel to another in milliseconds, but many of them may receive an *off* signal. In principle this works like a sequence of motion picture frames that the brain perceives as continuous film; some researchers believe that in reality the effect is not quite the same. "We find it hard to imagine that the Nucleus implant has been as effective in patients as claimed, purely from the standpoint that the patients are receiving one channel of extracted information at a time rather than simultaneous stimulation across the speech frequency range," says Rebscher. "If the implant used a different processing strategy and could multiplex and update all 22 channels fast enough, it would produce a more natural pattern of stimulation and give the patient more speech cues. But it's too slow to give the patient enough information to perceive sound in a multichannel way."

The Nucleus implant has a specially designed integrated circuit with 1000 transistors that analyzes speech and pulls out its key elements. Using EPROM—erasable, programmable, read-only memory—researchers can use hospital computers to program the chip according to the individual patient's needs. This includes measuring the threshold and maximum comfortable levels for stimulation on each electrode, determining how many hertz to use on each, and if necessary reversing the order of stimulation on some electrode pairs.

The Future of Cochlear Implants

The success of these and many other cochlear implants depends on, among other factors, how well they work, how safe they prove to be, and what kinds of psychological as well as auditory benefits they provide.

PERFORMANCE

As we have noted, it is difficult to make meaningful comparisons between implants available now. "I believe that there will be no 'best' device, just what is most appropriate for the patient, but right now there is no way to determine that," says Berliner.

Trying to find a way to do so, Dr. Gantz's group has compared four of the major implants. They have thoroughly tested 14 patients: seven with single-channel and seven with multichannel implants. Four of the patients have the 3M-House device, five have the Nucleus implant, and two have the Ineraid (the rest have a single-channel intracochlear device developed by the Vienna researchers). All these implants significantly improve the recipients' ability to lip-read. The patients can distinguish differences in both fundamental frequency and prosody (the information that distinguishes syllables in speech). Performance among the implants differs in that the patients with the multichannel implants consistently understand 35% to 50% of speech, while those with single-channel implants do not. One Ineraid recipient understands 90% of speech without lip-reading.

Of course, these are extremely small numbers of patients from which to try to draw conclusions. "For all we know, we could have seven patients who use their multichannel implants well and seven who use their single-channel implants poorly," says Dr. Gantz. To complicate matters, researchers elsewhere have had different test results. The Vienna researchers have several patients with single-channel implants who understand open set speech; it is not clear whether the difference could be attributed to the patients themselves, the German language, or the manner of testing. The UCSF investigators have found that the 3M-House and the Nucleus devices perform similarly to the Vienna extracochlear device.

If valid, these latter findings would really call into question the efficacy of implants (which open an area continuous with the brain and central nervous system) if the same results could be obtained with much less risk of infection.

On the other hand, all these investigators may get different results once they have been able to test more than a few patients with each implant.

SAFETY

Designing a cochlear implant for maximum safety can require a compromise with its effectiveness, partly because of the severe limitations of space in the cochlea. Electrodes are likely to be less toxic if they have a large stimulation area. The larger the electrode, the less heat, chemical changes, or other detrimental interactions with cochlear fluid are likely to occur. Unfortunately, larger contacts placed farther apart contribute to current spread, which causes electrode interaction and degrades the signal perceived by the patient. These factors must be traded off against the amount of space necessary between the electrode pairs to prevent channel interaction.

We have noted that implantation or removal of a device might damage the inner ear, with subsequent degeneration of nerve fibers. Researchers do not yet know what effects, if any, cochlear damage would have on long-term use of an implant. Preliminary studies suggest that sound perception is not affected, but researchers speculate that a patient with cochlear damage would derive less benefit from an upgraded implant than he or she might otherwise have enjoyed.

Scientists are working on methods other than extracochlear implantation to prevent these problems. "With the possible exceptions of intracochlear damage reported by the Australian group that developed the Nucleus device and the risk associated with a chronic percutaneous plug in the Ineraid implant, the devices as implanted in adults all seem to be designed from a very good safety and reliability standpoint at the present time," says Rebscher. The lead containing the electrodes on the UCSF implant, for example, has well-controlled flexion and a vertical structure that keeps it from bending up and damaging the basilar membrane or its bony shelf. The House group has shortened their intracochlear lead for greater safety.

PSYCHOLOGY

No doubt exists about the value of cochlear implants in general in relieving the social isolation and depression that accompany deafness. Improvement of hearing, however, requires adjustment on the part of the patient and his or her family. A change in roles within a family or the recognition of the implant's limitations may increase rather than decrease the psychological and emotional complications of deafness.

Researchers are extremely careful not to raise the expectations of potential patients or the general public about what implants can achieve. "We do our best to prevent patients from hoping for too much, but an experimental program like ours is likely to cause more disappointment than would occur with a purely clinical program," says Rebscher. "We're at the forefront of research, and each of our patients tested with the percutaneous cable hopes to have a breakthrough in speech understanding. It's difficult for them to reach a plateau and discover that although they have made great progress, their hearing will be limited." The investigators prepare for these possibilities by scheduling psychological interviews for the patients throughout the testing periods.

"A cochlear implant will most help the patient who wants to hear and recognizes that sound, even imperfect sound, is more important than anything," says Consuelo. "Those not motivated to hear might find it very frustrating to put up with the bulk or the cords of the apparatus and the inability to hear normally. As for me, all I can think of is what I would have lost during the past few years if I had not had my implant: the learning, the experiences, and just the awareness of sound itself."

Sometimes these issues are less important to the patient than they are to his or her family. An implant recipient may be delighted to hear anything, but a spouse or parent may be disappointed that normal hearing has not been restored. Researchers worry in particular that parents, grasping at any hope, will subject their children to high-risk procedures in an attempt to help them hear. "We have not found that attitude," says Berliner. "We educate parents and judge their motivations carefully. They agonize over the decision of whether or not to have the implantation, and we explain to them that no one can predict its results."

Like their colleagues working with high-technology visual prostheses, researchers in cochlear implants must empathize with the people they are trying to help. "Most deaf people want their hearing restored," says David, "but there is also a congenitally deaf population that has formed its own culture. These people associate only with each other and don't want to be 'restored to normal,' much less to wear electrical devices." The play *Children of a Lesser God* addresses this issue through the character of a deaf woman who identifies with her disability so strongly that she refuses to lip-read or speak; instead she expects other people to learn to sign so that they can communicate with her.

Potential risks and complications are more tolerable to those who remember that only a decade ago cochlear implants were considered totally unacceptable. The medical profession has evolved from arguing about whether implantation should occur at all to arguing over which implant is best and which recipients are most likely to benefit from them.

"Cochlear implants are extremely crude, so it's exciting that patients hear with them even as well as they do," says Dr. Gantz. The devices may prove to be primarily a high-technology alternative to other aids for the handicapped, but they are certainly here to stay.

The Tactile Auditory Substitution System

Another up-and-coming alternative for the deaf is a cousin of the TVSS—the tactile auditory substitution system (TASS). The TASS was developed by Frank

Saunders, Ph.D., Senior Scientist at the Smith-Kettlewell Institute of Visual Sciences. Trained as a psychologist, Saunders became interested in sensory devices 16 years ago, when he was called to Smith-Kettlewell to evaluate the TVSS. Using the concept of electrotactile stimulation to aid the deaf, Saunders created a device modeled on the natural ear. As we have seen, the cochlea contains a membrane covered with hair cells that is curled into a spiral. Saunders and his group unwound the spiral: they made a belt of 16 electrical stimulators in a row arranged in sequence from low to high frequency.

The belt is made of vinyl. The user's waist is moistened with warm water, and the perspiration produced by the belt keeps the area moist and sticky for good electrical connection. A lightweight battery pack hangs from the belt, and a microphone plugs into it. The wearer can use a switch to adjust the intensity of stimulation to the most comfortable level.

The TASS is designed for simplicity. The microphone connects to a bank of filters, each of which listens for a particular pitch. When a filter detects its correct pitch, it turns on the corresponding stimulator with an intensity that reflects the sound's volume. The word *shoe,* for instance, creates two stimuli: the *sh* at the high frequency end, and the *oo* at the low frequency end. A sound like *aah* stimulates from the center of the belt; a siren runs up and down its length. The device has electronics that emphasize speech in contrast to other environmental sounds.

"The primary goal of our device is to enhance communication," says Saunders. "It does so in three ways—by conveying sound in the environment, by conveying information about the speech of others, and by providing feedback on the wearer's own speech." Saunders does not see the TASS as a sensory substitution device, just as the TVSS subjects did not imagine that mechanism as a substitute for sight. "There's nothing magical about touch," he says. "The TASS is not communicating hearing over nerves in the skin, but it *is* providing information about sound. This information must be interpreted just like Braille or sign language."

Although not intended to mimic or replace hearing, the TASS does augment lipreading or other hearing aids by providing additional information that the brain learns to use. The deaf person must learn the pattern for each sound until the stimulation becomes "natural."

As an augmentation device, the TASS provides information that is not available from other hearing aids. It offers clues that cannot be picked up from lipreading, and vice versa. By itself, lipreading takes tremendous work, concentration, and energy; even the best lip-readers are only 60% accurate most of the time. While lipreading can tell the deaf person the difference between *key* and *tea,* for instance, the TASS can show the difference between visually indistinguishable sounds like *Sue* and *zoo* or *heed* and *hid.* If speech is enunciated slowly, the TASS can pick up its important characteristics.

The TASS is particularly useful in teaching congenitally deaf children to talk. Having learned as much as he or she can about the mouth positions and breathing required to make a sound, the child can then use feedback from the belt, imitating the teacher until the tactile stimuli from both their pronunciations are identical. Saunders reports that deaf children are as fascinated by the tactile vibrations and the insight they gain from them as they would be by actual sounds.

A researcher in the field has remarked that cochlear implants and tactile aids like the TASS have more to do with each other than either does with normal hearing. Each type of device tries to provide information through unusual stimuli, whether on the auditory nerve or on the skin. Cochlear implants have the advantage of being more conveniently wearable and providing a sensation of hearing that has psychological benefits. "But I think that our device can communicate as much information as any existing implant without the risk and cost of surgical procedures and without interfering with any hearing that is there," says Saunders.

Saunders is starting FDA-approved clinical trials with other centers and hopes to market the TASS through his company, Tacticon Corp., as soon as possible.

SECTION

VI

REPAIRED NERVOUS SYSTEM

20

Contacting the Muscles

Sam Khawam has taken his show on the road. At professional meetings across the country Sam climbs a few steps to a stage, walks to a podium, speaks briefly, descends from the stage and returns to his seat—a wheelchair. Sam is paralyzed from the chest down.

In the gait laboratory at the VA Medical Center in Cleveland Sam demonstrates his ability. In his right hand are two switches attached to a finger cuff. These are wired to a black box hanging from his belt that has an alphanumerical display across the top. Expertly tapping the switches while the crystal flashes in response, Sam lurches to his feet and grasps a nearby stair railing for balance. Supporting himself with his arms, he climbs three steps, walks across a small platform, and descends to the floor. His wheeled walker has been placed at the bottom of the stairs; Sam leans on it and strolls across the laboratory with Rudi Kobetic, M.S., the chief biomedical engineer, wheeling his chair behind him. He moves through a doorway into a small office and clunks to a stop at an unoccupied desk. Maneuvering around him in his own wheelchair, Dennis Johnson arrives at another desk and shakes his head in mock irritation. "Sam," he says, "You're always cramping my style."

Dennis remembers nothing about his fall through a skylight in February 1980 while at work on a construction site. The 27-year-old electronics engineer from Stow, Ohio, woke up a week later to find himself paralyzed and numb from the navel down. Spinal cord injuries often make such sudden, permanent changes in the lives of the victims, who are usually young men. An estimated 200,000 Americans are confined to wheelchairs because of spinal cord injuries; 10,000 others join them each year. About half are left paraplegic and half quadriplegic. Until researchers perform the miracle of reversing damage to the central nervous system, functional neuromuscular stimulation (FNS) of the kind that Sam and Dennis are using holds the most promise for improving their lives. The goal of scientists working with FNS is to provide safe, functional walking and grasping for patients unable to perform these movements themselves.

The Central Nervous System

The brain and spinal cord are composed of nerve tissue that extends out through the vertebrae in 31 pairs of branches. Starting from the neck these branches are divided into the cervical (C), thoracic (T), lumbar (L), and sacral (S) regions. The classifications serve as a shorthand for describing spinal cord injuries: Sam, who has a T8 injury, is paralyzed below the eighth thoracic branch, while Dennis, with a T11 injury, is paralyzed below the eleventh.

For years researchers have experimented with electrical stimulation of either the nerves or their target muscles to correct conditions like scoliosis, relieve neuromuscular disorders, or restore functions. In Chapter 21 we will see how spasticity, convulsions, and other symptoms may be relieved by stimulating the cervical spine. Electricity has been applied to the brain, spinal cord, or peripheral nerves to try to alleviate pain. Stimulation of the medullary cone, the tip of the spinal cord, can help a paralyzed person increase his or her bladder control. When the spinal cord has been damaged, resulting in paraplegia or quadriplegia, stimulation of the nerves themselves can help restore movement.

The brain and spinal cord act together to supply movement and sensation. In some cases the cord reacts immediately even before the stimulus reaches the brain. When the cord has been damaged, commands from the brain cannot reach the peripheral nerves that produce movement; impulses from the peripheral nerves cannot reach the brain for interpretation. The purpose of FNS is to take over for the damaged spinal cord.

Lower Extremity FNS

A widely publicized demonstration of FNS took place a few years ago at Wright State University in Dayton, Ohio. Nan Davis, a 22-year-old student paralyzed from the rib cage down in an automobile accident, took six steps using an external FNS system developed by Jerrold Petrofsky, Ph.D., Executive Director of the National Center for Rehabilitation Engineering. The system was cumbersome at best: Nan was supported by a harness; she used parallel bars for balance, and she was wired to a laboratory computer that sent signals to about 30 sensors and electrodes taped to her legs. Since then the computer has been made portable, but the system remains external, as do those of several other centers here and abroad that are doing this research.

Meanwhile, Sam, Dennis, and their fellow research subjects are experimenting with state of the art technology involving partially implantable systems. The research program on restoration of walking to paraplegic people at the VA Medical Center has six divisions: patient, metabolism, modeling, closed loop, sensor, and electrode development. "As far as we're able to tell, we have the most advanced program in the country, mainly because of the concentration of experts we have in this one area, and the fact that some of them have been at it for 20 years," says E.B. Marsolais, M.D., Ph.D., the program's director.

Like other research centers the Cleveland VA aspires to total implantation, but before its staff proceeds to human trials they must perfect the system. "I look forward to the implantable system because it will make showering and other tasks a lot easier," says Dennis. "But they can do only so many operations because of scar tissue, and they want the system right before they implant it." The FNS system will become part of the patient's body, and he will expect it to work reliably and safely. Meanwhile, much remains to be done.

THE FNS SYSTEM

Initially the Cleveland researchers worked with stroke victims who were unable to walk. They then discovered that totally paralyzed subjects made better models. "With partial paralysis, or with a stroke victim who has partial control, I can't tell what the patient is doing and what I'm doing," says Dr. Marsolais. "Totally paralyzed subjects enable me to define the effects of our system. Spinal cord patients also tend to be otherwise healthy, dynamic, vibrant, young, interested people whose brains are totally intact. This is very important for a project like ours in which the subject is a crucial member of the team."

To enter the program, subjects must meet definite criteria. They must have been injured between levels T4 and T11. Subjects injured above T4 can develop shortness of breath and other physiological problems; those injured below T11 have damaged their spinal cords at levels that need to be used. The peripheral nerves below the level of damage must be intact so as to carry electrical impulses to move the muscles. The subjects must be able to walk 10 feet using crutches with the assistance of one person. They should also be physically fit and have no problems with excess weight, damaged joints, or severe skin diseases.

The current FNS system consists of implanted electrodes along with an external stimulator and finger cuff switches, all of which are wired together.

Each Teflon-insulated, stainless steel wire electrode is placed in a hypodermic needle for injection into the muscle near the target nerve, where a small barb anchors it. After placement the needle is removed, and the wires are gathered at the skin surface and soldered to small connector sites that are taped in place. Each subject has one connector site located toward the inside of each thigh and on each inner calf. Wires plugged into these connector sites run to the stimulator and to the switches that control it.

The stimulators are built by Dennis, who has been an employee of the VA as well as a research subject since August 1983. Each stimulator consists of a battery-powered microprocessor that translates the switch commands into electrical signals; these in turn fire the nerves necessary for desired movements. In the laboratory the system can also be plugged into a large floor computer for experiments.

Sam has two or three electrodes implanted in each of 12 muscles in each leg. The maximum number of electrodes in the present system is 64, with a 32-channel stimulator controlling hips, knees, and ankles. Each electrode has a redundant twin, because they usually last less than a year. All the electrodes are numbered and tested weekly for electrical characteristics and monthly for force in each subject. A failed electrode is removed, and selected ones are scrutinized with scanning electron microscopy before disposal.

Part of improving the current FNS system will involve implanting electrodes in more muscles. "If a patient had active pushoff and hip extension, for instance, those muscles would bring his body forward, and his arms wouldn't have to do it," says Kobetic. With a totally implantable system, the team will be able to save channels by placing the electrodes closer to muscles they want to use but cannot reach by injection. "In some cases we are stimulating three muscles when better electrode placement would enable us to use just one," says Kobetic. Implantation will also help eliminate electrode shift, which requires extensive surgical procedures.

A subject can begin walking once a preprogrammed sequence of stimulation has been customized for him and placed into his portable computer. He is then ready to use his switches. The *ready* switch is a safety feature that keeps him from accidentally turning himself on. The *go* switch releases this safety feature for action.

To begin, the subject turns on the stimulator; the crystal shows *STND*, the prompt for standing. To proceed, the subject presses the *go* switch and stands. Then he can press *ready* twice to begin walking. Before pressing *go* for the first step, he can press *ready* twice again; the menu will then offer him a choice between up stairs, down stairs, level walk, or sit. When he has chosen a mode he is offered the option of starting with his right or left leg. Selection causes the chosen leg to bend while the other stiffens to support him. Once the subject starts moving an automatic system takes over, so by repeatedly pressing *ready-go* he can move his legs forward alternately. It takes about 3 months for a subject to learn to walk.

The exercise mode goes through the entire set of movements while the subject is seated. The subjects are allowed to take their systems home from the laboratory only to exercise and to stand briefly, keeping records of all their movements.

FNS IN USE

Sam, a 25-year-old industrial engineer from Beirut, Lebanon, was paralyzed from a gunshot wound in March 1982. After hearing about FNS at a convention of the Spinal Cord Society,* he went straight from the hospital into the VA program. Now, 2½ years later, he is the program veteran, devoting his full time to evaluation and development of the FNS system.

Sam can walk at a rate of 0.3 meter per second. (Average walking speed is 1 meter per second; since this figure was computed from walkers in New York City, Sam is doing pretty well compared to the rest of the country.) He can stand with 99% of his weight on his legs and walk with 80% of his weight on

*An organization devoted to finding cures for people with spinal cord injuries.

them. Using a walker, he has gone 400 feet, although he is still very slow and unstable on crutches.

Sam and Dennis are part of a group of six men who are about to be joined by a woman subject. All six walk, one of them with crutches. Three of them, including Dennis, can climb stairs. "I can't really do more than the others," Sam says. "It's like sports; not everybody can do the same amount, and this requires a lot of work."

At home the subjects use FNS for reaching or other tasks that they cannot perform in wheelchairs. "The system isn't that comfortable for functional use yet," Sam says. "It takes energy that I could use better in the wheelchair." Sam looks forward to its improvement. "If I didn't have faith, I wouldn't be here," he says. "Every 2 or 3 months we have some kind of dramatic breakthrough. That keeps me going. I think it's a great system, and I'm optimistic that some day we'll be able to use it functionally."

The researchers define "functional" in terms of everyday use. FNS recipients will need wheelchairs for long distances, like more than a 30-minute walk at their speed. They will not be able to walk over rough terrain like forest paths or beach sand. The main reason for these restrictions is to protect them from injuring themselves because they cannot feel. "Even if we perfected our system, it would be limited because of the elements we're not replacing, like sensation," says Dr. Marsolais. "To be truly functional, the system must be designed within these limitations."

The subjects can, however, get in and out of chairs, use the bathroom, and squeeze into small spaces. "I hope that this system will enable people to park in a lot and walk to their destination, or go to a restaurant that's down a flight of stairs without having to be carried," says Kobetic. "It's somewhat like the amount of function a person would have with one broken leg."

By popular demand, dancing has been placed on the agenda of FNS refinements. (Dr. Marsolais insists that one patient has already gone dancing without permission, but no one in the laboratory will confirm this claim.) Another high priority is restoration of the physical motions of sexual function, which are currently absent from all of the subjects. Combined with a penile prosthesis or more specialized FNS, the Cleveland system may help patients perform actively. "This will enable them to enjoy sex more fully, and it's extremely important for them psychologically," says Dr. Marsolais.

Because the subjects cannot feel strains, sprains, and the like, their safety is a major concern in the FNS program. "So far we've had no fractures, but I know that our subjects will get hurt," says Dr. Marsolais. "They would destroy themselves from injuries if we did not carefully try to foresee problems and avoid them." FNS recipients will always use crutches except possibly when walking short distances at home. They will also wear ankle braces made of lightweight polypropylene. These will have sensors to detect the amount and position of pressure on the feet; this information will be relayed both to the computer and probably also to electrodes implanted in the recipient's back. By stimulating these electrodes in a code, the system may be able to relay information about the menu or activity without the use of a crystal display.

Sensory feedback will also protect the recipient while standing or stuck. The computer will adjust the stimuli to save energy while the recipient stands; it will use only the muscle force necessary to keep the body erect. Subjects can stand in the laboratory for over an hour, but eventually an automatic shutoff will be added to keep them from extending that time much longer. "This too must be individually programmed, but within broad limits—a paraplegic person can't remain in one position for more than 2 hours, no matter what," says Dr. Marsolais.

The present microprocessor has no way of knowing whether the subject's foot has become jammed on an obstacle. The researchers have made a prototype closed loop system that gives it the locations of the knees and ankles so that if necessary the system can boost the foot. An alarm for abnormal pressures will help to protect the feet.

The team wants also to protect FNS recipients from ice, ramps, and other treacherous surfaces that may cause them to fall. The researchers are using computer modeling to determine how to help their subjects fall reasonably safely and get up again. The subject's muscles must respond to a fall dynamically and athletically so that they absorb energy. Programming this cushioning fall requires the best in both rehabilitation and engineering to make the most helpful, least adverse adjustments.

"Safety is a crucial factor," says Dr. Marsolais. "It's a reality of life, not a defect of the system. These subjects must recognize that they are not totally normal and that to benefit from the system they must follow certain rules. If they don't, they will pay some big prices."

OTHER APPLICATIONS OF FNS

While providing functional walking for their paraplegic subjects, the Cleveland researchers hope eventually to apply FNS more broadly. The microprocessor can not only relay interrupted messages from the normal brain, of course; it can also send information that would otherwise have come from a damaged brain. This substitution could benefit some of the 2 million stroke victims in the country, whose numbers increase by about 400,000 each year. Other candidates for FNS include victims of multiple sclerosis, head injury, and cerebral palsy. The system can overcome abnormal postures in cerebral palsy, for instance, but it poses problems of how to maintain the electrode sites and allow for growth. "We've made coils that elongate as a child grows," says Dr. Marsolais, "but we still aren't sure how long-term, constant stimulation might affect growth and function, and we haven't designed systems to last for 50 years."

FNS is in the research stage; the VA patients are doing preliminary rather than FDA-approved clinical trials. The team has a final implantable version ready for animal testing. "It shouldn't be terribly long before we implant the system in a human," says Dr. Marsolais, but he defines this time period as the "foreseeable" or "realistic" future. No one in the laboratory will give even a guesstimate as to when the system will be ready. "The problems are by no means solved here or anywhere else," says Dr. Marsolais. "The time it takes will depend on the level of functioning we can achieve. I'm not sure that paraplegia victims would accept a system like this if it required an athlete who would undergo rigorous training and extensive surgery," says Kobetic.

Despite this conservatism on the part of the investigators, the VA already plans to set up FNS centers because it will take each one years of development to implement the final system. The Cleveland staff meanwhile encourages mental and physical preparation for FNS. Clinical physicians need to think in terms of their patients being able to walk again. They should avoid doing rhizotomies (a surgical procedure that relieves spasticity by cutting nerves) or similar operations that could permanently damage a patient. If a patient breaks a bone, it should be reset just as carefully as if he could walk normally. Paraplegic people should keep themselves in shape to use FNS. "All the rules they learn in rehabilitation have become much more important than ever," says Dr. Marsolais. "They should stay trim, exercise for aerobics and muscle strength, stand with braces to maintain strong bones, and do joint range movements."

They should also be "patient patients." "Our bionic men are still limping, but they're getting there," says Dr. Marsolais. "To be acceptable to patients, FNS must provide more function than its inconveniences and really be able to compete with the wheelchair. We're being conservative—we want to do as much as we can do well, and we're learning a lot as we do more." "So much research is going on that paraplegic persons should have a lot of hope," says Sam. "Look for FNS in the future."

Upper Extremity FNS

A paraplegic person using FNS to walk gives a very dramatic presentation. His movements result from powerful forces that could hurt him badly in case the system failed. Leg movements therefore require precision in both the technology and the implementation of FNS. As we have seen with artificial limbs, however, it can be *relatively* easy to restore *some* walking function compared to grasping function. The legs have fewer degrees of freedom than the arms, and their movements can be approximated with simpler supports.

Creating an FNS system to help quadriplegic people grasp is therefore quite different from helping their paraplegic counterparts to walk. Hand movements are more subtle than leg movements and usually require very little force. Although the casual observer may consider it less impressive than walking, grasping with a paralyzed hand is a delicate procedure in which small errors have a much greater effect. Hand movements pose difficult problems in coordination that have important implications for quadriplegic persons. "People confined to wheelchairs can be tremendously productive, if only they can use their hands," says P. Hunter Peckham, Ph.D., Associate Professor of Biomedical Engineering and Director of the Case Western Reserve University Rehabilitation Engineering Program at Metropolitan General–Highland View Hospital and VA Medical Center in Cleveland. "We think that it's extremely important to develop all the available techniques to restore motion to the upper extremities; FNS is one of them."

Several years ago the Cleveland FNS program split into upper and lower extremity groups to tackle these different problems. The hand program, which began in 1972 following research done in the 1960s, has involved a total of some 30 to 35 subjects. "Since 1982 we've actually fit clinical systems on about 15 of them," says Peckham. "We have eight subjects in the program right now, and to our best estimate five of these are daily or frequent users of our system."

Like the walking program, the grasping program seeks to restore motor (and possibly sensory) functions to victims of spinal cord injuries. These are C5- or C6-injured patients (those paralyzed below the fifth and sixth cervical levels, who comprise nearly 70% of quadriplegia victims). They can move their shoulders and flex (but not extend) their elbows. Some can flex their wrists; C6-injured patients can extend their wrists and have limited sensations in their hands and forearms. Neither group can grasp objects; individuals use whatever residual function they have to manipulate.

Quadriplegic subjects in the hand program have met the same criteria as their peers in the leg program. They undergo motor and sensory evaluations, then have electrodes implanted in their forearms. Unlike paraplegic persons they cannot prepare themselves to use FNS, so they spend their first 2 months exercising and having their systems recalibrated as their muscles grow stronger. They develop movement patterns like palmar prehension (tip-pinch), in which the tips of the thumb and finger come together, and lateral prehension (key-pinch), in which the thumb is brought alongside the index finger. After about 3½ months they are ready to use their customized systems.

THE GRASPING SYSTEM

Like the walking system, FNS for grasping has three components. Instead of using switches, quadriplegic subjects give commands with a two-axis device: a transducer taped to the chest, connected by a thin rod to a sensor taped to the shoulder. Moving the shoulder forward and back regulates the grasp of the opposite hand; quickly moving the shoulder up or down locks the command to maintain the grip. This mechanism is different from that of the walking system in that it is a proportional control: rather than triggering off a sequence of movements like those involved in walking, the subject has control over his muscular force.

Which hand is designated for FNS depends on the range of motion in the opposite shoulder; if both shoulders move equally well, the subject uses what had been his dominant hand. The shoulder commands are sent to a four-channel processor-stimulator that translates them into electrical signals. Electrodes are implanted in the forearm muscles with the same hypodermic injections used for the legs. The wires are gathered into a connector site on either the back or the front of the forearm, depending on which muscles are to be stimulated.

Because of the transducer design and the shoulder movements needed to grasp, it would be difficult to make the present system into a bilateral one. The muscles would have to be coordinated; the hands would have to be controlled by wrist or cortical signals rather than shoulder movements, and the recipients still would not be able to feel what they were doing. Because they could not use both hands equally well, one would remain dominant. "These problems can't be solved just by stimulating muscles," says Peckham. "We're dealing with integrated sensory motor functions, and those considerations are just as important as the stimulation itself, as they are in the lower extremity system."

Lowell Bailey, a 27-year-old computer programmer who lives outside Akron, Ohio, has been confined to a wheelchair since July 1983, when he had a diving accident. He was treated in Highland View Hospital, met Peckham in the hall one day, and entered the FNS program in April 1984. "It gave me an outlet during my rehabilitation," he says. "I could come in for a few hours each day and diversify my time by thinking about and participating in the program."

Lowell's accident shattered his C5 vertebrae and damaged his spinal cord. Paralyzed from the chest down, he has about 30% function in his arms, including biceps and wrist extension. He has almost total sensation in his thumbs and less in his fingers and inner forearms.

Lowell usually has three electrodes in each side of his right forearm; he has had as many as eight in his forearm, plus five experimental ones in his triceps. Twice a week a family member removes the bandages, cleans the wires, and changes the connector. This takes half an hour; just having someone put on the system takes several minutes. Since it must be removed before he can be helped into a car, Lowell uses his system only three to four days a week. He believes, however, that it is worth having. "The advantages definitely outweigh the hassles of putting it on," he says. "It just becomes part of you, like putting on your socks." At night he puts on a resting splint and sets the stimulator in an exercise mode. While he sleeps, his hand exercises and rests for half-hour intervals.

FNS enables Lowell to write, to position a comb or toothbrush, and to pick up or switch utensils while eating. "At breakfast I can open a carton of milk, pour it on cereal, eat the cereal, put down the spoon, and pick up a fork to eat something else," he says. "This makes for more independence and crisper cereal. I can pick up a full can of Coca-Cola; without the system, I'd end up wearing it instead of drinking it."

Once Lowell gets a van for transportation, he will be able to wear his system all day and go back to work. "The FNS will definitely enable me to return to computer programming," he says. "I can write, use a pen to type, and use a mouse or other adaptable features to program." This prospect is one of several pleasant surprises. "I didn't expect as much from the system as it has enabled me to do," he says. "It doesn't open up tremendous opportunities for me, but it does give me a little more independence."

FNS IN PROGRESS

Researchers working on FNS systems make hard day-to-day decisions about what to improve next—technology or clinical application. Like other researchers in bionics, they want to achieve the most function with the greatest simplicity, at least for the time being. "Right now we're trying to implement techniques that will eventually lead to more complex implantable systems," says Peckham. "Meanwhile we try to develop minimally invasive systems that resemble our ultimate goal as closely as possible." This means taking the temporary percutaneous grasping system as far as it can go in precisely stimulating and coordinating the actions of many different muscles.

The researchers expect, then, that the implantable grasping system will be better clinically but about the same functionally. The implantable stimulator, for instance, will be an eight-channel circuit inside a polymer-coated titanium package. Although it has four channels, the present percutaneous stimulator works as a pseudo-eight-channel device. "We designate four channels—the minimum needed for each grasp—for either tip- or key-pinch because the patients can't use both at once," says Peckham. "We have to design the implantable system differently but with the same result. We are, however, using more channels with the percutaneous system to add elbow control." Peckham's group would prefer to implant the entire system but may have to settle for the stimulator and electrodes, using radio frequency to access and supply them with power from an external transducer.

Like his colleagues working on lower extremity FNS, Peckham cannot say when the implantable system will be ready for human use, but it is presently undergoing animal evaluation. "We hope it won't be long," he says.

Other research in progress will be applicable to either the percutaneous or an implantable system. One project is a closed loop system to compensate for muscle property changes like fatigue; this is being designed by Patrick Crago, Ph.D., Assistant Professor of Biomedical Engineering in the Case Western Reserve Rehabilitation Engineering Program. "I can hold a pen for up to an hour without fatigue, but repeating a task for half an hour straight would tire my shoulder," says Lowell. "Fatigue time is different for everybody." As an FNS recipient's muscles tire, they produce less force, and the grasp movement pattern may change. The closed loop system works on the same principle as the cruise control of a car, which maintains a certain speed regardless of driving conditions. It allows for fatigue by increasing the electrical stimulation so that the same command will continue to elicit the equivalent force and position. Since a whole muscle is not needed to grasp, the recipient has some leeway between closed loop stimulation and exhaustion.

The upper extremity system might someday incorporate a sensory stimulation technique being developed for the hand by Ronald R. Riso, Ph.D., Senior Research Associate in the Department of Biomedical Engineering at Case Western Reserve University. Transducers stuck on (and later implanted in) the fingers will send messages either to the muscle stimulator or to a second one. The stimulator will signal five electrodes implanted in a row in the upper arm or back. The recipient will be able to tell how large and heavy an object he is grasping by position and pulse speed. The heavier the object, the faster the pulses; the larger the object, the further along the electrode row the stimulation will occur. Another method might be to send this information from the shoulder movement, which gives the commands, rather than from the hand movement, which executes them. A modified version of the technique might be applicable to the lower extremity system.

The Cleveland group has also evaluated other approaches to restoring grasp sensation, including the use of special gloves. Some researchers have tried making gloves with strain gauges, for example, that transfer signals to a part of the body with sensation. One such glove, developed by Collins and his group at Smith-Kettlewell, has a pressure-temperature sensor on each fingertip. Designed for a leprosy patient, the glove is connected to a vibrotactile array on the patient's forehead, which still has sensation. Within an hour of wearing it, one subject could discriminate the weight, shape, edges, and even textures of objects. He noted that he particularly enjoyed the sensation of touching his girlfriend. (The researchers are also designing a boot for the millions of diabetic patients whose feet have damaged nerves and blood vessels.)

Andrew Schoenberg, Ph.D., Research Associate Professor of Bioengineering at the University of Utah, and his colleagues got the idea of trying sonar to replace sensation: they transmitted ultrasound into the environment and even tried sending it through the skin to bounce signals off bones.

A more successful application of this principle has involved the use of a piezoelectrical film, polyvinylidene fluoride (PVF_2), which could be made into a glove. When PVF_2 is exposed to the vibrating pressure of sound, it emits a corresponding vibratory electrical charge. In this mode PVF_2 is used as a sonic receiver (microphone). Conversely, when a vibratory charge is applied to PVF_2, it vibrates mechanically and produces a sound. In this mode it is called a transmitter (speaker).

Schoenberg's group sandwiches a 1-millimeter thick layer of silicone rubber between two layers of PVF_2. They excite the outer layer (transmitter) with a pulse producing a corresponding sound that travels across the rubber to the PVF_2 receiver. The travel time of this pulse varies with the thickness of the rubber. By repeating the sound pulse and travel time measurement 1000 times per second, the group obtains a nearly continuous measurement of compression of the rubber layer. This compression signal, which varies when the glove touches something, can then be used to modulate a touch or sound that the wearer can understand.

A glove made of PVF_2 could be used to measure temperature because it is sensitive to heat as well as pressure, but it would be difficult to decouple the signals. The glove could also provide feedback from a prosthetic hook. Unfortunately, as is so often the case, Schoenberg's group lost their funding for this project.

"One of the most important things happening now is that neural prostheses are becoming much more selective and more closely simulate the ways in which the normal nervous system used to work before the damage occurred," says F. Terry Hambrecht, M.D., Head of the Neural Prosthesis Program at the National Institute of Neurological and Communicative Disorders and Stroke. "This involves more and smaller electrodes to make finer connections with the nervous system." However sophisticated it becomes, FNS will be most effectively applied along with surgical techniques for restoring function to victims of paralysis. A person who can pull up his or her wrist, for instance, can have a tendon transfer, in which one of the two wrist extensor muscles is surgically reattached to move the thumb or a finger. FNS can then further increase the function of this improved grip. "The real power of FNS is how it fits into the whole standard armamentarium for restoring neuromuscular functions," says Peckham. "Stimulating paralyzed muscles will become a really viable clinical technique within this larger context."

Nerve Regeneration

Although electrical stimulation will greatly decrease the disabilities of paralysis, everyone concerned would like to see the ultimate in movement therapy—nerve regeneration.

Some researchers have tried using bridges of electrical wires to close the gaps caused by damaged nerves in the spinal cord. Others have noted that while parts of the central nervous system will not regenerate, peripheral nerves will. Some element in their environment rather than the makeup of the peripheral nerves themselves may give them this ability. In one experiment central nerve cells in rats regenerated when peripheral nerve sheaths were placed around them, although it remains to be seen whether the new nerves can carry signals. At least one manufacturer is testing plastic sheaths made of bioresorbable molecular compounds that may help severed nerves to grow. Nerve sheaths could also release growth factors or stimulate regeneration with weak electrical currents.

David J. Edell, Ph.D., and his colleagues at Harvard University and MIT are developing implantable computer chips that can be placed near damaged nerves to pick up their electrical signals and transmit them to the appropriate part of the body. These may someday be used to power artificial limbs, make keyless word processors that work from decoded neural impulses, and bridge damaged areas of the spinal cord.

"I don't think that anybody has had real success with nerve implants," says

Lyman, who has worked on cuffs with specially designed surfaces that prevent ingrowth and neuroma formation at nerve junctures. "But data are building up in the form of little interesting bits and pieces of results. At some point, someone will figure out how to move the investigation to the next stage." Beyond then, a day may well come when the wheelchair becomes just a museum piece. We are about to meet one young woman who would happily donate hers.

21

Stimulating the Nerves

When Carmen Scozzari was a 7-year-old California girl, something terrible began happening to her. It started with her frequently turning in her left ankle and falling. "I was told it was growing pains; then I got a reputation as a klutz," she says. "I began getting Charlie horses without having run and writer's cramp without having written." Excruciating muscle spasms would grip first her left leg, then her right one, then her right hand. The condition was misdiagnosed as juvenile rheumatoid arthritis.

By the time Carmen reached her teens, the disease was affecting opposite arms and legs on different days. "I went into a period of deep depression," she says. "I was horrified by watching this slowly happen and knowing that my body was turning against me."

It was not until she was 25, after years spent taking dozens of tests and medications, that Carmen was correctly diagnosed. A neurologist told her that she had dystonia musculorum deformans, a complex syndrome of neuromuscular symptoms. Dystonia causes garbled messages to be sent from the brain through the nerves to the muscles, which spasm into distorted positions for hours at a time, then suddenly leave exhausted limbs flopping. "Try clenching your fist as hard as possible as long as you can bear the pain," says Carmen. "Now imagine not being able to unclench it. That's what it's like. When the neurologist explained what dystonia was, I almost passed out in his office."

Carmen graduated from college in 1977; although later confined to a wheelchair by her severe disability she managed to reattend a different school in 1980. By then she wore heavy steel braces on her wrists and legs, along with a back brace that resembled a parachute harness. "I looked just like the Statue of Liberty during its restoration, surrounded by all that scaffolding," she says. At night Carmen switched to plastic braces; sometimes her muscle spasms would pop their rivets, or she would fall out of bed with her leg splints. The fingers of her right hand were splayed. She had developed torticollis, a neck muscle spasm that pulled her head to the left; she clenched her teeth so hard that she needed a bite plate to protect her tongue. Her right eye had developed nystagmus: the eyeball drifted horizontally and caused double vision and severe migraines, so she wore an eye patch.

Carmen had also lost control of her urinary sphincter and had to wear a catheter, which she particularly hated. "Every three to four weeks I had to have the urologist change the catheter, a procedure made worse by my leg braces," she says. "Catheters are usually worn by paraplegia victims, and the staff couldn't understand that I could feel what they were doing down there." She also felt the "memorable" irritation of bladder infections and the embarrassment of once having her collection bag spill urine on a classroom floor.

Eventually the only parts of Carmen's body left free of the dystonia were her left eye, shoulder, elbow, and last three fingers. At times she thought of suicide. "I kept wondering, 'Why me?'" she says. "I was angry at the disease and at myself. I had to live at home and hated the fact that the dystonia was making my life choices for me."

In 1982 while visiting the UCLA Neurology Clinic, Carmen accidentally learned of a surgical procedure for motor disorders developed by Joseph M. Waltz, M.D., Director of the Department of Neurological Surgery at St. Barnabas Hospital in the Bronx. Dr. Waltz had relieved the motor dysfunction of several types of neurological disorders by placing four electrodes along each patient's cervical spine and stimulating the spinal cord with a small external transmitter. Carmen gave the matter considerable thought, then made arrangements to fly east and have the implantation.

Three years later, Carmen remembers the details vividly. Her operation took place on Thursday, July 29, 1982. Stimulation was begun one day later. On Sunday Carmen was eating lunch in her hospital room while engrossed in a television program. Suddenly she noticed that out of long-forgotten habit she had picked up her knife and fork and was using them both. Extremely excited, she tried turning her head to the right—and succeeded. When Dr. Waltz came in on his rounds, Carmen twirled a pen in her fingers, snapped them, turned her head and asked, "Is this what I'm supposed to be doing?" Indeed it was.

During her next several weeks of physical therapy in the hospital, Carmen developed the habit of tossing her braces one by one into her empty wheelchair. "That chair started to look like a relic from Lourdes," she says. Two weeks after surgery, the staff delighted her by removing the urinary catheter. Together with her technicians, Carmen planned a surprise for her father and Dr. Waltz: the next time both men visited, she climbed out of bed, walked over to them, turned around and strolled back. She feels that the cliché is the only way to describe the result—"there wasn't a dry eye in the room."

At the end of 6 weeks Vince Barton, the hospital photographer, did his last videotape of Carmen's progress, including a shot of her standing next to the wheelchair piled with braces. "When I saw that tape of myself walking normally, I was shocked," she says. "What had happened finally sank in. I had rediscovered myself."

Carmen returned to Los Angeles and school but found it difficult to concentrate. "I kept thinking about Dr. Waltz and the other patients and felt that I had to go back and work with them," she says. The opportunity to return came unexpectedly because 6 months after her operation, Carmen began a total relapse. Back went the braces and the catheter as one by one her symptoms slowly reappeared. On returning to St. Barnabas, however, Carmen got to relive her miracle—the problem was a broken antenna wire. By having the faulty part replaced and repeating her therapy program, she recovered all her progress.

Today, at 31, Carmen works as a research associate to Dr. Waltz, explaining the stimulation system to new patients and doing follow-ups on the others. She has only minimal symptoms of dystonia but remains mildly disabled by osteoarthritis in her left hip and right shoulder. "I play the guitar again; I feed myself; I can write quickly; I have no bladder problems," she says. "Some of my medications have been decreased or eliminated. I have to watch myself because of the hip, but I could certainly run for my life if a fire broke out."

Carmen's ordeal has left her with patience, a sense of humor, and a self-acknowledged guilt trip. "Now I say, 'Why me?' because I don't know why I got so well," she says. "I love the work I'm doing—I can't believe I get paid for it!" Carmen offers other patients the unique perspective of someone who went from good health through a nightmare of disability and back again. "I'm glad I stayed around," she says, "to see this happen."

Spinal Cord Stimulation

The photography laboratory at St. Barnabas has a library containing before and after videotapes of each patient who has had this operation. They include the woman whose torticollis was so painful that Vince had to stop the camera and catch her before she fainted; now she can move her head comfortably. An attorney with dystonia was so crippled by his curved back and shoulders that he had to stop work; now he shows no symptons and has opened his own law office. A teenager with Friedreich's ataxia who could not write her name or walk without support is now a confident wife and mother.

Other tapes are less dramatic, but the majority of these patients are noticeably improved. Cerebral palsy patients, who show the best response, find that their spasticity and drooling decrease and their painful spasms cease. Many can sit up without straps or supports and hold their heads erect. Their speech and swallowing improve, as do their posture, walking, balance, and urinary control. For the first time some can feed themselves, brush their teeth, and switch appliances on and off. Many feel more relaxed and alert; their schoolwork improves, and their activities increase.

In addition to its applications for paralysis, FNS has been used successfully to correct scoliosis by contracting muscles to straighten the spine. We have seen that it might overcome abnormal postures in neurological disorders by stimulating muscles antagonistic to those that are clenched. This has already been accomplished with spinal cord stimulation. Unlike FNS, which contracts muscles to restore movement, spinal cord stimulation acts on nerve fibers to decrease spasms, relieve pain, and improve neurological and motor functions. FNS has been used mainly for spinal cord injuries, while spinal cord stimulation has effectively relieved symptoms in disorders of the motor system, including cerebral palsy, dystonia, torticollis, spinal cord injury, degenerative diseases, and posttraumatic brain injury.

MOTOR FUNCTIONS

Like the FNS systems, the spinal cord stimulation system is relatively simple to implant. The patient lies face down on the operating table. Dr. Waltz inserts a needle into the patient's back; through it he passes a lead up the spinal column to find a clear path to the cervical vertebrae. He then removes the lead and substitutes a fine catheter containing four electrodes. Using a fluoroscope, he guides the catheter through the epidural space until the electrodes are placed between the second and fourth cervical vertebrae. The catheter containing the connective wires is run subcutaneously down the back and over to the left or right side, opposite the dominant hand. There Dr. Waltz implants a computerized receiver the size of a watch. The site is protected, hidden, easy for the patient (and surgeon) to reach, and firm enough to support the antenna for a good interface.

The external part of the system consists of a small radio-frequency transmitter and a flat, doughnut-shaped antenna, which is taped to the skin over the receiver. The transmitter can be set to any combination of 18 electrode settings, polarity, and monophasic or biphasic stimulation ranging from 25 to 1500 hertz; the receiver has over 300 gates that convert coded radiofrequency signals into minute electrical impulses. Each patient must be carefully tested to find the best combination of the spinal cord level stimulated and the frequency and polarity of the stimulation. Unlike FNS patients, however, they do not need to readjust their systems once they start to improve. "We adjust only when a patient with a degenerative condition happens to lose the progress gained," says Dr. Waltz. "Then we try to recapture it."

Dr. Waltz starts the patients on 24-hour stimulation; most follow the strict protocol and use their systems constantly. Some patients find, however, that they can go all day or all night without stimulation. Those who stop using the system might find that it takes several days to a few weeks for the symptoms to return. Several patients who have recovered from spasmodic torticollis have been able to discontinue stimulation completely without relapse.

Since 1975 Dr. Waltz has used spinal cord stimulation on over 650 patients with motor system disorders. The number of cerebral palsy patients improved is now 91%; those with dystonia, torticollis, or spinal cord injury, 75%; posttraumatic brain injury, 73%; poststroke, 57%; degenerative diseases, 71%; and epilepsy, 62%. Dr. Waltz rates them according to functional improvement: ability to perform functions for the first time (markedly improved), significantly better function (moderately improved), and slightly better function (mildly improved). Of the cerebral palsy patients, 84% have improved markedly to moderately, as have 77% of the torticollis patients and 70% of the dystonia patients. Cerebral palsy seems most amenable to treatment because its symptoms can be stimulated more specifically than those of epilepsy, for instance, which can originate anywhere and be much harder to reach.

Prediction of how an individual patient will respond to treatment is not possible. Generally the stimulation arrests the progress of conditions like dystonia and torticollis. Within anywhere from 6 months to 3 years patients reach a plateau of improvement; usually this happens faster to those with torticollis and dystonia. Patients with degenerative diseases like multiple sclerosis may still expect to deteriorate, although symptoms appear more slowly than they would otherwise.

Dr. Waltz believes that the system may work by stimulating nervous tissue directly, indirectly by means of neurotransmitters, or both. To improve symptoms all over the body, a diffuse mechanism, probably involving the ascending or descending reticular formation, must be at work. The reticular formation is a long conduit of nerve fibers in the spinal column that carries a hodgepodge of information about position in space, sensation, motor tone, and coordination between the brain, cerebellum, and peripheral nerves. "It's the only part of the cord that communicates reciprocally from the brain all the way to the lower extremities," says Dr. Waltz, "and the only means by which we could have the far-reaching effects that we see."

Stimulation for neurological disorders changes the nerve transmission by modulating the abnormal electrical impules. The use of different stimulation parameters for spinal cord injury, on the other hand, enhances rather than alters transmission. "I wouldn't know which direction to take to change transmission for spinal cord patients," says Dr. Waltz. Muscles respond to a threshold stimulus; below this level no response is seen, while above it no increase in contraction occurs. "I postulated that the impulses coming through in spinal cord injury were subthreshold—too low an amplitude to be functional," he says. "With the stimulation, we can raise the amplitude to threshold so that the muscles can contract."

Dr. Waltz had previously worked with the stimulation system developed by the late Irving S. Cooper, M.D., Ph.D., for use on the cerebellum, which controls muscular coordination. At first Dr. Waltz saw improvements in spastic conditions with this device. With more patients, however, results became inconsistent and too poor to warrant the extremely risky operation required. About the same time (the early 1970s) Dr. Waltz was using spinal cord stimulation to treat pain; one of his patients who has multiple sclerosis showed improvement in motor function. "The cerebellar stimulator that was getting similar results had to be working through the tracts of the spinal cord," says Dr. Waltz. "I felt that there was better access to these nerve tracts through direct stimulation of the cord and began working on that. I stopped the cerebellar stimulation in 1976 and haven't done any since."

Initially Dr. Waltz got promising results by implanting electrodes in the upper thoracic area as had been done for pain. Since the nerve tracts are more concentrated and accessible in the cervical area, however, he moved the electrodes to various levels there and got improved responses. Analysis of the first 100 patients showed that the C2 to C4 area was a "magic region" within which responses markedly improved but still varied from patient to patient. By developing and patenting a four-electrode system, which covered this area and could selectively stimulate various levels, Dr. Waltz improved the results even more dramatically.

The original system that Dr. Waltz developed was limited because he had to make all the combinations for the different electrode settings by hand using percutaneous leads. He found that he had about a 2-week period within which to individualize the system before the patient could develop an infection. This was particularly difficult for torticollis and dystonia patients, with whom it generally takes longer to find the right electrode combinations. Working with electrical engineers from Neuromed, Dr. Waltz developed a computerized transmitter and receiver that allow all this testing to be done with the receiver already implanted and no worries about time or infection. "I'd love to have more electrodes," says Dr. Waltz. "Six of them would increase the combinations by a factor of thirty times. The manufacturer, Neuromed, can provide six or eight electrodes if they enlarge the receiver and transmitter by only about 10%—perhaps less, due to advances in microchips."

Total implantation, however, is still theoretical. One problem is finding a battery powerful enough to supply extremely high frequencies. If not equipped with a recharger that the patient could use while sleeping, the battery would have to be replaced every 6 months. Another difficulty is that implantation would prevent the patient from controlling the amplitude of stimulation. As patients move around, the intensity of their stimulation changes, so they constantly monitor it. New technology will someday make it possible to adjust the stimulation through the skin. "The crucially important thing was to improve the technique so that we could implant the electrodes with little risk or discomfort to the patient," says Dr. Waltz. "Right now the system works well with the minor inconvenience of having to wear the transmitter. Total implantation would not improve its therapeutic aspects, and we can't get into a situation in which we're hell bent for leather trying to maximize convenience." This is actually a trade-off, since Dr. Waltz has seen patients in wheelchairs refuse the operation because they did not want to bother with a transmitter.

According to the most recent statistics from the Office of Scientific and Health Reports at the Department of Health, Education and Welfare, an estimated 750,000 Americans have cerebral palsy, 3 million have head or spinal cord injuries, and 1 million have neuromuscular disorders like dystonia and multiple sclerosis. Medical bills for each of these patients average $5000 per year. With spinal cord stimulation, which costs $13,000, Dr. Waltz conservatively estimates that it would be possible to functionally improve 125,000 cerebral palsy patients (considerably reducing their $625 million costs a year), 500,000 head or spinal cord injury patients ($2½ billion a year) and 150,000 patients with neuromuscular disorders ($750 million a year).

Now if some enlightened funding agency can see fit to advance Dr. Waltz some money against these estimates, perhaps he can set up a large-scale research program.

PAIN RELIEF

For nearly 2000 years physicians have known that applying electricity to the head, feet, or other parts of the body could relieve pain; the ancients used eels and torpedo fish for this purpose. Although not fully understood, the mechanism of pain relief has been the subject of two hypotheses. The gate control theory suggests that adding electrical stimulation overloads the nervous system and blocks the pain signals from being carried to the brain. Another idea is that electricity triggers nerve impulses that release endorphins, the natural opiates that switch off pain signals. However it works, this method was first used in 1967 with the implantable dorsal spinal cord stimulator. By the 1970s the procedure had become acceptable and had received FDA approval.

Like Dr. Waltz's system, for which they were the prototypes, spinal cord pain relief systems have external transmitters and antennae. These send radiofrequency signals to implanted receivers, which translate them into electrical impulses conveyed to electrodes. Besides improving motor function, these stimulators seem also to decrease peripheral vascular disease symptoms by increasing blood flow to the extremities. Because they use far lower frequencies than Dr. Waltz's system, spinal cord stimulators for pain can be fully implanted; their batteries last for years, and they can be programmed much like heart pacemakers.

Other pain relief systems involve implanting the electrodes deep within the brain. Scientists have worked on similar "brain pacemakers" that stimulate the pleasure areas of the brain, thereby relieving some of the unpleasant symptoms of mental illness.

For peripheral nerve stimulation for pain, patients can choose between any of several transcutaneous electrical nerve stimulation (TENS) devices or implantable systems. The latter use the standard radiofrequency technique, but instead of being lined up in a catheter, the electrodes are imbedded in a silicone rubber cuff wrapped around the target nerve.

Some of these pain relief systems may become closed loop and adjust their stimulation levels to the amount of endorphins being released. In recent years, however, the application of neurostimulation for pain relief has become questionable. None of the systems relieves pain completely, and some patients are not helped at all. Those who do benefit, however, show dramatic improvement without the side effects of operations or analgesics.

URINATION

Paraplegic and quadriplegic patients often develop problems with their micturition reflex, which controls urination. Carmen has vividly described the unpleasantness of having a catheter; chronic use of one can damage the urinary tract and kidneys. By stimulating the medullary cone or the bladder wall itself with an implanted system, the paralyzed patient can once again control urination.

Peripheral Nerve Stimulation

In some cases electrical stimulation works better when applied not to the central nerves but rather to those in the problem area. A good example is the diaphragm pacer, which is being overhauled by researchers in Cleveland.

Often overlooked in respiratory therapy, the diaphragm is the breathing muscle that makes the chest cavity bigger and sucks air into the lungs like a bellows. It is powered by two phrenic nerves, which extend from the C3 to C5 levels of the spinal cord down into either side of the chest.

During the polio epidemics of the 1940s and 1950s Harvard University researchers put electrodes on patients' necks to stimulate their breathing. Since the late 1960s the work of William W.L. Glenn, M.D., Director of Cardiothoracic Surgery at Yale University, has made phrenic nerve stimulation into a viable technique. Dr. Glenn surgically placed electrodes on the nerves themselves above the C3 level and used radiofrequency to stimulate them. Later he moved the electrodes near the ends of the phrenic nerves on the chest.

Dr. Glenn's work freed high quadriplegic persons from ventilators. These individuals cannot breathe on their own; in most countries they are still being left to die. One of Dr. Glenn's earliest patients was able to open his own small business; another completed law school in Texas. An estimated 100 high quadriplegic persons a year nationwide could benefit from the system.

Another class of patients have disorders affecting the respiratory centers in the base of the brain or the areas of the spinal cord that receive or send the signal. These people breathe inadequately. They could benefit from respiratory assist if it were offered easily and with mininum risk. Putting them on respirators would cause their breathing muscles to atrophy; electrical stimulation, on the other hand, would help condition their diaphragm muscles. Phrenic nerve stimulators are potentially hazardous, however, because the surgery is painstaking (the nerves are delicate and easily damaged) and often requires thoracotomy (opening of the chest). Most surgeons are unfamiliar with the procedure, and the equipment is expensive. Damage to either nerve in a nonquadriplegic patient could leave him or her worse off than before.

This problem has interested Thomas Mortimer, Ph.D., Professor of Biomedical Engineering; Michael L. Nochomovitz, M.D., Assistant Professor of Medicine and Biomedical Engineering; and David K. Peterson, M.S., all of Case Western Reserve University. This group works in Cleveland, home of percutaneous neuromuscular stimulation, and they decided to implant electrodes into the diaphragm muscle itself. They imagine this procedure as an alternative to Dr. Glenn's system, one that is less invasive, safer, and easy to use temporarily.

Working with laboratory animals, Dr. Nochomovitz and his colleagues (including surgeon Thomas J. Stellato, M.D.) have used a laparoscope to implant two electrodes, one near each phrenic nerve. In humans this procedure could conceivably be performed with local anesthesia. So far, stimulating the diaphragm muscle has had the same effect as Dr. Glenn's method of stimulating the nerves. The Cleveland group also wants to close the loop on the system. "Signals for breathing go to other muscles," says Dr. Nochomovitz. "We've geared the stimulation of a dog's diaphragm to its own signal in the muscles of the nostril. These muscles could signal the stimulator when to activate the diaphragm."

Unlike Dr. Glenn's system, the Cleveland stimulator will use percutaneous power until it is adapted for radiofrequency. "We're still worrying out the conceptual, physiological, and functional questions," says Dr. Nochomovitz. Unlike the standard system, which has a small application, the new system could interest major companies capable of converting it to microprocessor technology.

Besides helping those patients mentioned, the Cleveland system could also be used by children with hypoventilation syndromes and people with some obstructive airway diseases. "It's hard to know just how wide the application would be because we haven't tested all the possibilities yet," says Dr. Nochomovitz. "We think, however, that as an assist and conditioning device, our system could help thousands of people a year." The group hopes to conduct initial tests on brain dead patients following chronic experiments in animals.

In general, it will be interesting to see which procedure works better—stimulating the muscles, the nerves, or either, depending on the particular condition being treated. While researchers try different treatments for various neuromuscular disorders and injuries, they are at the very least gaining valuable insight into the workings of the nervous system.

SECTION

VII

ENHANCED SELF-ESTEEM

22

Saving Face

Like arms and legs, artificial features are custom-designed prostheses. Along with reconstructive surgery, they are crucial in restoring a person's appearance to normal. Facial deformities usually bring to mind cosmetic surgery, but *reconstructive surgery* is the correct term for their treatment. Cosmetic surgery improves features that are already within the broad category of "normal." Reconstructive surgery, on the other hand, restores to normal what had been missing congenitally or destroyed by disease or trauma. A person whose nose was fixed to straighten a hump had cosmetic surgery; one whose nose was repaired after removal of a malignant tumor had reconstructive surgery. Even insurance companies recognize this distinction.

In reconstructive surgery, bionic parts treat not just the physical but also the psychological effects of disease or trauma, whether the victim's embarrassment is due to cosmetic problems or functional ones. As we will see later, these psychological forces are so powerful that widespread education about reconstructive surgery is leading to earlier diagnoses of cancer that may well save lives.

An estimated 350,000 people incur facial injuries each year from automobile accidents or fires. One in 750 babies is born with a cleft palate or some other type of facial defect. Rumor has it that the incidence of oral and facial cancer is rising, although statistically it has been stable at 3% of the population since 1978. Certainly some of the skin cancers are becoming more common, and they can destroy large portions of the face.

Oral and facial cancers can be treated by radiation, chemotherapy, or surgical removal of the tumors. The defects remaining from cancer treatment or those caused by accidents can be surgically reconstructed, rehabilitated by protheses, or both. The size, location, and nature of the defect; its effects on function or speech; the type and amount of reconstructive surgery required; the possible recurrence of disease; and the patient's age, health, and attitude all help determine which approach is best.

Prosthetic reconstruction of the head and neck is part of the field of prosthodontics, which is divided into three subspecialties. The first two, fixed and removable prosthodontics, deal entirely with intraoral prosthetic reconstruction. The third and newest subspecialty is maxillofacial prosthetics, practitioners of which use artificial materials for anatomical, functional, and cosmetic reconstruction of missing or defective areas of the jaws and face. Although easier to pronounce than the alternative, *maxillofacial prosthetics* is a bit of a misnomer because the mandible (lower jaw) is involved more frequently and harder to reconstruct than the maxilla (upper jaw).

Advances in Dentistry

Dentists are notable among health care professionals for the hard work they do to put themselves out of business. Besides performing routine procedures and trying to educate the public (90% of adults develop gum disease, almost entirely from negligence), dentists busily invent better ways to replace damaged or lost oral structures.

Bone grafts for the lower jaw can be shaped by a mandibular device developed by Dr. Leake's group. It is made of Osteomesh, a polyethylene terephthalate mesh fabric impregnated with polyurethane elastomer. The semirigid mesh is heat cured in the shape of the mandible and can easily be trimmed with scissors. Once a mandibular tumor has been removed, the implant is placed to bridge the space from which the bone has been cut away. Bone graft particles taken from the hip are then packed tightly into the tray. Osteomesh can also be used in restoration of the skull or other facial bones.

We have seen that new materials used to replace bone are being applied to restorative dentistry. Those serving a cosmetic as well as functional use include composite resins that can be bonded to teeth. A resin bond looks natural, preserves the tooth, and can be applied quickly, cheaply, and with minimal drilling. Unlike porcelain and metal caps, newer ceramic crowns fit well and look extremely natural, insulate the teeth, do not block x-rays, and are nonirritating and durable.

Different types of dentures are also being improved. Characterized dentures, made to reflect their owner's age, sex, and personality, can have stains, fillings, gum line recessions, and crooked placement so they will match the other teeth. Overdentures last at least twice as long as regular ones because they are attached to two remaining teeth, which help stimulate the jawbone and prevent its deterioration. Etched cast restorations, of which the most famous is the "Maryland bridge," require no capping of natural teeth to hold them in place. Instead they are attached to abutment teeth that have been slightly etched in acid to make their surfaces adhere better to the bridge wings. The result is cheaper and more comfortable than standard bridges. Etched cast restorations can also be used as retainers to stabilize mobile teeth and maintain them after orthodontia.

Some 200,000 Americans have dental implants of various kinds; the number rises annually by 25,000. Tooth roots made of HA or other materials are popped into fresh extraction sites to prevent resorption and provide stable bases for dentures. Implants ranging from stakes to blades to cylindrical honeycombs are being tested or used. Endosteal implants made of titanium are put through a chin incision and tapped directly into the jawbone; they have posts that protrude through the gums to anchor dentures. Subperiosteal implants are similar but rest beneath the gums on top of the alveolar ridge. Another type of implant, made of alumina, is screwed into the jawbone; after the tissues have healed, the screw is attached to a post that anchors a permanent bridge. Some removable dentures have projections that snap into tiny pockets created in the gum tissue.

A new implant, the Tissue Integrated Prosthesis (TIP) invented in Sweden, shows promise for attaching not just dental prostheses but others as well. Designed to be imbedded in the bone, where it integrates with hard tissue, a specially designed TIP has anchored bone-conduction hearing aids. It has also been used to reconstruct damaged or diseased joints in the long bones and may prove an effective attachment for facial prostheses and artificial limbs.

These and other innovations imply that one of these subspecialties will be swallowed up by the other as more replacement dental structures become permanent. "Oral disease is rampant," says Thomas R. Cowper, D.D.S., Staff Maxillofacial Prosthodontist in the Section of Dentistry and Maxillofacial Prosthetics in the Department of Plastic and Reconstructive Surgery at the Cleveland Clinic Foundation. "Millions of people have no teeth, and many have problems with lower dentures. Successful implants for fixed bridges would completely revolutionize prosthetic dentistry."

Maxillofacial Prosthetics

Throughout history everyone has wanted to look, speak, eat, and drink just like everybody else. For this reason, artificial eyes, ear, noses, and other prostheses have been traced back to the Etruscans and ancient Chinese and Indians (who penalized adulterers by chopping off their noses). Records show that a sixteenth-century French surgeon tried using a prosthesis to close a palatal defect. For the most part, however, it was dentists who used prostheses to restore self-esteem. Maxillofacial prosthetics is part of dentistry rather than plastic surgery because historically dentists had experience in making specific replacement parts. "Dentistry was poised toward prostheses because unlike surgeons, dentists were familiar with the materials and techniques used to replace missing segments of the head, neck, and oral cavity. The extension to creating a nose or a denture to replace a lost part of the face or jaws was a rather easy one to make," says John Beumer III, D.D.S., M.S., Director of the Maxillofacial Prosthetics Hospital Dentistry Group at UCLA. "The field got its real start during World War II, when drafted dentists saw many facial and oral injuries," says Dr. Cowper. "They became involved with aesthetics and started introducing plastics and other materials that had not been used before."

Maxillofacial prosthetics is not a well-known field. The American Academy of Maxillofacial Prosthetics has about 100 active members, of whom perhaps 50 have steady day-to-day practices. Other physicians, especially those in smaller cities, often do not think to recommend reconstructive surgery or prosthetic features to their patients. Even dental students do not learn much about the discipline unless they take an elective course or wander into a guest lecture. Those who become hooked must complete 4 years of dental school followed by 2 years of prosthodontics and 1 year of maxillofacial prosthetics. Prosthodontists learn to design, sculpt, and fit prostheses according to the needs of individual patients. Although the patient categories show different percentages from clinic to clinic, most prosthodontists treat cancer victims, especially those with oral cancers of the jawbone or soft palate. Their other patients have congenital deformities (especially cleft lips and palates) or defects resulting from accidents.

PROSTHESES

To make a facial prosthesis, the prosthodontist makes an impression of the defect using a dental stone like hydrocolloid. A wax model of the prosthesis is made on this stone cast. The wax pattern is used to make a mold from which the final silicone prosthesis is cast. This is tinted to the colors of the patient's skin and can be further shaded to match a tan or sunburn. The prosthodontist adds details like pores, freckles, lashes, and brows made from the patient's own hair. The edges of the prosthesis are feathered to blend with the skin, and sometimes glasses are used to hide its outlines.

One of the most easily constructed and realistic prostheses is the artificial eye. The prosthesis is colored to match the other eye and has extremely fine red thread run through it to simulate blood vessels. An ophthalmologist takes an implant, sets the eye muscles to it, and encloses it with tissue. This tissue moves with the other eye, so once fitted to the implant, the artificial eye moves too. It is removable but can remain in place for long periods of time.

An artificial ear can be either a prosthesis attached to the head or a skin graft scaffold that may be anchored with Dacron mesh.

An operation on the upper jaw can cause the patient's speech to become nasal, hollow, or unintelligible; he or she may also have difficulty swallowing. A child whose cleft palate has been surgically repaired is often left with a short soft palate and nasal speech. Correct speech can be quickly restored with a pharyngeal speech aid. An adult with a palatal defect left from a cancer operation instead needs an obturator attached to a partial or full denture.

The most common operation on the lower jaw is the partial mandibular resection, which leaves only one functioning jaw joint. If enough teeth are left, a flange prosthesis is used to realign the bite; the patient may also need dentures. When both joints remain intact, the surgeon can use a bone implant for reconstruction.

Prostheses are attached to the face with skin adhesives or implants. The adhesives do not hurt or damage skin; they hold the prosthesis in place even while its owner is bathing, and they allow it to be removed regularly for cleaning. "Adhesives are really quite effective for most patients, especially if the prosthesis has been properly designed," says Dr. Beumer. Alternative methods have included snaps, precision attachments, imbedded magnets, gold rings, and more recently, osteointegrated implants.

PROBLEMS WITH PROSTHESES

Nature of defect. A facial or intraoral prosthesis may or may not effectively disguise a defect. The most successful prostheses are those confined to one body part—an eye, ear, or nose. Conventional complete dentures pose few problems. The upper jaw, a static piece of bone, can also be restored reasonably well, allowing the patient to speak, chew, swallow, and look normal. These defects tend to have less impact on the patient and his or her family than more extensive ones.

Disguising even a minor defect, however, can contribute greatly to a patient's comfort. "Harold Lindberg,"* a 77-year-old Los Angeles musician, had an operation for skin cancer in 1972 that removed the right side of the tip of his nose. He covered the defect with an adhesive strip until a year later, when he had unsuccessful plastic surgery. Harold's surgeon then suggested that he consider getting a prosthesis. "I had no option but to try it," Harold says. "I had a difficult situation, but I still perform onstage and needed to do the best I could about my appearance."

Now Harold has a silicone prosthesis that restores the shape of his nose. "I always have it on when I go out—my wife is the only person who has seen me without it," he says. "I feel much more natural with it on. It stays on all night or when I wash my face—I can't take chances with its not being secure."

Harold has worn his prosthesis for 11 years. "It's just like any other permanent nuisance—it requires infinite patience and good attitude," he says. "But it's certainly worth it. I don't have to worry about not being attractive, and the people around me accept it as just a part of me."

The least successful prostheses are those intended to cover large facio-oral defects—the loss of a nose, cheek, lip, and upper jaw, for instance. "The more extensive the ablative surgery and the larger the prosthesis, the harder it is to do and the less successful it is both functionally and cosmetically," says Dr. Cowper. "The most difficult case is that of the patient who has lost parts of both the upper and lower jaw—I've had three of them," says Dr. Beumer.

Cancer of the mouth, tongue, or lower jaw can make restoration very difficult. The jawbone itself can be restored, but not the tongue. The patient will need mandibular guidance therapy—exercises and appliances to train the jaw muscles to correct its position—followed by speech therapy. "The tongue makes incredibly fast, complicated movements that affect speech, chewing, swallowing, and saliva control," says Dr. Beumer. "We can make the tongue look normal, but most patients probably won't be able to speak effectively or

---

*The two maxillofacial prosthesis patients mentioned in this chapter have asked that their names be changed.

eat normally without drooling. They might not be able to work, so they and their families will have many adjustments to make."

Materials. Another difficulty involves the materials used to make prostheses. "We have a greater choice of materials now than in the past," says Dr. Beumer, "but we are not yet entirely satisfied with the physical properties of all of them." Prosthodontists have tried acrylic resins, acrylic and vinyl copolymers, latex, polyurethane, and silicone elastomers. The best materials available for external prostheses are just not lifelike. The preferred material is silicone, which is extremely durable but has a surface texture that is difficult to control. Prosthodontists cannot create as much detail as they would like, and unlike the skin, the silicone does not stretch when the person changes facial expressions. "Various centers are working on new materials, and everyone, including us, is basically trying out whatever their chemists come up with," says Salvatore J. Esposito, D.M.D., Head of the Section of Dentistry and Maxillofacial Prosthetics in the Department of Plastic and Reconstructive Surgery at the Cleveland Clinic Foundation.

Implanted prostheses need not be realistic; instead their surfaces must be biocompatible. Standard implants made of silicone and other materials are available for reconstructing or cosmetically improving the shapes of the head, forehead, ears, cheeks, cheekbones, nose, upper lip, chin, and Adam's apple (as well as the arms, breasts, nipples, buttocks, and calves). Besides posing the usual risks of infection, implants for the face and jaws can create enough pressure to cause bone resorption. At least one case has been reported in which a silicone implant placed in front of the chin eroded the labial plate, wandered through the lower jaw and popped through the floor of the mouth surrounded by a connective tissue capsule.

Some of the new bone replacement materials may help solve this problem. We have noted that HTR is one of the new dental materials being used to augment the alveolar ridge; it also shows some promise for use in maxillofacial reconstruction. Dr. Leake has used solid HTR as an implant for a boy with congenital absence of the condyles (articular projections of bone on either side of the lower jaw). A large piece of acrylic placed in his chin to lengthen the jaw had become infected and was extruded. "The patient must be followed," says Dr. Leake, "but so far the HTR has been inert and seems to be working very well."

One especially controversial procedure is the use of liquid silicone for facial injections. Popularized in the 1940s for breast and other soft tissue augmentation, liquid silicone was found to cause dangerous side effects. By the 1960s the medical establishment had halted its use to enlarge breasts. The FDA has not approved silicone injections for cosmetic improvement of the face, but since the agency regulates only interstate commerce, it cannot stop physicians from using liquid silicone obtained within their own states.

Over a period of several weeks the silicone is injected just under the skin in microdroplets, which are walled off by collagen in tiny cysts. By lifting the skin, the cysts smooth out its defects. Injected in too large an amount, the silicone can migrate or cause pain, discoloration, and bumpiness. "The success of these injections depends on who is doing them and how well they are done," says Dr. Beumer. "Some licensed investigators have done dramatically good work, but others have caused deformities."

Surgical procedures. Another problem facing prosthodontists is that head and neck surgeons, like those who amputate limbs, often operate without considering that the body part they remove might be prosthetically replaced. "Everyone has priorities, and psychologically the operation and restoration are at different ends of the spectrum," says James Smith, D.D.S., Oral and Maxillofacial Surgeon in the Section of Dentistry and Maxillofacial Prosthetics in the Department of Plastic and Reconstructive Surgery at the Cleveland Clinic Foundation. "The surgeon considers the operation successful if he or she removes all the cancer, regardless of whether the site can be prosthetically restored." Hundreds more surgeons than prosthodontists are in practice, and when the operation proceeds without prior consultation, the patient benefits far less. "The surgeon-prosthodontist interaction is probably one of the more critical things that happens before and during surgery," says Dr. Beumer. Most of the time the prosthodontist need not be in the operating room unless procedures have been left ambiguous or a prosthesis needs manipulation, but his or her involvement at an early stage makes later rehabilitation much easier. Like prospective amputees, persons facing head and neck surgery should insist on such a consultation.

The Cleveland Clinic Foundation is one of the few institutions in which several disciplines are lumped together in the Department of Plastic and Reconstructive Surgery. "Someone once said that the surgeon gives life, and the prosthodontist gives quality of life. Here that distinction doesn't exist. We do pre- as well as postoperative care of head and neck cancer patients," says Dr. Esposito. "We have a lot of interplay and consultation, and we don't hesitate to suggest surgical improvement of a patient's defect if that will help," says Dr. Cowper. This group practice approach permits specialties to overlap in a patient's treatment: depending on the defect, a patient may need a head and neck surgeon, a plastic surgeon, a radiation therapist, a maxillofacial prosthodontist, an oral maxillofacial surgeon, and a speech pathologist.

One area in which interdisciplinary teamwork is crucially important is the lifetime treatment of patients with cleft lip–cleft palate. A newborn baby with a cleft lip–cleft palate may or may not need a prosthesis immediately; when used, the prosthesis gives the baby a surface to feed against and keeps the jaw arches properly aligned. When the baby is approximately 3 months old, its cleft lip will be closed surgically; the palate is closed in a second procedure when it is 6 to 18 months old depending on the surgeon's preference. "Surgery plays the primary role in the initial treatment of the cleft palate," says James Zins, M.D., Plastic and Reconstructive Surgeon in the Department of Plastic and Reconstructive Surgery at the Cleveland Clinic Foundation. "The prosthesis helps growth along and makes rehabilitation easier." Once the palate is closed, a prosthesis is no longer needed, although the child may wear a speech prosthesis later on if he or she has teeth missing. "Well done surgery gives the patient speech as normal as possible," says Dr. Zins. "We're concerned with speech first and appearance second, because the lay public immediately associates improper speech with subintelligence."

Despite various difficult and complicating factors, the prosthodontist's work may well become easier and more successful as new materials and techniques make their way into the field. Computers are being used to simulate cosmetic and reconstructive surgery on the screen, allowing the surgeon to examine alternatives in advance to determine which procedure will give the best results. Like other bionic parts, facial and oral prostheses may be more efficiently and better made with CAD/CAM. The surgeon will call up a computer image, color in a defect, then use the data describing its exact dimensions to sculpt the prosthesis automatically. The computer will control the machine tools that create the model or prosthetic part.

THE PSYCHOLOGY OF PROSTHETICS

Maxillofacial prosthetics involves considerable emotional stress for both patients and prosthodontists. The excision of benign tumors usually causes disability that is cosmetic rather than functional, and although most patients are elderly they are understandably upset by the results. Like a mastectomy, facial cancer surgery is disfiguring and raises all kinds of psychological issues. Physicians generally lack training in psychosocial issues and information about facial prostheses, so a cancer patient may be treated insensitively and left ignorant that he or she can be reconstructed. Upon learning of their predicament most patients initially feel shocked, depressed, even guilty; they want to deny that their problem exists. "Patients and their families must deal with life-threatening diseases and permanent disabilities. Most people come to grips with the outcome fairly appropriately, depending on the nature of the tumor or defect," says Dr. Beumer. "Everyone wants to walk out of our offices looking fabulous, and that presents a problem," says Dr. Esposito. "We can't really play up the results because they'll be disappointed, and I think we are our own worst critics."

A recent article in a dental journal compares the responses of prosthodontists, patients, and artists to various prosthetic reconstructions. The prosthodontists were the most critical, the patients the least. "I think that a patient may initially agree with us that the prosthesis looks artificial, but as time goes on he or she adapts to it psychologically and sees it very differently," says Dr. Cowper. "The patient either thinks that it looks better than it really does or else totally rejects it." Prosthodontists are hard on themselves partly because they can make intraoral prostheses like fixed bridges and dentures look so natural that even other dentists are fooled at first glance. "The facial prostheses are so much less lifelike because of their materials that they really suffer by comparison," says Dr. Esposito. "We say that they could be better because they're not perfect."

These factors mean that a prosthodontist's work elicits his or her full range of emotions. The prosthodontist lives and to a certain extent dies with the patients. The patients are closer and more dependent than they are with other physicians; they show anger and appreciation more readily. It is very difficult for a prosthodontist to meet a patient with a large facial defect who is expecting miracles, even more so to have completed a prosthesis that the patient has not seen. "We have the patients take their prostheses and put them on at home by themselves," says Dr. Cowper, "although that may be a cop-out on our parts because it's so terribly hard to see their disappointment when it occurs."

The Cleveland Clinic staff also spends time counseling families on what they and patients should expect before the prosthesis is unveiled. "Often a patient isn't terribly pleased with the prosthesis, but if his or her family and the staff here are all really positive, that greatly helps its success," says Dr. Esposito. A happy patient is a real joy to the prosthodontist. "The rewards, the pleasure and satisfaction we get are greater than those of any other field of dentistry I know of," says Dr. Beumer.

"June Watson," a 76-year-old retired court reporter from Los Angeles, lost her nose to cancer 10 years ago. She covered the defect with gauze and tape until Dr. Beumer saw her in an elevator at UCLA one day and suggested that he could help her. Most people do not know that June now has an artificial nose. She needs to wear glasses, and the rest of her face is mottled from radiation treatments, so the nose is not obviously artificial. "I would advise anybody who needed one to get a prosthesis," she says. "A nose is certainly better than a piece of gauze."

Prosthetic features are a poor imitation of nature and will remain so until better materials for them are developed. Meanwhile, they allow people who would otherwise avoid being seen to go out and socialize comfortably.

23

Carefully Selected Contours

Like prosthetic features, breast implants are important partly because of their tremendous psychological clout. This holds true whether the implants are used for cosmetic surgery or for reconstruction. Implants intended to enlarge small breasts are considered cosmetic surgery. As the song "Dance Ten, Looks Three" from *A Chorus Line* illustrates, larger breasts (and matching derrière) can make all the difference in the world of show business. Women in other fields may be dissatisfied with breasts they consider too small or poorly shaped. Perhaps their breasts do not match well or have changed contour because of weight gain, dieting, or pregnancy. Although *not* a "deformity" that needs to be restored to "normal," small breasts can limit a woman's sense of self-esteem. Breast implants will not work miracles, but they can indirectly improve the quality of a woman's life.

We have already noted the distinction between cosmetic and reconstructive surgery; breast implants after cancer surgery are reconstructive. As men feel less than whole if they are impotent, women feel abnormal when they have lost a breast to cancer. No wonder—their body symmetry and balance are askew because a part of themselves is gone. Regardless of the type of mastectomy she has undergone, however, just about any woman can be reconstructed. By encouraging women to examine their breasts and report suspicious findings promptly, the prospect of breast reconstruction is likely to save lives.

As of summer 1984, 1,100,000 women had breast implants. About 75% of these were cosmetic implants, the rest reconstructive. An estimated 40,000 reconstructions are performed each year, and that rate is rising very quickly as more women become aware of this option.

The earliest record of breast reconstruction dates back to 1896, but no implants were available then. The search for a satisfactory implant for cosmetic enlargement accelerated in the 1940s. Researchers tried all kinds of substances for breast implants: paraffin wax; various synthetic plastics; Dacron; nylon; polyethylene; and different sponges, foams, and shredded materials. All of these implants caused catastrophic infections or were extruded or otherwise rejected.

In the 1950s researchers tried pumping up breasts with liquid silicone injections, which created all kinds of difficulties. A major problem was that the body forms capsules around each droplet or pocket of silicone that make the breasts look and feel lumpy and cannot easily be distinguished from cancer. As a result, many women had to have their enlarged breasts removed by mastectomy.

Types of Breast Implants

These complications ended in 1961 with the development of the silicone implant similar to the many types available today. The standard is the gel implant, a seamless silicone envelope filled with silicone gel. The two silicones are identical except for their molecular structure: the shell is composed of larger molecules, the gel of smaller ones. The implant is quite inert, lighter than water, and as dense as natural breast tissue, so it looks and feels real.

Another type, the inflatable implant, is a silicone shell that is injected through a valve with saline solution at the time the prosthesis is placed. This implant can be inserted through a small incision; its size is adjustable, and it has a natural feel. It can leak, however, and has therefore lost popularity.

The bilumen (sometimes called the gel-saline) implant has a shell full of silicone gel enclosed within an outer shell into which saline solution can be injected through a valve. Like the inflatable type, this implant can be adjusted to a particular size, and its structure may help prevent silicone droplets from bleeding through the shell. It might also minimize encapsulation: the surgeon can either overinflate, then deflate the implant (to loosen the capsule) or add cortisone or steroids to the saline solution (to slow capsule formation). Like the inflatable implant, however, the bilumen can leak saline solution, although the results are not as embarrassing because the breast retains most of its volume.

Another type of gel-saline implant, prefilled with silicone gel, has a valve through which saline solution is added to form little bubbles. Like the bilumen type, this implant allows for variable sizes without risk of saline solution leakage, but it is not very popular.

Some researchers have reportedly glued two implants together and implanted them with the smaller one in front. The purpose is to enlarge the breast sufficiently without the sagging or flattening of one big implant.

Two newer implants have recently joined this group. One is a temporary expander implant. The empty implant is attached to a small tube and filling reservoir that are placed beneath the skin next to it. If the chest wall is very tight after a mastectomy, this device can be filled gradually to stretch the skin until it can accommodate a standard implant of the proper size.

Another new implant is a silicone one covered with a thin layer of polyurethane foam sponge rubber. This type seems less likely than the others to harden from encapsulation, possibly because fibrous tissue is separated and disorganized as it grows into the foam. Its disadvantages are that the breasts can change appearance over time and the foam often disappears—and no one knows where it goes or in what form.

Implant Surgery Procedures

Three types of surgical incisions for cosmetic implants are in use. The most popular is the inframammary, in which the incision is made at midpoint just above the breast's lower fold. A periareolar incision is made in the lower border of the areola to hide the scar more effectively. A transaxillary incision, made just below the hairline under the arm, is also well hidden. All three procedures, especially the periareolar incision, may cause temporary numbness around the nipple.

In performing reconstruction, most surgeons cut through the mastectomy incision unless they are supplementing the implant with a flap from the back or abdomen. Some women, especially those who have had radical mastectomies, may need this kind of autograft to replace skin and muscle removed from the chest. If the flap is taken from the abdomen, the surgeon may install plastic mesh there to prevent a hernia.

Some patients may need an implant to replace missing muscle as well as a breast implant. Standard silicone elastomer implants come in several sizes and can be cut to the specifications of the pectoral muscle. Most implant manufacturers also supply contour defect molding kits used to fill hollows in the chest wall. The surgeon sends a mold of the patient's chest to the supplier, who returns a custom-made silicone prosthesis. These are sometimes attached with Dacron tabs.

Like his colleagues in maxillofacial prosthetics, Richard V. Dowden, M.D., Head of the Section of Breast Surgery in the Department of Plastic and Reconstructive Surgery at the Cleveland Clinic Foundation, consults with the cancer surgeon beforehand but makes no suggestions about procedures. "The surgeon should not modify the surgery in a way that will make reconstruction easier," he says. "The main thing is to do everything necessary to remove the cancer, and let me worry about reconstruction. Even when reconstruction is difficult, I have the comfort of knowing that the patient had the most thorough treatment of the cancer."

Risks of Implant Surgery

Implantation is like other operations in posing some risk (less than 1%) of hemorrhage or infection. Microscopic droplets of silicone gel may bleed through implants, but no evidence exists that the material causes cancer. As we will see in a later chapter, the FDA has recently decided to regulate breast implants more strictly. Surgeons have characterized this decision as a last-minute scare that at best seems unwarranted and at worst may offset some of the lifesaving benefits of implants by scaring the public away from them.

Every breast implant is surrounded by scar tissue. Encapsulation is not defined as a complication, perhaps partly because it has certain benefits. The capsule keeps the implant from moving, may protect it from injury, and may keep the cohesive silicone gel in place if the shell should rupture. Women have gone for years without knowing that one of their implants was broken.

On the other hand, active contraction of the capsule can cause hardening—the number one problem with breast implants. Estimates on the incidence of hardening range from 25% to 33% of patients. "Our experience has been that one in 20 women requires removal and replacement of the implant, while 15% of reconstructed women and 20% of cosmetically enlarged women have some degree of hardening," says Dr. Dowden. "To combat this, we have patients do compression exercises three to four time a day for 6 months to a year to stretch the capsule. That takes care of it pretty well." When hardening does occur, the physician can manually squeeze the breast to break the capsule open. This usually works but could cause hematomas or displacement, distortion, or rupture of the implant.

Women often ask about "rejection" of breast implants. True rejection (the body's intolerance of an implant) is extremely rare. As noted earlier, however, infection may occur; if it is not treated, the implant might be forced out through a thin area of the skin. A woman with this problem might think that her body is "rejecting" the implant when in fact it is just responding to infection in its usual manner.

A woman might worry less about potential complications than about what effect, if any, breast implants could have on cancer detection. Some physicians argue that by pushing the breast forward, cosmetic implants can make cancer easier to detect; others claim that it is difficult to tell whether a lump is an extension of old capsular scar tissue or a tumor. Either viewpoint could potentially be correct.

With reconstruction, the situation is more clear-cut. The implants do not cause cancer to recur or change the patient's medical status. They are placed behind the muscle. "The majority of recurrences in the chest are in the skin or between the skin and the muscle, in front of the implant where they can easily be felt," says Dr. Dowden. "Theoretically there's a chance of recurrence behind or along the implant, and like deep cancers in large breasts this would need to be examined in a certain way." Through the back of the implant the physician can examine the entire front surface of the chest wall with the enhanced sensitivity afforded by the slippery shell. Mammograms and CAT scans can also be done on a reconstructed breast.

Ultimately the detection of recurrence in the chest may not alter the outlook for the patient. "One very important fact is that cancer that returns in a woman's chest is almost always associated with its recurrence elsewhere, and this metastasis really determines the eventual outcome," says Dr. Dowden. "If a chest recurrence is left untreated, the patient will survive an average of 32 months; if it is treated, she'll survive an average of 4 months longer. Yes, we do want to find recurrences, but statistically it won't make much difference with respect to the long-term prognosis."

Her prognosis should have no bearing on a woman's decision about reconstruction. "I would reconstruct a patient even if she had only a few years left," says Dr. Dowden. "The age is passing when surgeons said, 'Run along now, honey, we'll tell you what to do about your health.' Women are entitled to live the rest of their lives with two breasts if they so choose."

Consequences of Reconstruction

The most important fact about breast reconstruction is the improvement it is expected to make in survival statistics. It is legitimate to fear losing a body part, but many women have been brainwashed by society into believing that it is better to be dead than lose a breast. For the first time, women can have a positive rather than a negative motivation to do breast self-examination. The earlier they find a lump, the greater the chances that they will need just a lumpectomy or partial mastectomy, the simpler the reconstruction, and the sooner they will be back to normal. Women who know about reconstruction tend to seek help sooner, are less devastated by the thought of mastectomy, and have much more peace of mind.

A mastectomy patient who is not reconstructed is reminded every day that she looks abnormal, that she has had cancer, and that she may die. It is very difficult to escape from this burden, which saps a woman's energy tremendously. The main benefit from reconstruction is that it frees her mind so that she can use her energy to live. "I was reminded every day that I had cancer, and now I forget. Reconstruction really helps a woman shed her fear of cancer," says Susan Kaskey, who has had both breasts reconstructed. "Reconstruction removed my preoccupation with possibly losing the other breast, because now I know that I can replace it," says Sophie Maher, who has had one breast reconstructed and has passed the 5-year survival mark. Often mastectomy causes subtle but deep mental changes, and restoring a woman's self-image and self-confidence seems to have huge benefits. "My patients' husbands and boyfriends come in and say, 'I had no feeling one way or the other about the breast, but I sure do like the fact that her personality has been restored,' " says Dr. Dowden.

On the other hand, it is important to note that having breast implants for whatever reason will change a woman's relationships only insofar as it changes her personality. It will not dramatically improve relationships that were troubled before the operation. Reconstructed breasts do not look entirely normal, and success is often determined by how a woman looks in a bikini rather than in the nude. No patient should schedule cosmetic or reconstructive surgery with false expectations.

Susan, a 37-year-old secretary from Berea, Ohio, had a modified radical mastectomy with lymph node removal on one side and a subcutaneous mastectomy as a preventive measure on the other. "I had spoken with a woman from Reach to Recovery* and was able to accept the surgery knowing that I could be reconstructed," she says. The mastectomies were followed by a year of aggressive prophylactic chemotherapy. "The surgery was nothing, but the chemotherapy was hell," says Susan. "The prospect of reconstruction was always the light at the end of the tunnel."

Susan's reconstruction took place in August 1983. "This was my first positive surgery, and I was absolutely thrilled with the results," she says. "I woke up after surgery and couldn't see my feet! I was really excited." The following July, Susan had her nipples reconstructed. "My breasts look and feel great, and I'm proud of them," she says. "I have two children and had been starting to sag a bit, so they look better than before."

Except for a few numb spots, particularly under her armpits, Susan has skin sensation all across her mastectomy sites. Her breasts look different to her and have a few pencil-thin scars, but she considers them very acceptable. "It's important for women to know that they should look for cancer, get it out in time and get reconstructed," she says. "It's not vanity. If you had your arm removed, it would be foolish not to put it back if you could, because it's just a normal part of your body. Instead of feeling different and strange, I feel really self-confident now."

Sophie, a 43-year-old former secretary-manager from Olmstead Township, Ohio, now works as a volunteer for the Reach to Recovery and RENU† cancer information groups. Sophie used to do breast self-examination in the shower but not while lying down; it was her physician who found a lump in her breast when she was 36 and put her in the hospital for a biopsy. She woke up from the operation to find her breast gone. "My initial response to reconstruction was, 'Are you kidding me? You just ripped off part of my chest, and now you're going to put it back?' I put the idea aside, because right after surgery no one is in a hurry to jump back up on the gurney and go in for more," she says.

Sophie's physician suggested that before being reconstructed, she wait out the 2-year period during which chest recurrences might appear. About 2½ years after losing her breast, she had implant surgery. "I nearly changed my mind at the last minute, but it was really the best decision I have ever made for myself in my life," she says. "I was ecstatic right from the start, even though it took 2 to 3 months for my skin to stretch enough to take an implant of the right size." Her nipple was reconstructed 6 months later, and 4 years afterward she had what she calls "a new front end alignment"—the implant was repositioned and enlarged.

"I had been married only 3 years when I lost the breast, and I felt guilty as well as uncomfortable because it seemed to me that I looked so different," says

---

*A national information group.
†An acronym for Reconstruction Education for National Understanding.

Sophie. "Reconstruction is never going to replace my natural breast, but it sure comes close. I don't know anybody who is sorry that she was reconstructed. It's fantastic that this is available and that breast cancer no longer has to be the end of the world."

Like the creation of prosthetic features, breast implant surgery is upbeat work with enormous rewards for the physician as well as the patient. The same holds true for implants that restore certain bodily functions.

24

Private Prostheses

Psychologically it is important not only to look normal but also to behave appropriately. The more intimate the behavior, the more traumatic are any deviations, especially with respect to performance in the bedroom and the bathroom. The ability to control bowel movements and urination is one of the first significant departures from infancy, while sexual maturity confirms the onset of adulthood. Loss of control over either toilet or sexual functions can be devastating to a person's sense of well-being. For this reason, courageous people who have overcome either problem are unusually open about discussing personal details in the hope of educating and helping others.

Urinary Prostheses

THE ARTIFICIAL SPHINCTER

Experts estimate that 8 million Americans of all ages are incontinent, or unable to control their urination. Incontinence is a complication of all kinds of disorders, including spina bifida (a birth defect that affects the spinal cord), radical prostatectomy or transurethral resection (prostate surgeries), spinal cord or pelvic injury, congenitally malformed bladder or urethra, neurological disorders, diabetes, multiple sclerosis, poliomyelitis, infections, and aging. Women, especially those who have had several children, may leak urine while exercising, laughing, coughing, or sneezing.

In most cases incontinence can be reversed by surgery or medication or controlled by adult diapers, collection bags, catheters, or clamps. These control methods, however, can be troublesome, embarrassing, and for some victims, ineffective or dangerous. Incontinent people may become depressed and withdrawn; they may be afraid to go out and unable to work or attend school. "Incontinence is a terrible thing," says Irv Berwick, a film director in his mid-sixties from Encino, California. "I was working when I became incontinent, and it was very embarrassing," says Ken Mason, a 59-year-old retired engineer and private pilot from Ventura, California. "I had to use the bathroom really often and would wet my clothes. We took a trip to China during which I tried using adult diapers, but they didn't control the leakage." "When we were out sightseeing on the bus, there was no way for Ken to use a bathroom," says his wife, Betty, a 59-year-old retired grocery checker. "One time he got up, and his bus seat was all wet."

In 1972 F. Brantley Scott, M.D., of the Department of Urology at the Baylor College of Medicine in Houston, performed the first implantation of an artificial urinary sphincter he had developed. Since then, different versions of the device have enabled over 5000 incontinent people to resume normal lives. The latest model, the AMS Sphincter 800, went on the market in 1983; the manufacturer is American Medical Systems of Minnetonka, Minnesota.

The kidneys constantly produce urine, which flows down the ureters into the bladder. The sphincter muscle, which encircles the urethra, squeezes it closed to store urine and releases it when the person wishes to urinate. The artificial sphincter performs the same function.

The prosthesis consists of a silicone elastomer cuff, balloon, and pump connected by silicone tubing. The balloon is placed in the abdomen near the bladder, the pump in a male's scrotum or a female's labia or vaginal lips. When implanted, the components contain either a contrast solution that shows up on x-ray films or a saline solution if the patient is allergic to iodine. The cuff, a balloon the size and shape of a Band-Aid, is implanted around either the urethra or the bladder neck. When filled with fluid, the cuff presses the area shut so that urine cannot pass. Both the cuff and the amount of pressure generated by the balloon are adjusted during surgery to fit the patient's needs.

To urinate, the wearer squeezes the pump, which transfers the solution from the cuff to the balloon. As the cuff loosens, the urethra or bladder neck opens so that urine can flow out. As soon as the pump is released, the cuff slowly begins to refill from the balloon. The wearer has a few minutes in which to urinate before the reinflated cuff again staunches the flow. "My prosthesis usually takes 4 to 5 minutes to refill," says Irv. "When the leakage stops, I know that the cuff is reinflated." Implant recipients say that they can wait about 3 hours between visits to the bathroom. If they delay too long, the urinary pressure will bypass the sphincter and cause leakage. This safety measure prevents the pressure from damaging the bladder or kidneys.

A button on the pump can be used to deactivate the device and lock the cuff open. This feature is extremely useful if the patient has surgery and needs a temporary catheter. A study by David Barrett, M.D., of the Mayo Clinic indicates that a patient can safely lock the cuff to perform self-catheterization. "Some of our patients can't empty their bladders completely. The lock permits them to do self-catheterization three or four times a day and use the sphincter the rest of the time," says Gary E. Leach, M.D., Director of the Urodynamics Laboratory at the Kaiser Foundation Hospital in Los Angeles and Assistant Clinical Professor of Urology at UCLA.

Dr. Scott's group, however, does not approve of self-catheterization with the artificial sphincter. "We make certain before implantation that the patient can empty the bladder spontaneously and completely once the sphincter is in place," says Irving J. Fishman, M.D., F.R.C.S.(C), Assistant Professor of Urology at the Baylor College of Medicine and an associate of Dr. Scott's.

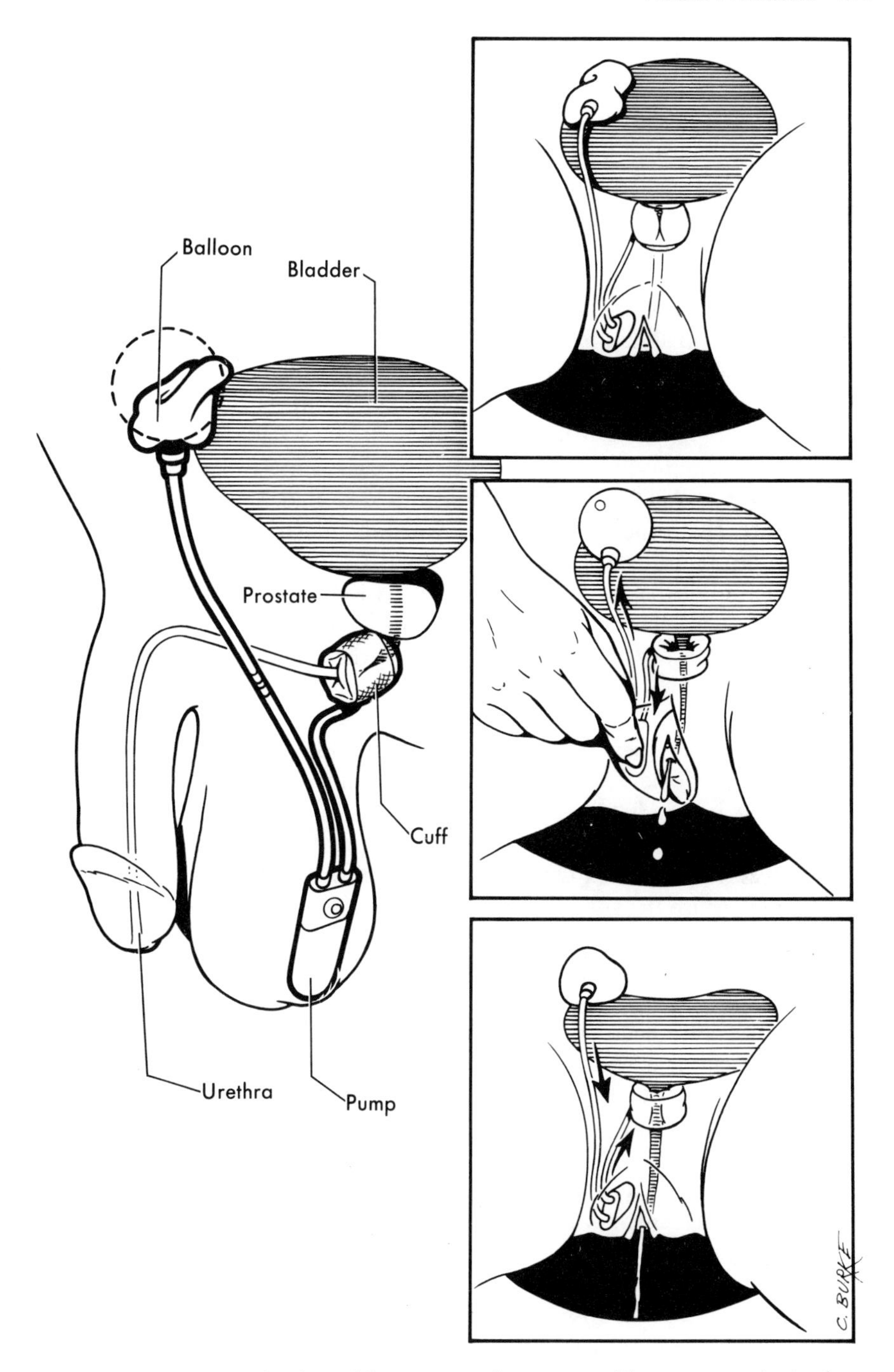

The AMS Sphincter 800 implanted in a man and a woman. The woman urinates by squeezing the pump implanted in the labium to activate the device.

"Intermittent catheterization could be dangerous for sphincter patients because it might injure the urethra and disrupt the mucosa overlying the cuff. The latter poses a high risk of infection and loss of the sphincter."

Catheterization definitely should not occur while the sphincter is activated. "The lock is a really good safety feature," says "Mark Harrison" (who asked to remain anonymous), a 50-year-old engineer from Texas. Mark has been paraplegic since an accident in 1962. "If I were injured, recovering from surgery or even just really sick, I'd have to be catheterized, and without the lock that could do all kinds of harm," he says. Mark carries a card in his wallet stating that he has the sphincter; Ken wears a Medic-Alert bracelet. "If he has an accident and is taken to an emergency room, it alerts the staff not to catheterize him but to call Dr. Leach instead," says Betty.

Ken developed cancer of the prostate 4 years ago. His radiation treatments weakened the tension in his sphincter muscle so that a year later he became incontinent. The problem worsened when he developed Shy-Drager syndrome, a rare neurological disorder that may be a contributing factor.

Neither adult diapers nor other methods Ken used to control the leakage worked very well. Dr. Leach had mentioned the artificial sphincter to Ken, who learned from tests that he was an appropriate candidate for implantation. "I was all for it," he says. "I'm anxious to try anything new, especially if I think it will help."

After spending a few days in the hospital for his surgery, Ken recuperated at home for 6 weeks so that he could heal fully before the sphincter was activated. He has used it for a year with no problems and has been tremendously happy with it.

For security Ken uses Depend pads, which resemble sanitary napkins. "I wear them because the sphincter takes longer to reactivate than I want to spend urinating and because if I strain very hard and have tension in my stomach muscles, I might leak a little bit," he says. "The pad controls leakage whether I'm active during the day or relaxed or sleeping at night."

Ken and Betty have both benefitted greatly from the implant. "It doesn't bother me at all; it makes a really positive difference in my life, and Betty no longer has to wash clothes continually," says Ken. "I'm very happy because I knew that Ken was miserable while incontinent, and he spent half his life in the bathroom," say Betty. "Now it's 2 or 3 hours before he feels the pressure to use the bathroom, so we can lead a normal life. We go to church, and out to dinner, and he wears nice suits, with no more odor or embarrassment."

Mark found that using the sphincter and taking prophylactic doses of oral antibiotics helped clear up a chronic *Pseudomonas* infection that he had endured for the 18 years that he wore catheters. "Since the sphincter leaks when I bounce around, getting in and out of cars, for instance, I usually wear an external catheter," he says. "I have a penile implant that lets me roll the catheter on smoothly over an erection; it never leaks, and I'm able to do business traveling all the time without any trouble."

Patients considering the artificial urinary sphincter must have adequate presurgical evaluation of their lower urinary tract function, not only because other treatments may end incontinence but also because implantation of the sphincter without satisfactory bladder function can cause horrible results.

The sphincter is appropriate only for very specific sources of incontinence like radical prostatectomy or perhaps pelvic fracture or urethral surgery. Patients must have the capacity and dexterity to use the device as well as adequate bladder capacity at low pressure (high pressure in the bladder can cause leakage, erosion of the cuff into the urethra, or damage to the kidneys). "Most of my implantations are performed on men incontinent from prostate surgery," says Dr. Leach. "The manufacturer estimates that 15% of implant recipients are women, but I can usually help them be dry with some other type of surgery."

Several hundred children with spina bifida have had the sphincter implanted by Dr. Scott's group. "These patients have had a 75% success rate in achieving urinary continence," says Dr. Fishman. "The psychological effects of implantation on these children are tremendous—instead of being social outcasts, they are suddenly accepted by their peers. Their dramatically improved self-image is reflected in both their social and their scholastic achievements." Yet like other implants, sphincters may pose some risks for children in particular. "Nobody really knows what will happen in the long term if the sphincter is implanted in a growing child," says Dr. Leach. "Although the age range of potential recipients is wide, the sphincter should always be considered a last resort."

Irv became incontinent after prostate surgery. Another physician introduced him to Dr. Leach, who told him about the artificial sphincter. "My response was, 'Let's go!'—and I'm very happy that we did," he says. Aside from noting the time to remember to urinate, Irv finds use of the sphincter to be a normal experience in most respects. "I had to make a few adjustments," he says. "I can't sit on the edge of a hard chair because that presses against the area I squeeze in order to urinate. I also have to be careful about bending over to tie my shoes, for instance, because that might cause leakage."

Irv also has a penile prosthesis, but he appreciates the sphincter more. "I enjoy sex, of course, but if I had to choose between the sphincter and the penile implant, I would certainly choose the sphincter," he says. "It's a hell of a lot better than wearing a rubber bag around my thigh. The implant has done a tremendous amount for my life and my attitude, and I'm delighted with it."

Dr. Leach recommends the sphincter enthusiastically to patients he feels are appropriate candidates for it. "I tell them that if it doesn't work, they can hardly be worse off than they are," he says. "The 70% to 80% of patients for whom it is successful are unbelievably happy. The difference in their lives before and after the surgery is totally dramatic." One of Dr. Leach's patients, an 18-year-old man, quit school when he became incontinent; after his sphincter implantation, he resumed his studies. Another went home after his surgery, gathered all his collection devices, pads, and diapers into a big pile in his yard, and built a bonfire.

REPLACEMENT BLADDERS

Each year an estimated 40,000 cases of bladder cancer are diagnosed in the United States. One fourth of bladder cancer victims, as well as some patients with genital (especially cervical) cancer, will require eventual removal of the bladder. Most patients have an ileostomy, a surgical procedure in which a section of the intestine is used to reroute urine from the kidneys to a plastic bag worn outside the body. Unpleasant in several respects, an ileostomy can also cause a number of complications. Scientists are therefore trying to improve on this method by creating artificial bladders.

Bioresorbable Bladders. Several years ago researchers at the Cleveland Clinic Foundation developed what might be called a hybrid bioreresorbable bladder. "It was intended to replace a part of the natural bladder, to enlarge it," says Andrew Novick, M.D., Head of Renal Transplantation at the clinic. The team removed pericardium from cows and treated it with acetic anhydride. The chemical modified the structure of the membrane to make it bioresorbable, but the researchers were unable to time the resorption to coincide with regrowth of the natural bladder tissue. "We tested this approach on six or seven humans," says Dr. Novick, "and while it caused no serious problems, it just didn't seem to enlarge the bladders all that much. It met with some success but didn't provide the definitive solution to the problem."

The Baylor researchers have worked extensively with both fresh and glutaraldehyde-treated human amniotic membranes as bioresorbable grafts for reconstructing blood vessels and the bladder, urethra, and ureters. In experiments on dogs, they have removed the dome of the bladder and covered the defect with a patch of amnion. Within a month, normally functioning bladders had regenerated; within 6 months, all layers of the bladders had regrown. "We saw no evidence of stone formation or leakage, and even more fascinating, there was no rejection of the human tissue," says Dr. Fishman. "We are presently instituting clinical trials and anticipate no rejection of the graft by human patients."

Artificial Bladders. The Department of Artificial Organs at the Cleveland Clinic Foundation is now working on a total artificial urinary system including a bladder and ureters made of silicone. "So far we've had problems with connections, valves, calcification and other obstructions, and bacterial infections," says Malchesky. "But we're making good progress."

To work effectively an artificial bladder would need tight connections to the ureters, with valves to protect the kidneys from the high pressure and bacterial infections that occasionally occur in the natural bladder. Its materials would need to preclude calcification and withstand pressure and stretching.

The drainage system would have to remove all urine to avoid the development of stones and infections. The patient would want to control urination and would need a feedback system to inform him or her when the bladder was full.

Jacob Kline, Ph.D., Chairman of the Department of Biomedical Engineering; Eugene C. Eckstein, Ph.D., Associate Professor of Biomedical Engineering; and Norman L. Block, M.D., Professor of Urology and Biomedical Engineering at the University of Miami, have made encouraging progress toward solving these problems. They have developed an artificial bladder made of polyether urethane (for mechanical strength) coated with hydrogel (to avoid calcification). Shaped like a domed soup bowl, the device is designed to minimize bending and pressure that could wear out these materials. A flexible diaphragm attached to the bowl expands away from this base as the bladder fills with urine. The device will be available in various sizes and will range in capacity from 500 to 600 milliliters.

The bladder is implanted in the pelvis after the biological bladder is excised. It empties by the combination of gravity with pressure of the abdominal organs and muscular tension in the abdominal wall. Drainage channels assure that it empties completely. The bladder is sutured to the natural ureters through a ureteral nipple attachment. An artificial urethra with a valve attached to the prosthesis provides a medium for voiding. The urethral valve is connected to a small bulb implanted beneath the skin in the pelvic area and activated when the bulb is compressed. "Initially the wearer will activate the valve to urinate four times a day," says Kline. "Later we may include sensors allowing him or her to urinate on demand." Meanwhile the wearer would be warned of a full bladder by small amounts of urine leakage. The system is designed to be fail-safe.

"We've tested components of the bladder in dogs over the past 8 years," says Kline. "Some of the initial ureters made from hydrogel have been implanted for as long as 3 years." In addition to working on the problems encountered by the Cleveland Clinic group, the Miami team is trying to improve their materials and find the best means of implanting the artificial urethra. They are also negotiating with manufacturers who might be willing to subsidize further research and support the clinical investigations required by the FDA.

Penile Prostheses

An erection not only represents sexual fertility, it also identifies its owner in his own eyes as a full man. A man suffering from impotence will find all areas of his life affected by the disorder, as its victims will attest. "Sex is a great tranquilizer, and being impotent means that your anxieties show up in your work, your family relationships, and your moods," says Jerry Melberg, a 42-year-old self-employed computer systems consultant from Anoka, Minnesota. The man's sexual partner may feel angry, hostile, and humiliated by his failure to perform;

the couple may have trouble discussing their feelings about the situation. "I wasn't prepared for impotence and wondered what was causing it," says Jerry. "It's very frustrating, and I kept making all kinds of excuses for it. My wife and I confronted it, but I imagine that it can cause all kinds of problems if there's a communications gap." Impotence often develops slowly, preying on a man's mind. "Our biggest problem was that we both got really depressed about it," says Bobby Roberts, a 41-year-old pest controller for the Pensacola Naval Air Station near his home in Cantonment, Florida.

At least 10 million Americans, or one out of eight men, are impotent. The problem is so widespread that in 1983 Bruce and Eileen Mackenzie of Chevy Chase, Maryland, started Impotents Anonymous and I-ANON, which now have 10 chapters nationwide.

The more complicated a piece of machinery, the more ways in which it can break down. The same holds true for the delicate hydraulic system that creates an erection. Two erectile chambers filled with spongy tissue run the length of the penile shaft. When the penis is flaccid these chambers are collapsed. During sexual arousal, blood rushes in and distends the chambers; valves in the penile veins close to prevent its flowing out. The resulting erection has been caused by erotic thoughts or tactile stimulation combined with certain chemicals like testosterone and neurotransmitters. With all these physiological and psychological stimuli factored in, it is hardly surprising that the desired response may be difficult to achieve.

It used to be thought that most impotence was psychological. Now researchers know that at least half the cases of impotence have any of several physiological origins. Medications, vascular conditions, illnesses, injury, or surgery can all contribute to impotence. The list of culprits includes alcohol and nicotine as well as prescription drugs for hypertension, depression, psychosis, abnormal serum cholesterol, weight reduction, ulcers, stomach disease, and cancer. Erections depend on blood flow, so atherosclerosis, Leriche's syndrome, sickle cell anemia, and hypertension can cause impotence. (One third of men who do *not* take medication for high blood pressure, however, likewise become impotent.)

Illnesses that cause impotence include diabetes and kidney disease (half of men with either disease are impotent), multiple sclerosis, and hormonal imbalances. Injury to the brain, spinal cord, or penis as well as surgery on the aorta, iliac arteries, prostate, colon, bladder, or penis can damage nerves or impede circulation, resulting in impotence. Of course, organically caused impotence is aggravated by the very psychological trauma it gives rise to.

Fortunately, various types of penile prostheses are able to restore potency and provide satisfactory sex lives to men who might otherwise be past hope. Like the artificial sphincter, a penile prosthesis should be implanted only after thorough physiological and psychological tests have eliminated other treatment possibilities. Tests of blood chemistry, nerve functions, hormone levels, and presence or absence of erections during sleep will inform the urologist whether the impotence is an organic problem that can be treated easily.

If implantation is called for, the patient must be free of infection and capable of operating the prosthesis. The physician must inform him fully of all risks and complications and modify any unrealistic expectations he may have. The patient should also be evaluated by a psychiatrist to uncover significant psychological problems and eliminate the possibility that other therapy might be appropriate. He and his wife should both meet with the urologist as well. "Some couples have major marital difficulties and should have counseling before the husband has the implantation," says Dr. Leach. "Nothing is more tragic than the reported cases in which divorce occurs after implantation because the situation was not well evaluated beforehand."

Psychologically impotent men can have a prosthesis implanted under appropriate circumstances. "We do implants on psychologically impotent patients," says Dr. Leach, "but only after an extremely thorough trial of psychotherapy has led one or more psychiatrists to state that they have exhausted the alternatives and the implant is a last resort."

The three types of penile prostheses are the semirigid, which comes in several models, the inflatable, and a new type that combines features of the other two.

SEMIRIGID PENILE PROSTHESES

A man with a semirigid penile prosthesis has a constant erection; some prostheses have hinges or flexible construction that allow the penis to be tucked against the body and concealed by clothing. The original semirigid prosthesis, the Small-Carrion, is now used only infrequently. More popular are the Flexi-Rod implant, which has a hinge like a gooseneck lamp, and the Jonas prosthesis, which has a silver wire core that bends at any point in any direction.

Implantation of a semirigid prosthesis is a relatively simple operation that takes perhaps 20 to 30 minutes under local anesthesia. If a week or so goes by without the recipient developing an infection, he will almost certainly not experience mechanical problems. The semirigid prostheses are also much less expensive than the inflatable model.

The major disadvantage with semirigid prostheses is the permanent erection, which is obvious whenever the man is nude. Wearing jockey shorts, a jock strap, or tight elastic underwear enables a man to conceal the implant under boxer-style shorts or bathing suits. Older versions of the Flexi-Rod bent during intercourse, but the newest model, which is much more rigid, has practically eliminated this problem.

After losing his wife, Irv joined a social organization for widows and widowers, began dating one of the other members, and opted to have a penile implant. "Dr. Leach and I decided that the Flexi-Rod would be best for me because I already have an implant in my scrotum and didn't want the additional potential complications of an inflatable prosthesis," he says. "I have about two thirds or three fourths of an erection. Occasionally it can cause a problem with bending, buckling, or slipping, so its success would depend on how cooperative a man's partner is. It's not something a bed hopper would want. Were it not for my sphincter implant, I would have chosen the inflatable type, but under the circumstances I can recommend the Flexi-Rod very highly."

Men with free choice often feel that the semirigid implants would be too uncomfortable and embarrassing. "They don't make the penis as rigid as I would like, and I wouldn't want a partial erection all the time," says Jerry. "I decided against a semirigid implant because I just couldn't picture myself spending the rest of my life with one," says Bobby. "I think that men are choosing them because they don't know about the inflatable type of prosthesis."

THE INFLATABLE PENILE PROSTHESIS

The most natural penile prosthesis available is the Inflatable Penile Prosthesis, created by Dr. Scott and first implanted in 1973. The idea came to Dr. Scott one day when he inflated an unrolled cuff on an artificial sphincter and noticed that it became long and stiff. The inflatable implant mimics the natural erectile system by operating hydraulically. It consists of two soft silicone rubber tubes that are implanted in the erectile chambers. These are connected by silicone elastomer tubing to a reservoir of saline or opaque solution, which is implanted in the lower abdomen, and a small bulb-shaped pump placed in the scrotum. When the wearer or his partner presses the scrotum, the pump forces the fluid into the tubes, which expand and distend the erectile chambers. To lose his erection, the wearer squeezes a valve on the pump, and the fluid returns to the reservoir.

More than 19,000 men have received the Inflatable Penile Prosthesis. They range from a 19-year-old motorcycle accident victim to an 85-year-old man; the median age range is 40 to 69. Over 600 medical institutions around the world now perform the surgery. Urologists report a 90% success rate; patients can have natural intercourse, including orgasm and ejaculation, 6 to 8 weeks after implantation. Several children have been fathered by men who were impotent before their surgery. Two of the happiest parents are Bobby and Miki Roberts, whose 2-year-old son, Bobby Jr., was the fourteenth American child to be fathered by an implant recipient.

The Robertses had been married a year when Bobby developed diabetes. They bought several books on the illness and would later recall indignantly that none of them mentioned that impotence was a frequent complication. During their first several years of marriage Bobby and Miki tried to have children, but although all their fertility tests were normal, they had no success. Then Bobby started to become impotent. "We talked about it and thought at first that it was just the way that men are supposed to go downhill sexually while women are going uphill," says Miki, 36, who works in contract sales of building materials. "I didn't believe Bobby at first when he said that the doctor had told him that diabetes had caused his impotence. We were just stunned."

Bobby offered Miki a divorce, which she refused. "I had a new job and just threw myself full force into that, resigning myself to a sexless marriage," she says. Meanwhile Bobby was being irritated by people ignorant of his impotence

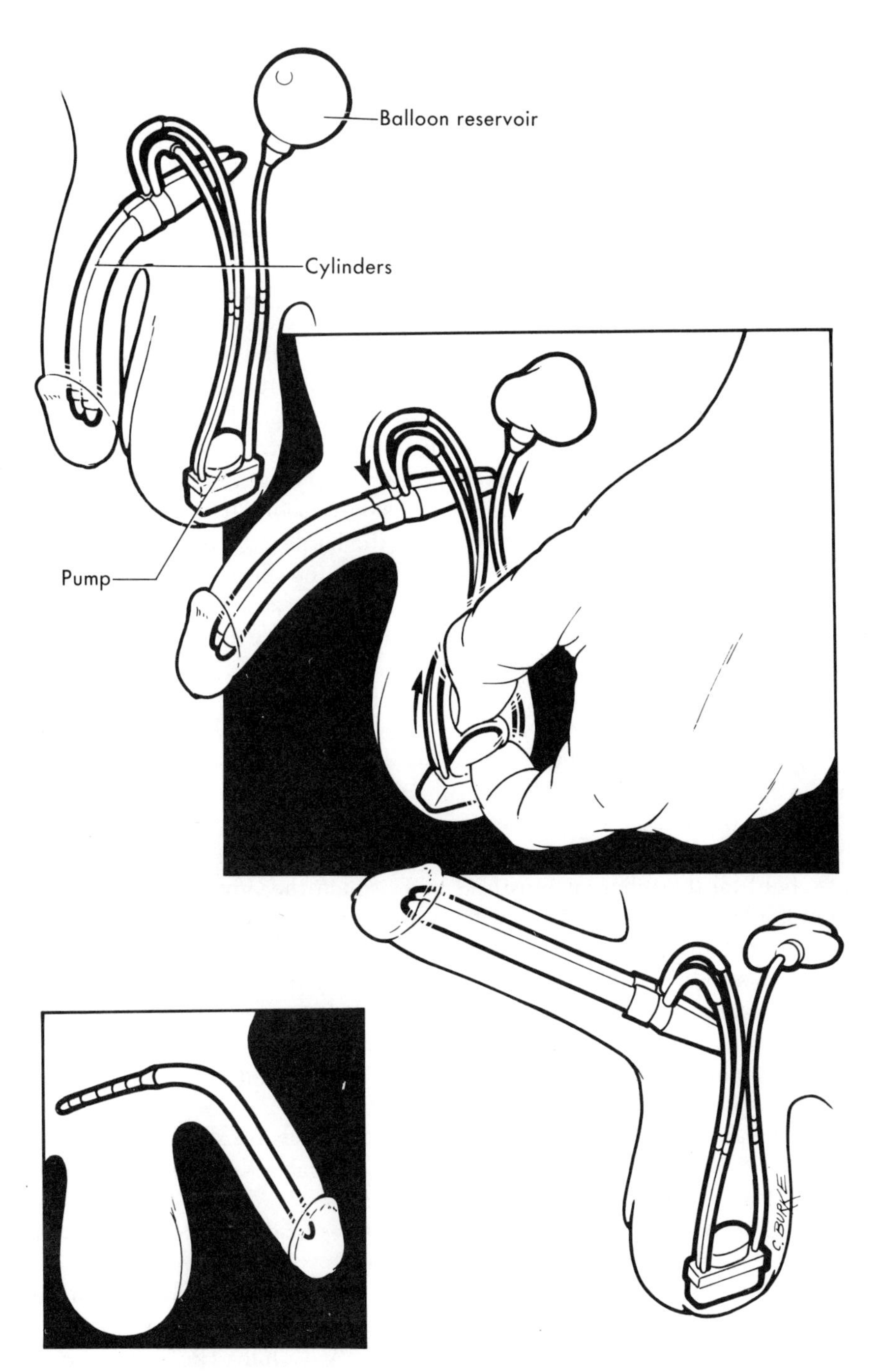

The Inflatable Penile Prosthesis being activated by the wearer to achieve an erection; a semirigid penile prosthesis.

who kept teasing him about not having children. "We were getting older; I was impotent, and we thought we'd never have kids," he says.

Then Bobby happened across an article by Dr. Scott about the Inflatable Penile Prosthesis. "I asked my diabetes doctor about it," he says, "but like most doctors, he hadn't heard about it. Doctors are still telling impotent men to live with their hard luck." But his physician also recommended, however, that Bobby have the surgery. "Bobby asked me what I thought about it," says Miki, "and I told him not to have it on my account, because years ago I had made up my mind to stick with him for better or worse." Having made his own decision, Bobby arranged with Dr. Scott to have the surgery done in Houston.

Like most inflatable implant recipients, Bobby spent a week in the hospital and 5 weeks recuperating at home. In the 4½ years since then, he has had no problems at all with the implant. The difficult part was readjusting to having sex again and deciding whether or not to continue trying for children. "I was ambivalent at that point because Bobby was a 38-year-old diabetic patient, and I was 32," says Miki. "The age factor made it the hardest decision I've ever had to make." Both Robertses were delighted when they learned that Miki was finally pregnant. "I haven't regretted my decision, because we have a precious baby," she says. "I'm grateful every day that Dr. Scott developed the implant, because our baby is the most important part of our lives."

The Robertses report having described the implant to several impotent men who were on the verge of suicide. "They told us that we don't know how much we did for them," says Miki. "Any impotent man of any age should consider having the implant—it's a shame that more people don't know about it."

Dr. Leach, of course, thoroughly educates both patients and medical students about all the different prostheses. "I explain the pros and cons of each type of implant and try to let the patient make the decision, mainly because I think that the inflatable may cause mechanical problems," he says. "The failure rate is supposed to be 15% to 20%, but I tell patients that eventually we may be operating again on everybody. We can't predict the future of a prosthesis implanted in a 30-year-old man." Although surgery for such repairs might not be troublesome, the initial surgery is quite painful and usually requires a general or spinal anesthetic.

The head of the penis, which is not inflated like the shaft, remains soft during erections. "This is just about the only observable difference in an erection resulting from an inflatable implant," says Mark, who obtained one after hearing about it through the media. "It makes no difference to me." Most implant recipients do not consider it a problem.

Although familiar with some of these disadvantages, Jerry is also enthusiastic about his inflatable prosthesis. Like Bobby, Jerry developed diabetes soon after his marriage to Jan, who works with him. He too was ignorant of the connection between his illness and the impotence that began about 4 years ago. "A registered nurse from the VA hospital came to talk at our diabetes group meeting about sexual problems caused by the disease," says Jerry. "She mentioned implants but didn't make clear that they were appropriate for many men, not just the handicapped, and that they were generally available." Again like Bobby, Jerry found out about the Inflatable Penile Prosthesis quite by accident. "A year and a half after that meeting I had lunch with a friend of mine who works for American Medical Systems, which manufactures the device," he says. "I asked him what the company did, and when he told me, I explained that I was impotent and got more information about the prosthesis." About a month later, Jerry went to the Mayo Clinic to have his implantation.

After enduring his first surgery, Jerry had to have the procedure repeated. "I wasn't prepared for the degree of pain I felt from the operation," he says. "I was instructed to pump up the device a few times a day right from the start, and that didn't help any." A year ago Jerry's prosthesis was replaced because the pump in the original one had failed. "That operation wasn't bad at all," he says, "and my full recovery time was half what it had been with the first implantation."

Other drawbacks Jerry mentions are that a single man who did not want his partner to know about the prosthesis might find the pumping and release mechanism awkward to disguise. Sometimes the prosthesis is not comfortable. "If I leave it pumped up too hard and use it an awful lot, I get a little irritation at the tip of my penis," he says, "but that might happen without the implant." "Some patients complain of various aches and pains in the area," says Dr. Leach. "A lot of it depends on how vigorous they are during sex."

The first time he tried the prosthesis for intercourse, Jerry was dubious—still a little sore, and not quite sure how well it would work. He quickly became enthusiastic about it. "I'm extremely happy with my prosthesis," he says. "It's a super product, and I really push it because it's so much like a normal erection that sex is better than ever before. I have no anxieties, because I can control my rigidity; I can perform anytime; I can even ejaculate early, and it doesn't really matter because I stay erect as long as I want. Sex is a big part of my life, and I would have been stupid not to have had the implantation."

Like Jerry, Mark had to have his inflatable prosthesis repaired twice since its implantation with the sphincter in 1980. In each case the cylinders gave out. "I had doubts about the initial surgery but absolutely none about the repairs," Mark says. "I'd drop everything and go to Houston to have Dr. Fishman fix the prosthesis. The procedure is quick; I don't need anesthesia, and in general the prosthesis seems to be very stable." Mark notes that other men whose implant cylinders fail should be prepared for the diuretic effects of the contrast solution. "You may want to take it easy for a day because your kidneys will be working overtime," he says, "but it won't be long before you urinate it all out."

Although he is paralyzed from the chest down, Mark was aware for a few months after his surgery that two foreign objects had been placed in his body. "It takes a while for an implant to become part of you," he says. "I think that women with breast implants would feel the same way. Once you get used to them, though, there's no doubt about their advantages."

SELF-CONTAINED PENILE PROSTHESES

Surgitek–Medical Engineering Corp. and American Medical Systems have recently brought out new penile prostheses that use slightly different mechanisms to combine features of the semirigid and inflatable types. The American Medical Systems implant, called the Hydroflex, looks like a pair of semirigid rods but actually consists of two inflatable cylinders with a new hydraulic mechanism. The reservoir and pump are both within each of the cylinders. Squeezing the front cylinder tips transfers fluid from the reservoirs at their bases through passageways to the head of the penis, then through inflation chambers in the middle of each cylinder. The cylinders become stiff, with some degree of expansion. The wearer deflates them by pressing the cylinders behind their front tips.

The self-contained prosthesis may be an option for a man who wants neither a semirigid type nor the more complicated, expensive surgery required for an inflatable implant. It can be implanted as an outpatient procedure. Surgeons who conducted the early clinical trials on the device report that patients consider it a reasonable alternative, but the long-term results are still unknown.

Like breast implants, penile implants are not suitable for patients who hope that a prosthesis will transform their lives. "I recommend penile implants very cautiously once we've ascertained that psychologically the patient is an appropriate candidate, that all other forms of treatment have not worked, and that the patient really understands what he's getting into and does not have unrealistic expectations of the surgery or the results he can expect afterwards," says Dr. Leach. "Many men think that they'll have stronger sexual desire, better performance, more enjoyment, and more appeal to the opposite sex. I try to make clear that the implant provides just a rigid penis for sex and has no effect on these other factors, although psychologically it certainly can be very beneficial."

Now that impotence is coming out of the closet, more men are learning that their condition may be treatable and if it is not, a penile prosthesis can help. Both physicians and patients need better education about penile implants. "There are 10 million men who are impotent, and half the doctors in Pensacola don't know what a prosthesis is!" says Miki. Bobby reports that most men who ask him for information about prostheses are afraid to have an implantation. "They've heard that with an implant they can't wear shorts or a bathing suit, or that their penises will hang out of their pants," he says. "I point out that they've got nothing to lose by trying an implant—that gets their attention, and they want to learn more."

The Robertses, the Melbergs, the Mackenzies, and others have committed themselves to educating others about impotence and the availability of penile implants. Readers who wish more information about incontinence, impotence, the artificial sphincter, or penile implants can contact a urologist or American Medical Systems, Consumer Information Department, Box 9, Minneapolis, MN 55440, 1-800-328-3881.

Other Genital Prostheses

A surgeon can reconstruct a penis that has been lost to cancer or injury by using a silicone elastomer prosthesis as a frame to support skin grafts. The implant is a half-cylinder with rows of holes (through which blood can flow to the grafts) and Dacron patches across its flat side (for tissue ingrowth). The device is attached to the pubis with two tails. The surgeon constructs a new urethra with a tubular skin flap, places the prosthesis over its upper half, and covers the entire structure with another skin flap.

Similarly, vaginal reconstruction can be performed with the use of a gel-filled silicone stent that is available in several sizes. The stent is wrapped in a skin graft and inserted into a tunnel cut between the urethra and the anus. When the skin graft has taken as the tunnel lining, the stent is removed.

Testicular prostheses similar in construction to gel-filled breast implants come in various sizes. They are implanted to reconstruct undescended testicles or those lost to cancer or injury.

Researchers have experimented with different methods of reversing vasectomies. One is to use silicone elastomer tubing to reconstruct the vas deferens. Another is to implant a sperm reservoir, a small silicone rubber cup in which sperm are collected. The physician inserts a needle into a flange on the cup to remove sperm for artificial insemination.

Although such genital prostheses are applicable far less often than are penile or urinary implants, they likewise contribute considerably to the quality of their recipients' lives.

SECTION

VIII

IN VIVO LABORATORIES AND OTHER BIONIC SPIN-OFFS

25

The $100,000 Man

The patient being readied for surgery at the USC Medical Center would be described today as pure beefcake, but a decade ago that term had not yet come into use. Although in his mid-thirties, and therefore over the hill by Hollywood standards, he looked tall and well-built, with light brown wavy hair, blue eyes with thick lashes, perfect features, and full lips slightly parted to reveal nice teeth just irregular enough to be natural. He was not tanned, but he looked healthy and seemed relaxed as he lay on the table, his left arm extended with an IV shunt in it.

Two anesthesiology residents stood by observing a third who bustled around the patient, taking his blood pressure from a cuff on his right arm, checking his pulse rate, and reading the meters, dials, and gauges on the machine standing nearby. The resident carefully fitted a mask over the handsome face and began administering oxygen. Raising the oxygen level in the patient's tissues would give him a margin of safety during those stages of the procedure when he would be briefly deprived of it.

After 5 minutes the anesthesiologist rechecked the blood pressure and pulse, then gave the patient a test injection of sodium pentothal through the IV portal. Noting no change in the readings, he injected another, larger dose of the drug. The patient's eyes blinked a few times, then closed. The next step was to inject succinylcholine, a muscle relaxant that paralyzes every part of the body except the heart. The muscles of the patient's chest rippled, indicating that it had taken effect. The patient had ceased to breathe. The two residents watched attentively.

The silence was broken by burping as the patient began vomiting. Apparently he had disobeyed instructions not to eat before surgery. A foul odor wafted from the table. The two residents shuffled their feet, grimaced, and glanced at each other. "What in hell did he eat—a rubber pizza?" one of them asked. The third resident did not respond. The patient's carbon dioxide levels, pulse, and blood pressure were rising; if not given oxygen as soon as possible, he would die.

The anesthesiologist removed the mask, grabbed a suction line, and quickly vacuumed the patient's throat. Satisfied that it was clear, he inserted a laryngoscope to keep the patient's tongue out of the way and reveal the opening to the trachea. He grabbed an endotracheal airway and deftly inserted it through the opening and into the trachea to a point just past the vocal cords.

This was followed by a bite block to protect the airway. The anesthesiologist's movements were smooth and skillful, but droplets of sweat beaded above his brows. Inside the patient's throat a small cuff was inflated, filling the space between the airway and the tracheal wall. The patient's lungs were now sealed off completely, protected from any gastric juices that might otherwise drain into them. The resident connected the airway to the anesthesia machine, turned on the oxygen and nitrous oxide, and began hand ventilation by squeezing an inflated bag.

His audience sighed with relief. The whole delicate, complicated emergency procedure had taken less than a minute. "*Nice* going," one of the observers said, the other murmuring in agreement. Neither had attempted to help.

From this point on, the anesthesiologist was actually breathing "for" the patient, carefully monitoring the mixture of gases that entered his lungs. He lifted his glance to the mirror on the wall opposite him. "Some doctor," he thought. "I look like a ghost."

Abruptly the patient went into what looked like mild convulsions. He was bucking: his trachea was reflexively trying to cough out the airway. This meant that he was not fully anesthetized. The resident reached to the machine and increased the nitrous oxide dosage.

The bucking stopped.

So did all the vital signs.

The anesthesiologist was finally caught off guard. "What?" he cried, staring at the uncooperative patient in outraged disbelief. "*WHAT?*" His audience looked surprised but remained silent.

A door next to the mirror opened, and a man's head poked out. "Sorry about that," he said. "I just realized how late it is, and I've got to get going." The head disappeared.

The two residents grinned at their companion, stretched, and began murmured small talk. The third, still stunned, glared at the corpse. Beyond the open door a machine chattered. The anesthesiologist recovered himself. "Will the printout show that you terminated?" he called into the room. "I'm noting it right now," the man's voice called back. "By the way," the head reappeared in the doorway, "your last dose of nitrous oxide was exactly right. That's listed." It disappeared again.

"Thanks," said the resident. He pulled up the sheet to cover the patient's face. "Lotta good it did him."

Before legally trained readers hurry off to consult Lexis or reach for their telephones, we must note that the deceased was no ordinary patient. He was, in fact, a Lazarus of anesthesiology, someone who could die repeatedly and be resuscitated with the push of a button. He was Sim One, a remarkably lifelike manikin who reacted to medical procedures just like a human being.

As noted earlier, the terms *bionic* and *bioengineering* can mean the application of engineering principles to medicine *or* that of biological principles to engineering. Sim (short for simulator) One is an example of the latter. His mechanical parts substitute for natural ones not to help a living person but to replace that person entirely for educational purposes.

Anesthesiologists perform an unobtrusive but crucially important function in surgery. "Anesthesia is basically controlled poisoning," says Peggy Wallace, Ph.D., Instructional Media Specialist at the USC School of Medicine. "It's an absolutely critical task to keep the patient at a level of unconsciousness so that the operation can proceed without pain but at the same time not kill him or her." Unlike surgeons, anesthesiologists cannot learn their skills by practicing on cadavers: the course of anesthesia is determined entirely by how the patient reacts to the drugs and gases administered, and cadaver tissues are different from living ones.

This difference is particularly noticeable to someone performing endotracheal intubation, the process that enables the anesthesiologist to control the administration of oxygen and anesthetic gases. As seen from the example, the endotracheal airway must be inserted quickly and skillfully, a feat that requires considerable practical training. Mistakes can seriously hurt the patient.

Conventional training in endotracheal intubation involves practice on a live patient with an instructor standing by, ready to shove the anesthesiology resident aside and take over in case of emergency. The creation of Sim One, however, made it possible for students to practice simple, complex, and emergency anesthesia procedures over and over. Having proved themselves proficient, the three residents we just met were allowed to make appointments to practice and observe endotracheal intubation with just a technician present to control Sim One's responses. They had reached this point after several practice sessions with an instructor evaluating their skills.

Simulation is obviously useful for teaching all kinds of medical procedures. Many static simulators are used in medical schools, but there are very few interactive ones like Sim One in existence, probably because of their cost. Two of Sim One's younger cousins are interactive in the sense of being bionic creatures that mimic human functions. At the University of Miami School of Medicine, "Harvey" can simulate 20 cardiovascular disease states. His heart sounds vary according to where on his body a stethoscope is applied; they are synchronized with an ECG so that the student can also see what he or she is hearing. His creators plan to endow Harvey with the ability to age, showing the results of chronic heart disease.

At the University of Kentucky, "Hera" repeatedly gave birth to a fetus that had heart sounds and an umbilical cord attached to a placenta. Students could palpate her abdomen to check on the progress of labor and reach inside her vagina to examine changes in her cervix. Hera delivered 71 times before her worn-out vaginal lining began to interfere with the birth process. Like Harvey's, her interactive capabilities were extremely limited compared to those of Sim One.

Sim One began in 1964 at a three-martini lunch. "The original idea came from a conversation between me and some nut who turned out to be a really active, imaginative, totally altruistic fellow," says Stephen Abrahamson, Ph.D., Director of the Division of Research in Medical Education at the USC School of Medicine. The "nut" was an engineer named Tullio Ronzoni, who came to Abrahamson's office to discuss the unusual idea of using computers in medical education. Having been given a runaround by other staff members at USC, Ronzoni told Abrahamson that if they had nothing to discuss, he was taking his idea to UCLA. "Oh, no," said Abrahamson, "you *can't* do *that*."

The two men came up with some preliminary ideas, then arranged a lunch meeting with the late J. Samuel Denson, M.D., of the Department of Anesthesiology, and some engineers from Aerojet General Corp. After several martinis, all the participants fell in love with the possibility of building a manikin to train anesthesiology residents.

The NIH loved the concept also but would not fund it. "I got angry and depressed, so I rewrote our proposal and submitted it cold and blind to the United States Office of Education," says Abrahamson. "In 1964 we were on the verge of Johnson's Great Society, and damned if they didn't approve the idea and give us $272,000."

When Sim One was unveiled in 1967, USC made a splash that rivaled the glories of its football team. Abrahamson enjoyed active support for the project from the assistant secretary's office in the Department of Health, Education and Welfare. In 1969, however, changes in the Washington administration resulted in a loss of funding. To increase his cost effectiveness, Sim One was modified so that he could be used to train other medical personnel.

Sim One's creators had been Abrahamson, Dr. Denson, and Ronzoni; A. Paul Clark, Leonard Taback, Harry Loberman, and Hank Perez of Aerojet General; and John Alt and Don Carter of the Sierra Engineering Co. Their creature was something to be proud of. In his final form, Sim One was 6 feet tall and weighed 195 pounds—so the engineers could fit the necessary equipment inside him.

Sim One lay on an operating table beneath which was the rest of the equipment that kept him functioning. His left arm was extended and prepared for IV injections, and his right arm at his side was fitted with a blood pressure cuff. Designed to take intramuscular injections, this arm jerked if the needle was off target. Blood could be extracted from it, and a vein could be catheterized to measure central venous pressure.

A stethoscope was taped to Sim One's chest over his heart. He had realistic plastic skin and classic features. His mouth opened and closed with power assist, and like his throat it contained all the correct anatomical structures. Four of his upper front teeth could pop out if too much pressure was applied to them. His eyes blinked. Their pupils constricted and dilated realistically, using small black rubber cones that were drawn back or pushed forward to flatten against the inner eyeballs.

Sim One breathed visibly: his chest rose and fell, and his abdomen swelled if inflated with air. He could imitate apneic and depressed as well as normal respiration. Each type of breathing could be assisted by mask and bag or by attachment to a ventilator. Sim One had a heartbeat with temporal and carotid pulses and blood pressure that could be measured using auscultatory or palpation methods. He could simulate circulatory failure or shock by showing appropriate changes in central venous or blood pressure, heart rate, and respiration, and he returned to normal when given fluids. He responded to oxygen and nitrous oxide administered through a mask or tube and to 10 drugs administered intravenously. All his responses occurred in real time, forcing the student to act as quickly as he or she could in an actual emergency.

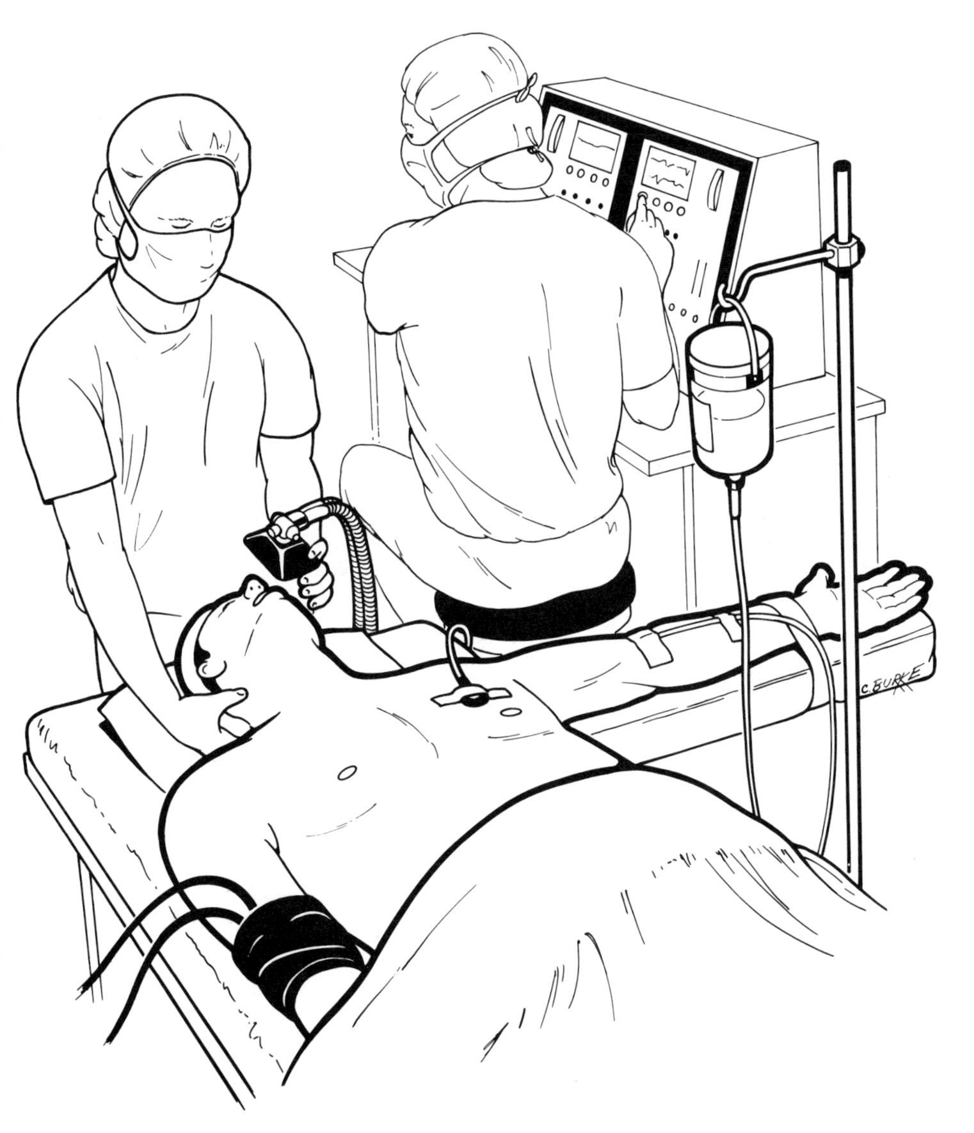

Sim One about to undergo induction of anesthesia.

The total Sim One system included the manikin and table, anesthesia machine, interface unit, computer, and instructor's console. Sim One was operated by a Honeywell minicomputer and recorder that registered all of his variables. A computer printout listed every detail of each procedure applied to him. All the anesthesia equipment (the machines, gas canisters, mask, bag, and blood pressure cuff) was modified so that the computer could determine exactly what the student was doing, read that against its standard, and make the manikin respond appropriately. "There were an infinite number of things the student could do to Sim One," says Wallace. "The computer was so finely tuned that it could respond to half a cc difference in drug injection. This closed loop system is state of the art even today."

The instructor's console, however, was really the source of all the fun. Between viewing the procedure through a two-way mirror and checking the console, the instructor could confirm details such as whether Sim One's lips had been pinched or his mask was fitted improperly. By pressing a hold button, he or she could put Sim One in suspended animation, then either continue a procedure or start over. With a mere twirl, push, or switch, the instructor could increase or decrease Sim One's pulse, respiratory rate, blood pressure, and jaw tension. He or she could induce vomiting, bucking, ventricular fibrillation, arrhythmia, cardiac arrest, right or left bronchus block, and laryngospasm—then bring Sim One right back to life.

In 1974 Abrahamson estimated that it would cost $100,000 to purchase a Sim One system if they were being produced. Sim One could now cost five to seven times that much, plus additional annual maintenance costs. He could well earn his keep, however, by teaching respirator application, intramuscular injection, recovery room care, pulse and respiration measurement, and endotracheal intubation in addition to the procedure for which he was created, induction of anesthesia. He could be used to train anesthesiology residents, interns, medical and nursing students, registered and licensed vocational nurses, inhalation therapists, and paramedics.

Sim One was built as a demonstration rather than a working model, but he lasted almost a decade. He began wearing out as funding got more and more scarce. The USC group had fulfilled their government contract of demonstrating the feasibility and cost effectiveness of Sim One and hoped someday to get money to fix him. Then something ridiculous happened.

Sim One was being housed in a barracks across the street from the Department of Medical Education. The hospital decided to take back the barracks and turn it into a headquarters for their chaplains. While Abrahamson searched for another place to store Sim One, workmen were sent in to refurbish the barracks. They tore down a wall, came through, and found Sim One lying on his table surrounded by hypodermic needles and little vials marked "sodium pentothal," "succinylcholine," and so forth. The workmen called security guards, who arrived at the scene and declared that needles and drugs could not be left lying around. Instead of contacting Abrahamson, a security officer called the hospital pharmacist, who said, "Bring all those things to me immediately, and I will destroy them." The paraphernalia were delivered. The pharmacist poured the contents of all the little drug vials (distilled water) down the sink and broke up the hypodermic needles.

Those needles had been magnetically coded. The IV portal into which they were inserted had a coil that recognized which drug and how much of it was being injected. The needles had been manufactured by a West German firm that no longer existed. "Need I say more?" asks Abrahamson. "Sim One hadn't worked well for the past year, so instead of raising hell, I just said that the mistake was too bad." The department had Sim One's left arm sawed off to make him more compact and stuck his carcass into a room in the basement. He can still be found there, lying in state among old, broken audiovisual equipment.

In 1975 a proposal for a Sim Two was approved by the federal government but never funded. Sim Two was intended to be a dramatic improvement over his predecessor, and today's microprocessor technology would make him even more spectacular. Imagine him as he was planned a decade ago, the ideal device for teaching emergency care. All the machinery and electronics are inside his body, so with his portable computer nearby, Sim Two can cause an emergency anywhere he can be carried and dropped. The student called to the scene is confronted with an unconscious patient and must identify the problem—again in real time. Sim Two might have any of several types of shock, different types of respiratory derangement or various arrhythmias, cardiac arrest, drug toxicity, abdominal or head trauma, diabetic coma, or the aftereffects of drowning or electrocution.

In addition to inheriting the talents of Sim One, Sim Two has skin, mucous membranes, and fingernails that change color. Depending on the amount of anesthesia he receives, his left fingers open and close with varying degrees of tension, the skin on his brow wrinkles, and his eyelids close rapidly or slowly. His eyes can rotate and move independent of their pupil dilation or constriction. If his head and jaw are improperly positioned after anesthesia is started, his tongue blocks the airway, a problem detected by a sensor in his esophagus. His trachea and either or both nostrils can be blocked. His mouth and throat secrete saliva and mucus and bleed if traumatized. During severe respiratory obstruction, the skin between his ribs retracts; his lungs and chest wall have various degrees of compliance. He responds to cardiopulmonary resuscitation and a larger number of drugs; the latter are coded by pH rather than by magnetic needles.

There are always disadvantages in trying to alter an established curriculum, especially if the new element is extremely sophisticated. A simulator is no exception. Its initial and maintenance costs are high. Instructors must be motivated and taught to use it, and it must work well to maintain their enthusiasm. If its mechanism is faulty, students could learn procedures incorrectly.

Nevertheless, there are excellent reasons to build a Sim Two. Anesthesiology residents trained on Sim One learned more thoroughly in half the time compared to a control group of residents trained conventionally. Sim One also had highly effective results in training medical personnel in other procedures for which he was adapted.

The students liked Sim One very much once they became accustomed to him. "They would come in uncomfortable, embarrassed and making jokes about being Dr. Frankenstein," says Wallace. "But as soon as they got into the procedure and the manikin responded in real time, their responses became equally authentic. They would get nervous; their hands would shake; and if the manikin came out of anesthesia, some of them would panic. And there would be no expert to push them out of the way and take over."

Sim One shows that bionics need not be applied directly to patients in order to benefit them. Given a choice, probably no one would elect to be used as a guinea pig to train students during surgery or in an emergency. Learning skills on live patients not only endangers them but also decreases students' respect and compassion. With a simulator, a student need not approach a live patient until he or she is thoroughly trained in the required procedure. The simulator is always available. Its responses to standard stimuli can be varied and its case made instructive or unusual. It is more interesting and involves students more closely than anything short of real life.

A simulator can save faculty and student time and ultimately decrease the costs of education. It can teach more students, especially since it is easily moved from place to place. It provides direct feedback plus a way to standardize complex tasks to evaluate student performances.

With a simulator, students learn from their mistakes more quickly. They can focus on a step and repeat it over and over. Real time can be compressed to show them the results of chronic conditions, but they can also work against it while handling emergencies. Instead of encountering emergencies before they are well trained to handle them, as often happens in real life, students can work up to increasingly complex situations. "It's like a pilot-training simulator—you get graduated experience, plus when you 'crash,' you just start over, and nobody dies," says Abrahamson. "I think that this value is absolutely enormous."

Sim One is remembered fondly by the Medical Education staff at USC. "We had a picnic with him," says Abrahamson. Although the carcass lies in the basement and its successor does not yet exist, Abrahamson believes that manikins will be part of the future in medical education. "There's no question in my mind, especially with improvements in computers, that we'll go toward simulation," he says. "I may not be the lucky one to find the money for development, but somebody will, and more power to him or her. It's a great way to go."

26

The Multiprobe and Company

Sim One has not been the only project that demonstrates how ingeniously biological principles can be applied to engineering. Other devices justify bionics as a legitimate field of endeavor that results in useful spin-offs as well as practical devices. Some intriguing examples of both results are discussed here.

An Artificial Nose

We have already seen researchers hard at work trying to make practical substitutes for sight and hearing. Scientists have been able to duplicate other senses as well, although not necessarily for use in the human body. A Japanese company has marketed a transducer for odor, and work on odor sensors was presented at the 1985 meeting of the American Society for Artificial Internal Organs by an Italian investigator. Collins likewise decided one day to make an American artificial nose.

Enjoying a place of honor in its own laboratory at Smith-Kettlewell, the nose looks more like ordinary machinery than like an appendage with which a human might identify. At its sniffing end is a small vacuum pump that draws in air. The air sample is run past three pairs of thermistors, each five thousandths of an inch in diameter. One of each pair is coated with a thin layer of material. As this material adsorbs an odorant, the condensation produces a slight temperature rise on the coated thermistor. By detecting such changes in fractions of thousandths of a degree, the device can measure that an odor has been adsorbed.

Collins discovered that the nose could possibly discriminate odors simply on this basis. Different materials are capable of various amounts and types of adsorption. "I used materials I thought might possibly be in the olfactory epithelium—a fatty material, a protein one, and a long chain simulating a carbohydrate," says Collins. This gave him three different adsorptions for each odor. "They formed a signature," he says. "The ratios of these to their sum is different for every material, giving me a unique means of identifying odors." Collins holds a patent on the method.

The signal output goes through a recorder to identify the scent. Each odor can be plotted in two dimensions and falls into a characteristic place. Interestingly enough, most of the odors that people find similar fall into well-defined groups (fruity, woody, herbal, spicy, and so on) when plotted on a graph. Collins has hooked up the system to a microcomputer that can speak the name of the odor: sniff, sniff—"rose." The system works as promptly and accurately as the human nose.

"In addition to replacing lost sensory capability, there are many other applications for a device that can sensitively discriminate odors," says Collins. The nose could be used in several industries to confirm the freshness and quality of foods, for instance, or to analyze raw materials and match batches of chemicals or perfumes. Police could use an artificial nose to screen for explosives and drugs at airports and to detect alcohol and drug intoxication. In a medical center it could aid in some types of diagnosis, while in the home it could alert residents to poisonous food, gas, or smoke. "To replace the sense of smell itself is really an esthetic consideration," says Collins. "You'd have to pick one of several possible ways to convey the information to the user—put it on the skin or use colors or sounds as analogs of odor." Collins suggests coding subtle differences in scents as a series of different notes and calling the results "smellodies."

The Artificial Mouth

Another device not meant for use in people that nevertheless serves a valuable purpose is the artificial mouth developed at the University of Minnesota, Minneapolis. Called ART, for artificial resynthesis technology, the mouth was built by William Douglas, D.D.S., Ph.D., Director of the Biomaterials Program in the School of Dentistry; Ralph DeLong, D.D.S., Ph.D., Research Associate in Biomaterials; and Neal Peterson of MTS Corp.

The artificial mouth was created when Dr. Douglas built a servohydraulic machine and got it to reproduce jaw movements. "I think it's still the current view that the movements of the jaw and the environment of the mouth are so complex that they can't possibly be resynthesized," he says. Dr. Douglas attributes this view to general ignorance of servohydraulics. A servomechanism is one with a feedback system for correcting errors; in the artificial mouth a hydraulic pump is controlled so that loads and positions can be maintained very accurately. "We made it for the pure fun of building an artificial mouth," says Dr. Douglas. "Then we had to come up with an environment—put the cheeks around it."

That environment consists of an acrylic chamber holding four fluid jets trained on a few teeth mounted in a rubber base. The system is connected to a computer that uses large amounts of software to process the resulting data.

The researchers can equip the mouth with natural or artificial teeth. Usually they test one upper and one lower tooth at a time, although they can put in a partial denture and simulate full-arch chewing. The machine is unique and state of the art in being able to duplicate the major movements of the lower jaw. The researchers can torture the teeth with any amount of pressure from 0 to 100 pounds. To simulate the whole oral environment, they can inject any fluid from water to natural or artificial saliva and use it to produce extreme temperature changes.

ART has great flexibility. It can produce long-term enamel deterioration by among other things controlling the symbiotic effects of corrosion, stress, and bacterial growth. The teeth can be subjected to a year's wear in a single day. The device starts by tracing the tooth surfaces in 50-micrometer steps both before and after the chewing period. Both tracings are then overlaid on the computer to determine the amount of wear. The researchers can look at a tooth from any angle and usually take about 40 profiles of its surface.

As noted earlier, dental materials are becoming more competitive and topical especially now that people are keeping their natural teeth longer. The artificial mouth can test various restorative materials and discriminate between them on the bases of their friction coefficients and wear rates.

Dr. Douglas, however, is more interested in its other applications. "We're trying to get smarter about occlusion: tooth shape, meeting, movement, grinding, friction, and forces," he says. "We're interested primarily in jaw movements and ultimately in analysis of the joint. Nobody has really brought high science to bear on temporomandibular joint problems even though they've become a real focus." Another important application has to do with clinical research in dentistry, which involves many patients and is difficult, long-term, expensive, and qualitative. "It's very difficult to carry out clinical measurements," says Dr. Douglas. "What most dentists, and doctors too, depend on are clinical impressions. We want to bring in a new element of quantitation, including changes in the shapes of biting surfaces, which nobody knows how to measure."

To carry out these plans, the group is developing a multistation ART that will handle four tooth setups at once rather than just one to test different materials simultaneously. The computer's capabilities are being expanded and will provide three-dimensional graphics of how a tooth is changing. "We'd like to master the problem of synthesizing the human masticatory cycle and function and also to become very good at clinical measurement," says Dr. Douglas. "If we can put those things together, then we can make some real strides in restorative dentistry."

The Multiprobe

It is hard to believe that a high technology product can be unique because it has been hand crafted. This is the case, however, with the multiprobe, a miniature laboratory whose microscopic components are new tools for measuring chemicals. The multiprobe principle combines two technologies (those of solid-state integrated microcircuitry and electrochemistry) and two size standards (those of pencil-sized ion-sensitive electrodes and pinhead-sized solid-state electronics). It is the work of several researchers here and abroad, including Imants Lauks, Ph.D., of the University of Pennsylvania; Kenneth Wise, Ph.D., of the University of Michigan; and Jiri Janata, Ph.D., Chairman of the Department of Bioengineering, and his group at the University of Utah.

The multiprobe is 1 millimeter wide. It can be run through a catheter the width of a spaghetti strand and inserted into a blood vessel or interstitial space. The probe relies on microscopic chips, 300 of which can fit on a 2-inch wafer, that carry ion-sensitive field effect transistors (ISFETS). Coated membrane patches are applied to these chips. Individual chemicals alter the electrical currents that travel through the transistors; these signals are amplified and processed by circuits on the chips. The data are collected and stored by a computer and may someday appear on the screens of physicians' desktop models.

The measuring principle of the multiprobe is similar to that of an ordinary ion-sensitive electrode, but it offers significant advantages. In the past, chemical analysis required separating fluids into their various components. The multiprobe eliminates that step because each of its membrane patches reacts to a single ion: sodium, chloride, potassium, hydrogen, calcium, or other substances. The present model can sense two ions; Janata's group has designed a probe that can measure eight. In principle, 10 to 20 membrane patches could fit on a single chip. The real breakthrough, however, is that the multiprobe requires only a very small measuring space and volume of sample.

The Utah group began developing the multiprobe in 1974. Looking for new areas of electrochemistry in which to work, Janata was receptive to the idea of combining that field with solid-state electronics. "Electrical engineers were making a fundamental error in stating that these probes could be used without reference electrodes," he says. "I knew that it couldn't be done, but by the time I had fully formulated my reasons several months later, we were so deep into and fascinated by this area that we just stayed here. It was absolutely crazy—we had no right to be doing this."

The multiprobe is intended for in vivo measurements (those taken within the body) rather than implantation. In vivo measurement occurs over a limited period of time under the close supervision of technical personnel, while an implant is a permanent addition to a person's everyday life. "I don't see the multiprobe becoming implantable, because it's an electromechanical device that works by closing an electrical circuit," says Janata. "If for any reason the circuit is interrupted, the information is immediately lost. It would be practically impossible to calibrate the probe in vivo."

If the circuit were part of a feedback loop for drug delivery, for instance, sending the actuator into a false operative mode could be lethal. For that reason, Janata is convinced that an in vivo sensor is unlikely to be used, as one might expect, to monitor glucose. While hoping to get faster and more accurate information from the multiprobe, he sees it as a trade-off compared to the self-calibration of standard laboratory equipment. "We must perform an act of faith," he says, "and believe that the probe will respond and the baseline remain the same during the relatively long period of time we would be measuring."

Researchers could have many opportunities to test the reliability of the multiprobe. Its main application would be in medicine, where small space and sample volumes are critical and where standard laboratory analysis is not always optimal. Blood drawn for analysis can easily change composition even before it reaches the laboratory; it is tested in discrete samples, and each procedure uses valuable time. With the multiprobe, however, rapidly changing blood chemistry levels of surgical or critical care patients could be monitored continuously with the long-term data stored in a computer. Physicians testing patients in general could compare their blood chemistry levels at different times, such as during exercise and at rest.

The multiprobe can be designed to measure substances other than ions. "We have made a probe for glucose and one for penicillin, more or less to demonstrate the principles," says Janata, "but there are other enzymatic systems that could be used in exactly the same mode." Enzymes could be incorporated into the sensing membranes. The multiprobe could also monitor electrolytes during dialysis to confirm that their balance has been restored by treatment. "Many materials, packaging, and safety issues must be addressed first," says Janata. "There's a long road from research to clinical trials."

Of course the multiprobe could be used more aggressively to measure chemicals outside the body. "Fluids like sweat, tears, or saliva, which are not available in large quantities, have been neglected as diagnostic sources," says Janata. Analysis of sweat, for example, can lead to early diagnosis of cystic fibrosis. A specialized dual probe sensitive to sodium and chloride ions can be applied to the skin of newborns less than 3 weeks old to detect the disease. This procedure is much faster, cheaper, more reliable, and more efficient than the standard test. The probe could also be used in the laboratory to analyze whole blood, plasma, urine, and other body fluids.

The multiprobe has broad nonmedical applications as well. In industry it could analyze electrical and chemical solutions and monitor them during production. The space program is obviously not a big market, but it does demand tiny, lightweight, energy-efficient devices; the multiprobe is almost tailor-made to monitor the metabolisms of astronauts. It can also contribute to chemometrics, a modern statistical technique that is developing very rapidly. Chemometrics allows statisticians to manipulate and use imperfect data by measuring various channels from different angles. Equations are solved in a multidimensional space. The multiprobe is ideal for chemometric information acquisition.

Even a speck of machinery like the multiprobe has room for its share of bugs. It poses problems of long-term stability, electrical safety, and biocompatibility. We already know that when blood touches anything foreign, clotting usually follows. If placed in the bloodstream, the probe could stimulate clot formation. Janata's group has tried placing it in a high blood flow area to minimize clotting. Using a very careful surgical procedure, they have also inserted the probe into tissues so that it measures from interstitial fluid without contacting the blood itself. "This problem has not been resolved," says Janata. "But changes that occur in blood from touching the multiprobe are still less than those that result from removing blood and sending it to a laboratory."

The major hurdle to be overcome is that of encapsulating the multiprobe. The chips themselves are made using high technology production methods at a cost of about 50¢ each. In Japan they are offered for $180 apiece fully encapsulated. Each chip combines two basically opposite areas, chemical sensing and electronics. The present multiprobe has two little windows with ion-sensitive membranes whose chemical treatment must be applied by hand under microscopes. To make matters worse, the membranes must be exposed to fluids to measure ions, while the rest of the chip consists of solid-state electronics that cannot get wet. The encapsulation process that would expose and protect appropriate sections of the chip should result from industrial, rather than academic, research and development. "These chips won't sell until industry takes an interest in them and develops automated encapsulation techniques to bridge this gap in their fabrication," says Janata.

While industry drags its feet, Janata's group is making multiprobe chips and distributing them to the public worldwide at a nonprofit price—$6.50 apiece, including shipping and handling. "We can produce 100,000 chips a year," he says. "We'd like to see more people using multiprobes in their labs for research."

The group is also working on multisensors for gases that could theoretically deliver as many as 600 pieces of information. The encapsulation problems are much simpler, and their selectivity would be useful in industrial control. "We're very optimistic that ion and gas sensors will lead to something important," says Janata. "Computer processing capacity is expanding, and physical and chemical sensors will give us new information to use."

Electric Injections

Several years ago Dr. Stephen had his eardrum anesthetized with lidocaine (Xylocaine) that was administered electrically. He was so impressed with the procedure, called iontophoresis, that he described it to Jacobsen, who began developing the technology for use in skin anesthesia. The result was a small portable unit that gives electric injections.

Iontophoresis is the electric passage of an ionized drug through intact skin. An electrode connected to the machine is glued on the skin, with a second electrode serving as ground so that the current loops in and out through the body. A drug is injected into a syringe entry port. When the electrons come along and hit the fluid, the drug is ionized and completes the circuit. It is driven through the sweat glands into the skin, leaving bacteria behind. So far researchers have been able to inject drugs 2 centimeters below the skin surface.

There are many applications for electric injections. Physicians treating the U.S. Olympic wrestling team and several of the National Football League teams have used iontophoresis to deliver steroids to inflamed tendons and joints. It can anesthetize the skin for minor surgery; incision and draining of abscesses, boils, and carbuncles; insertion of large IV cannulae; removal of warts and splinters; or excision of lesions. Using iontophoresis on postoperative patients results in greater pain relief with lower doses of morphine. It can help treat skin cancer and herpes and may even be used someday to deliver insulin.

Like regular injections, iontophoresis delivers drugs systemically, with the starting point getting the biggest hit. It eliminates the pain of needles and the risk of tissue damage or infection.

Artificial Gills

Since 1976 Joseph Bonaventura, Ph.D., Co-Director with his wife Celia Bonaventura, Ph.D., of the Duke University Marine Biomedical Center, has carried around an artificial gill in his pocket. The gill, a reddish brown, spongy material resembling seat cushion foam, is a mixture of Hypol, a polyurethane cured with water, and Bonaventura's own blood.

The plastic foam, called a Hemosponge, can remove oxygen from either air or water. The sponge is saturated with heme, the oxygen-binding ingredient in hemoglobin. When seawater is pumped through the sponge, the heme protein binds with the oxygen molecules in it, attracting them 60% to 80% as efficiently as it does in the bloodstream. The oxygen must then be released from the sponge by vacuum, electromechanical shock, increased acidity, or addition of another gas. The Bonaventuras are making Hemosponges from a different substance now being patented with which they anticipate using synthetic heme or a heme analog.

The researchers envision a diver wearing two Hemosponge tanks. One tank will discharge oxygen into a breathing bag mounted on the chest while the other is recharged with oxygen from the seawater. A Sodasorb canister will absorb carbon dioxide from the diver's exhaled air as it passes back into the breathing bag. "Along similar lines of research, we anticipate having an 'artificial fish' rather than just 'artificial gills,'" says Bonaventura, "in that it's a continuous cycling system that loads oxygen from the seawater and unloads it where it is used." The newest test version for divers supplies 0.25 liter of oxygen per minute; a working diver will need eight times that much.

Together with the Aquanautics Corp., the Duke researchers are exploring possible commercial applications of the Hemosponge in machinery, submarines, underwater communities, perishable products, and other areas. It may also be used in medical devices. "We see the Hemosponge technology as a means of oxygen enrichment of breathing air for patients with pulmonary diseases," says Bonaventura. "It would be placed inside a lightweight portable unit that the patient could wear." For scuba enthusiasts, it can meanwhile provide a subject for fantasies of extended vacations through the waters of the Great Barrier Reef.

Drug Delivery Pumps

As noted, devices like the Infusaid pump promise to revolutionize the management of diabetes. Implantable pumps have broader applications, however, that may improve treatment in all areas of medicine.

THE INFUSAID PUMP

Unlike the SPAD, which was invented to deliver insulin and is now being considered for use in chemotherapy, the Infusaid pump was initially created to deliver heparin. The Minnesota group's successful use of the pump for heparin therapy led them to move to other drugs best delivered continuously. Dr. Buchwald anticipates that one day the pump might be used to administer hypertensive and antirejection drugs; researchers at Dartmouth College think that it might be able to deliver a drug that could relieve some symptoms of Alzheimer's disease.

Chemotherapy. The pump has made a major contribution to chemotherapy, specifically for liver cancer. Conventional chemotherapy is systemic: toxic drugs are injected into the circulatory system to find and kill cancer cells. Unfortunately, these drugs also destroy normal cells and cause unpleasant side effects like nausea and hair loss. The pump, however, can be implanted near the diseased organ; a catheter placed in an artery targets a more powerful concentration of the drug to the cancer cells while its effects elsewhere are minimized. Patients spend less time and money in hospitals and feel better while receiving more effective treatment.

In 1977 the Minnesota researchers did the first human implantation of the pump for delivery of fluorodeoxyuridine (FUDR) to the patient's diseased liver. Most of the 10,000 patients who have received the Infusaid pump worldwide have used it for this purpose. One of them is Margaret Tollefson of Inglewood, California, a 60-year-old retired clerk administrator for the town's municipal court. Margaret has had surgery for colon cancer, which has since spread to her liver. For the past year she has had an Infusaid pump in her abdomen delivering chemotherapy directly into the hepatic artery.

The Infusaid pump comes in two models. Model 100 is used for diabetes or in other cases where one type of fluid is administered. Model 400 has a side port with a separate septum and no reservoir. Injecting fluid into the side port sends it through the catheter directly into the circulatory system. The reservoir in Margaret's pump infuses FUDR at a rate of 2 milliliters per day; the side port is used to deliver heparin, saline, or the new drug mitomycin C. Margaret's physician fills the pump every 18 days.

Considering her original prognosis, Margaret feels that she would not be alive without the pump. She can still perform many of her daily activities. "I can't clean house well or put on big dinners as I used to do," she says, "but I still go out to lunch regularly, or have my hair done, or go shopping." Several of her friends with cancer have suffered side effects of chemotherapy that Margaret has avoided. "All I've noticed is that I'm more tired after the pump is refilled," she says. "I don't get nauseous; I haven't lost any hair, and I haven't been in bed. I feel well despite the fact that FUDR is very powerful and my doctors didn't know what its effects would be combined with the other medication."

Two of Margaret's neighbors are receiving chemotherapy with Infusaid pumps. One, she says, is "doing just beautifully"; the other still plays tennis. "I would recommend it to anybody, even if it gives you only another year or so," she says. "My life has been extended, and I really think that I'm going to get well."*

Pain Therapy. Along with the side effects of chemotherapy, cancer patients must dread the prospect of terminal pain. Cancer itself presents researchers with an obvious challenge, but few patients can understand why physicians cannot at least alleviate its agony. Most cancer patients can be treated well with standard medications, but in some cases the physicians prescribe too conservatively, the pain escapes control, or the narcotic side effects are undesirable.

A drastic but standard remedy is a cordotomy, a surgical procedure in which a needle is inserted into the spinal cord and heated. The heat coagulates the cord and destroys the portion that carries pain impulses from the cancer site. A cordotomy performed on the right side of the cord relieves pain on the left side of the body and vice versa. The procedure does not work, however, for patients suffering pain in the midline pelvis or leg caused by genitourinary, peritoneal, rectal, or colon cancer. For these patients the Infusaid pump has been a major advance because it can deliver high levels of morphine to the nerves involved without causing side effects.

At the Cleveland Clinic Foundation, for instance, the patient is first tested with an external pump and a temporary catheter to see if local morphine delivery deadens the pain. Then he or she has the Infusaid pump implanted in the abdominal wall or in the chest if it would interfere with a colostomy or ileos-

*Margaret died on March 12, 1985. Her daughter Diane reports that she was mobile until a week before her death, when she entered a hospital and rapidly deteriorated into cardiac arrest. Margaret remained alert until the end.

tomy. The catheter is run around back, usually to the level of the second lumbar vertebra.

Physicians at the clinic have implanted 18 patients with the pump since 1982, the longest implantation for 21 months. All of these patients have reported at least a 50% reduction in their pain and are more alert. "I did our first implantation in a man literally bedridden from pain. Within a month he was home mowing his lawn," says Janet W. Bay, M.D., Head of the Section of Neurooncology in the Department of Neurological Surgery. "We wondered why his pump rate picked up so sharply; it was because he had been feeling well enough to lie out in the sun."

The clinic staff is doing psychiatric tests on pump patients to see if spinal morphine could negatively affect thinking or cause brain damage. They are also studying how pain relief affects depression. "Patients are extremely brave in facing death, but the debilitating effects of chronic pain on them and their families just can't be believed," says Dr. Bay. "Usually their cognitive abilities improve and their depression decreases after implantation. But sometimes the depression returns because the pain relief frees them to recognize that they haven't faced the prospect of dying." The clinic has staff members trained to help with such trauma.

Elaine Lasky, a 50-year-old woman from South Euclid, Ohio, has had a morphine pump for the pain of Crohn's disease, ulcerative colitis, and rectal cancer since October 1984. Elaine is a professor of speech pathology at Cleveland State University doing research in central auditory processing comprehension. Before her illness she had been active and energetic: running every day, lifting weights, carrying an academic work load, and raising a family for the past 25 years. This life-style prompted her surgeon to recommend the morphine pump. "My pain was intense, and I wanted my time left to be as worthwhile and as active as possible," Elaine says. "Although I'm usually extremely conservative when it comes to medical or surgical intervention, I jumped at the idea of the pump. And it's been just wonderful. I think that more people should know about it."

Like Elaine, some cancer patients find the pump's size troublesome because they are extremely thin. "It leans against the bone, and if I lie on my left side for too long, it can be uncomfortable," she says. "Otherwise I don't feel it. Some days I feel well enough even to run, but my doctors won't allow that with the pump."

Elaine's family has appreciated her new treatment. "My family said I sounded more like myself after the implantation than I had for the past 9 months," she says. "I sounded up and excited and alive as I hadn't before. My posture improved because I was no longer bent over with pain." Elaine has been able to start new activities and plan a final vacation with her family. They plan to spend 11 days in Israel, during which time her son attending medical school will refill her pump with the assistance of her daughter.

THE PROGRAMMABLE IMPLANTABLE MEDICATION SYSTEM

A more sophisticated cousin of the Infusaid is part of the Programmable Implantable Medication System (PIMS), invented by Robert E. Fischell, M.S., Assistant Head and Chief of Technology Transfer in the Space Department of the Johns Hopkins University Applied Physics Laboratory. The implantable part of the PIMS system is the Implantable Programmable Infusion Pump (IPIP), which releases computer-controlled minute amounts of medication into the body.

Like the Infusaid pump, the IPIP can be used to deliver insulin or various other medications. Unlike it, the IPIP can be programmed by the physician, using a Medication Programming System (MPS), or the patient, using a Patient's Programming Unit (PPU), to provide a variety of medication flow rates. The Infusaid pump is designed to provide a fixed, constant rate of infusion; a diabetic patient using PIMS can have any basal flow rate delivered into his or her body over a 24-hour period. The patient can also use the PPU to call up a shaped profile of his or her medication to provide the necessary insulin for any type of meal. The PIMS is currently being tested in animal trials at the Johns Hopkins University School of Medicine.

Fischell's group hopes that this alphabet soup of devices will be fortified by the eventual addition of SAMS, the Sensor Actuated Medication System. This implantable device would sense changes in body conditions, like a rise in blood pressure or the onset of an epileptic seizure, and release appropriate doses of medication.

In describing the PIMS, Fischell likes to point out the similarities between electronic medical implants and orbiting spacecraft, especially in their command, telemetry, and power systems; microminiaturization; and reliability. "Considering our ambitious hopes for the use of implantable microcomputer systems," he writes in a recent article, "it will be most interesting to observe what impact those systems will have on medicine by the end of this decade." It will likewise be interesting to see how such impacts on medicine will affect industry.

SECTION

IX

BIONIC BUSINESS

27

Conservatism or Exploitation?

It would seem from our discussion so far that the most crucial aspect of getting a bionic part to the public is arousing the interest of a manufacturer, that once on an assembly line, an implant or prosthesis would reach patients automatically and with comparatively little difficulty. Actually bionic parts being produced by private companies need promotion just like any other new products and may require marketing and distribution strategies uniquely suited to their unprecedented medical purposes. Successful bionic parts can create significant cost and marketing problems, as anyone familiar with the pacemaker industry or dialysis treatments can attest. The Jarvik-7 artificial heart in particular has raised two issues: the possibility of conflict of interest among some of its proponents and the nature of for-profit medical treatment. These issues are especially tricky because it is not always easy to define the point at which determination to save or improve lives ends and commerical exploitation begins.

Whether an artificial body part makes it from the bench to the patient is only partly a matter of its technology and the clinical success necessary for the FDA to approve it. Like other products, bionic parts need investment capital to ensure their development and sound marketing strategies to help them beat out their competition.

"Let's say that the ultimate market for cochlear implants is half a million deaf people," says Dr. Hirshorn. "The real market is that by the end of the decade, about 20,000 people will have received a cochlear implant. Implantation will be slowed down by the FDA, by the time it takes the medical community to accept it, and by the fact that few surgeons will be trained to do it. A product also needs to make large clinical leaps so that industry will become interested in it. The producing company needs the necessary technology; other relevant factors include whether other companies are allowed to share that technology, whether the results of basic research are being disseminated, and so on."

Bionics is big business. Marketing personnel do not sit idle while company engineers are busy inventing bionic parts; they are reviewing patent filings and medical literature and taking notes on the competition. Meanwhile, scouts are visiting research centers and conventions, meeting with engineers and physicians, and watching for devices that have potentially large markets.

Updated figures from a survey published in December 1982 by Creative Strategies Research International of San Jose, California indicate that in 1984 the revenues for seven of the largest selling implants totaled $2.042 billion worldwide. The company predicts that in 1986 this amount will rise to $2.944 billion and that by 1987 the pacemaker industry alone will have worldwide revenues of well over $2.5 billion.

These figures seem discrepant with the complaints voiced by researchers that industry often shows little interest in bionics. One explanation, of course, is that the numbers represent the best-sellers—their large markets are the reason they were manufactured in the first place. Another is that these amounts pale in comparison with those spent on bionics by foreign companies.

Bionics Research Abroad

A general conservatism about bionics has evolved in America over the past few decades. "We're seriously in trouble," says Dr. Nosé. "America is still the greatest nation with the highest technology, but if we keep questioning artificial hearts and other artificial organs and technique applications, we will become a second-class country. This happened in the automotive industry; it's happening in computers; we're already far behind in drugs; and it will happen in other medical technologies as well. I'm sure that we're going to lose our edge, and all of us will pay the penalty."

When Dr. Nosé emigrated from Japan 22 years ago, that country was poor, and the United States was wealthy, innovative, and optimistic. Now the United States is less enthusiastic about—or downright mistrustful of—much high technology in medicine. The nation seems to have forgotten that a healthy populace, with its brainpower and productivity, is our most important natural resource. Everyone worries about the cost of medical innovations like the artificial heart: the government is cautious about funding, and companies adopt only projects that will be marketable in 3 to 5 years rather than those that may prove beneficial and profitable in the long term. Even when patients are treated with the best medical technology available, they sue physicians for malpractice. These attitudes have so thoroughly permeated American society that few people remain objective enough to question them. No wonder the field of bionics remains sparsely populated.

Dr. Nosé, who maintains Japanese citizenship, is particularly conscious of and concerned about this deterioration in attitude. He points out that many of the other leaders in bionics research are emigrés like himself: Dr. Kolff, for example, is Dutch; Dr. Galletti is Swiss. "I don't think that there's a thoroughbred American WASP really excelling in the field," says Dr. Nosé. The attitudes of foreign countries toward medical technology help explain this situation. In France, for example, each deceased person is presumed to be an organ donor unless he or she carries a card indicating otherwise. "It makes an enormous difference," says Dr. Galletti, referring to his work on a hybrid pancreas. "Human insulin-producing tissue, for instance, which is extremely difficult to get in America, is available in France almost for the asking." Germany is another country that encourages research and has produced strong work in bionics. (Like the United States, Russia is falling behind.)

Our biggest threat, however, comes from Japan. Dr. Nosé formed the International Center for Artificial Organs and Transplantation specifically to provide better access to international ideas and technology; that includes those of the Japanese, who have unlimited curiosity, enthusiasm, and optimism when it comes to bionics research. Nobody there asks how much a project will cost or worries whether it is ethical to use bionic means to save lives; they instead ask how well a product works, and if they are satisfied, they develop it. One Japanese company is spending $40 million each year on research and development of kidney dialysis, plasmapheresis, and other uses for medical hardware; by contrast, the U.S. government spends less than $10 million annually for research in artificial kidneys and related devices. In the future another Japanese company may spend more than our government does to develop artificial hearts. These companies invest in worthwhile projects even if development is expected to take 20 to 25 years. "The American Society for Artificial Internal Organs [a group separate from the International Center] has slipping attendance at its meetings and a present membership of about 1400," says Dr. Nosé. "The Japanese society has 2000 members, many of whom think they're going to win the world."

In Japan the government, physicians, and companies all think that money that helps patients is well spent. The patients do not sue, because they believe that experimental procedures are over and above ordinary medical care, and they are grateful for them. (We have seen the same attitude here among the comparatively few patient pioneers who are testing investigational devices.)

"Americans are enormously provincial; we think that we are the center of the universe and control mankind, and that's not true," says Dr. Galletti. "Whether we want to participate in bionics research won't affect what other countries will do. Japan will do it anyway."

Cost Issues

The annual cost of dialysis in the United States is about $2 billion per year. Implanted in all the patients who need one, artificial hearts have been projected to cost at least $5 billion each year. Back in 1980 it took Medicare only 7 months to decide to discontinue coverage for heart transplantations; bureaucrats may decide that heart implantations are likewise too expensive to be covered.

Researchers and others interested in promoting artificial hearts and other bionic devices point out, however, that cost estimates are misleading. "The figures quoted for the artificial heart summarize the retail prices of all the services offered in connection with implantation," says David Jones, Chairman and Chief Executive Officer of Humana Inc. "Most of the hospital services and costs for implantation already exist, and all we incur are incremental costs—those that result specifically from artificial heart patients being here. Each service offered is using employees and equipment that are already present and paid for. I can't imagine how the University of Utah let this program get away if they likewise incurred only incremental costs. There are over 100 major hospitals in the country that could support the artificial heart program as easily as we do. All we brought to this particular situation is a certain imagination and boldness, if you will."

Even if the cost estimates were accurate, they would have to be set off against the cost-saving possibilities of artificial hearts. "The economic impact of these costs should be compared with the costs of long-term hospitalization of totally incapacitated patients and the likelihood of the return of highly trained and experienced people with a total artificial heart back in the work force," write Dr. Olsen and his colleagues in a 1981 article. "Cost is a cliché question," says Dr. Hill. "I don't think that $100,000 is a lot to invest in a 50-year-old man at the peak of his career. You wouldn't abandon your house because the kitchen needed fixing. I think that there's a good financial argument for blood pumps."

The same cost-benefit argument might be applied to any expensive bionic part. Implantable insulin pumps with sensors, for example, might initially be expensive, but if they reduce the incidence of diabetes complications, they could save the United States several billion dollars each year while earning huge profits for their manufacturers. Rehabilitative devices like myoelectric artificial limbs and FNS systems could have the same effect on a smaller scale.

"The cost question is one of the most curious ones I've heard about the artificial heart implantations," says Jones. "A researcher has a concept. That concept must then be proved—that's what Dr. DeVries is doing by implanting artificial hearts. If the concept works, then engineers miniaturize the device and make it affordable. The idea that the artificial heart will cost $250,000 is fanciful. Our society has plenty of resources to deal with this prospect. The artificial heart is potentially a weapon against the world's bigger killer, and if it proves efficacious, it will be mass-produced at far lower affordable cost."

"Every prototype of any device is expensive," says Dr. Nosé. "Later the cost comes down." The same is true of new treatments, as evidenced by the dialysis program, which began in 1961 at a cost of $25,000 per patient per year. In 1982, after 2 decades of rising health care prices, the cost was $24,725 per patient per year. The present billion-dollar annual cost results from the increase in the number of patients from four in 1961 (totaling $100,000) to 72,800 in 1982 (totaling $1.8 billion).

Researchers feel that society's reluctance to spend these amounts on bionic parts represents misplaced priorities. "Spending $200 million to date to develop an artificial heart is nothing—we gave more than that to poor countries over the last 25 years so that their presidents could keep mistresses. We need to spend 5 to 10 times as much as we do on medical devices because in the long run they will help us all," says Dr. Nosé. "Financial consequences can't be predicted," says Dr. Kolff. "People don't want to die—they want help, and we should help them if we can. To think that our society can't afford it is nonsense. We don't necessarily need more nuclear warheads; we need a more service-oriented society."

Society's argument that medical funds should be directed toward prevention suggests in some cases that existing treatments should be abandoned in favor of mere potentials to avoid disease. "Some argue that money spent on artificial hearts and assist devices should instead be spent on prevention of heart disease. If those who feel that way could save the patients who are dying now, I could accept their point of view. But it will be several years before we find remedies for arteriosclerosis, cardiomyopathy, and other diseases," says Dr. Harasaki. "There are already cases on record in which postoperative patients would have died without heart assist devices," says Kiraly.

It is interesting to note that in 1984, 3 years after publishing the article quoted earlier in this section, Dr. Olsen coauthored another article that discusses the cost implications of the Jarvik-7 artificial heart. The outlook of the second article is somewhat gloomier. "Realistically, the financial impact of employed recipients will probably be a minor influence in the overall cost of the total artificial heart," the researchers write, ". . . the artificial heart may become the first medical technology which does not reach its maximum application, because America just cannot afford it." It may be more accurate to say that America just does not wish to pay for it.

Reflecting the general conservatism, insurance companies are similarly reluctant to cover medical procedures involving artificial body parts. Coverage varies between states, companies, and policies. Insurance companies are still fighting with reconstructive surgeons over coverage of breast implants, for instance. A 1979 study of this issue by Dr. Dowden and his colleagues reports "a great discrepancy between the stated policies of the health insurance carriers and their actual practices" with respect to breast reconstruction. The companies studied claim to cover the procedure but then weasel their way out of payment by inadequate coverage and specific contractual exclusions.

American Medical Systems, Motion Control, and, no doubt, other companies come to the rescue of their products by informing their prospective users of the possibility of claim rejection. These manufacturers suggest strategic plans of attack on insurance carriers, starting with the attempt to obtain their prior written approval of the treatment. If that fails, say the manufacturers, the next step is to have the physician send a medical history along with a statement that the implant or prosthesis is medically necessary. If the policy excludes the device, ask the company to make an exception. If coverage is still denied, appeal right up to the top (the claim manager, medical director, or review board). Ask employers or union representatives for help.

At least those companies that have committed themselves to manufacturing artificial body parts are then ready to fight for their products.

The bottom line to cost issues, of course, is that patients desperate for longer or better lives will pay any amount to that end. Unfortunately this creates a perfect opportunity for unethical parties to indulge in commercial exploitation.

"Fraud, Waste, and Abuse"

The argument that taxpayers deserve to reap what they have sown in the artificial heart program in the form of Medicare-insured implantations can be countered with examples of insane applications of pacemakers and kidney dialysis.

THE PACEMAKER INDUSTRY

Medicare pays 80% to 90% of the cost of implanting a pacemaker and usually covers follow-up care as well. This means that pacemaker manufacturers can raise the prices of their products continually. Medtronic enjoys double the market share of either of its major competitors, Cordis and Intermedics, although advances in development have been shared by the three companies over the years. They and other companies vie for state of the art technology so as to upgrade pacemaker models and capture larger shares of the market. "Unnecessary complexity for a given function is costly," writes Dr. Buchwald in a 1983 article. "A vivid example before us today exists in the pacemaker industry, where many manufacturers have built-in [sic] a degree of complexity into their devices that has increased their costs but not their utility."

In 1982 the Senate Special Subcommittee on Aging completed their study entitled "Fraud, Waste and Abuse in the Medicare Pacemaker Industry," a report on which was made public in May 1985. The investigators had found the industry to be full of corruption. Physicians have accepted Caribbean or Colorado vacations, cash, and other bonuses for implanting certain pacemaker models in their patients. Terminal cancer patients with only a few months to live have received the most expensive types of pacemakers. The investigators also uncovered routine implantation of two pacemakers (the second a backup in case of failure) or implantations in patients who complained only of chest pains and were given no diagnostic tests.

The committee concluded that possibly half the money spent each year on 150,000 implantations and replacements of pacemakers might be wasted on unnecessary surgery. The amounts are considerable, since hospitals are billed five to eight times the cost of making the pacemakers and add their own markups of 50% to 150% before billing insurance carriers. The situation can be attributed to incompetent, unethical physicians, negligent government officials, and high-pressure salespeople.

While the Senate was studying the pacemaker industry, the Health Care Finance Administration, a branch of the Department of Health and Human Services that oversees Medicare, was independently finding the same conditions. In March 1984 they issued rules under which pacemaker implantation is justified. More specifically, they are developing guidelines to determine which illnesses require the implantation of the more expensive pacemakers. Hospitals or patients who do not follow the rules may have to pay the cost difference.

Over the next 4 years Medicare will also be phasing in a new prospective payment system covering what they call diagnostically related groups. Under this system, Medicare will reimburse fixed amounts for particular diagnoses that require pacemakers. The average hospital payment for pacemaker implantation will be around $7500; a hospital that charges a Medicare patient more than that will have to pay the difference, while those that charge less can keep it. In July 1984 Congress passed legislation requiring the Department of Health and Human Services to establish a national pacemaker registry to keep track of defective devices. This registry was supposed to have been in effect by January 1, 1985, but officials estimate that its implementation will take another year.

These changes have significant implications for the industry. No longer protected by government reimbursement, manufacturers must now attract customers legitimately with advanced product technology and sound value. They will have to market various pacemakers longer and find cheaper, less dramatically innovative ways to upgrade them. Research and development will slow under cutbacks, and some companies may have to diversify their products to maintain profit levels.

Although practices of marketing and implanting pacemakers have been scandalous, dialysis treatments are more often cited as a worrisome example of where potentially expensive programs like the artificial heart might be headed.

DIALYSIS TREATMENTS

In the early 1970s dialysis treatments for almost all patients were supervised by nonprofit organizations, and 40% of patients underwent dialysis at home. Kidney transplants were an up-and-coming alternative treatment for end-stage renal failure.

In 1972 Medicare began funding dialysis treatments under legislation that reimbursed center- or hospital-based dialysis without imposing cost ceilings. Almost immediately more patients started dialysis and far fewer dialyzed at home or had transplants. Profit-making dialysis centers increased in number to the point where they now treat about 40% of dialysis patients. The United States now has the largest number of dialysis patients in the world, about 85% of whom dialyze at hospitals or in centers.

From the beginning of its dialysis program, the Health Care Finance Administration had little reliable information on treatment costs. The agency therefore set reimbursements much too high and later had trouble cutting them back. In a 1983 article Dr. Kolff refers to a chart comparing the costs per patient per month for dialysis at different locations. The patient's Medicare pays $1430 per month for home dialysis, $1794 for center dialysis, and $4225 for dialysis in a hospital. For chronic dialysis services there are set technical and professional fees regardless of where the treatment is provided. This fee structure is intended as an incentive to use low-cost home dialysis. Stories pop up, however, about dialysis centers buying each of their physician owners a Mercedes-Benz and comatose patients being removed from nursing homes to be dialyzed.

On August 1, 1984 cost-cutting changes in Medicare reimbursements for dialysis went into effect. Now Medicare reimburses facilities prospectively at a set rate ranging from $118 to $138 per dialysis. (Dr. Kolff's chart lists the facility fees for home, center, and hospital dialysis, respectively, as $110, $138, and $325 per treatment.) The new rate applies to treatment at any location.

The cutback has had the desired effect. Facilities are recycling dialyzers, reorganizing staffs, and considering other cost-cutting measures. Suddenly home dialysis is popular again, and CAPD has never looked better. Kidney transplants, whose initial high costs drop sharply as time goes by, are being performed more often.

Not everyone is happy about the new ruling, however. Home dialysis is just not appropriate for all patients now being encouraged to use it. Patients fear that the Medicare cutbacks will lower reimbursements below their actual costs. Now that hospitals have discovered that disposable cartridges for dialysis machines can be flushed out and reused, companies that make them face a drop in profits. (Of course, those that manufacture CAPD equipment as well are not as badly hurt.)

Physicians are concerned about the Medicare cutbacks for legitimate as well as selfish reasons. "It's not a good idea to let a few disreputable physicians govern the legislation for the many," says Dr. Paganini. "The Medicare ruling is well-intentioned and may create some cost-effective moves, but I'd be very cautious about what it might lead to. Of course, we don't want to be totally controlled, but we need a very good quality assurance program to accompany a decrease in reimbursement.

"While there should be some form of cutback, and a cost reimbursement system can't go on forever, I wouldn't want to see us verge toward socialized medicine," he adds. "I am against socialized medicine not because of finances but because of the type of medicine it engenders. I was trained in Europe, where I saw it at its worst as well as its best, and even its best is not great. It's a crowd pleaser—but those who can pay and those who can't receive different levels of care." Physicians interested in cutting costs see the Medicare ruling as a step in the right direction but cannot predict how well it will ultimately succeed.

It is unfortunate that dramatically lifesaving devices like artificial kidneys and pacemakers have created these kinds of problems. Both devices have been used long and widely enough to be taken for granted, so that critics of innovative treatments can focus on their troublesome consequences rather than their enormous potential benefits. Since the artificial heart has not yet proven itself to be an equally necessary device, its potential for abuse can be similarly targeted. Isolated abuses of pacemaker implantation and dialysis treatment reflect on the industries as a whole and raise suspicions about all lifesaving bionic devices. Still in its start-up phase, the artificial heart has done likewise by arousing controversy about the business practices involved.

Bionic parts differ from other business products in that most of them perform unprecedented, very personal functions. Some artificial body parts are lifesavers that carry enormous psychological and emotional implications. Recognizing their importance and the need to avoid unthinkable choices about who should or should not live, the government opted to pay for treatments involving pacemakers and artificial kidneys. Now the government faces the same choice with respect to artificial hearts: contribute to their use (and possible misuse) or leave individuals to pay for their own.

It is not easy to resolve the problems posed by bionic devices. Lack of enthusiasm about artificial body parts is causing the United States to fall behind other countries eager to develop these medical innovations. Strong enthusiasm, on the other hand, makes bionic parts desirable and their potential recipients subject to exploitation. If the government does not fund the development and use of implants and prostheses, private companies will step in, bringing their own set of problems and temptations.

We are reminded that people healthy enough to study issues and make decisions cannot fully empathize with dying persons who want to live and be well at all costs. Sick individuals, however, are not in a position to recognize and protect themselves from commercial exploitation. Society needs to decide how best to distribute important health care resources like bionic parts. Perhaps we can look to Japan and other foreign competitors for ideas on how to run bionic business. If the United States does, in fact, begin to lag behind other countries, at least close observation may help avoid repetition of their mistakes.

28

The Business of the Jarvik-7 Artificial Heart

We have noted that as a business venture the Jarvik-7 artificial heart has raised two issues: that of possible conflict of interest and that of the appropriateness of medical experimentation at a for-profit institution.

Conflict of Interest?

SYMBION AND THE UNIVERSITY OF UTAH

Although large companies are increasingly moving into the field, many bionic parts are still produced by small private firms. After all, one way around the funding problem is to set up a company to make a product. This is why in 1976 Dr. Kolff founded Kolff Associates to produce artificial hearts. The company was started with too little capital and incompetently run by a board of directors who tended to squabble over issues. Six years later Dr. Kolff, who disagreed with the majority view of how the company should proceed, lost management control. Dr. Jarvik was elected president at a shareholders' meeting, and the board of directors renamed the company Kolff Medical. Now known as Symbion, Inc. (a conflation of "symbiosis" and "bionic"), the company produces artificial hearts, ventricular assist devices, and Ineraid implants.

What bothers observers is that Dr. DeVries has sat on the board of directors of Symbion and has held stock in the company, as have Humana and the University of Utah, where he has tested the Jarvik-7 artificial heart on humans.

Dr. Kolff set up his company somewhat prematurely as a safety measure in case his NIH grants for artificial heart research were not renewed. Initially the company's greatest asset was its relationship with the University of Utah. The availability of university facilities intrigued venture-capital firms while it provided government-funded research and development. Dr. Kolff supplied materials and hired scientists as company consultants who continued to work in their own laboratories.

Dr. DeVries and Humana are now more distanced from Symbion, but as previously noted, the company is still involved with the university. "Symbion has two relations with the University of Utah," says Dr. Jarvik. "We have license agreements on the technology for heart and hearing devices—exclusive rights to the patents and know-how on those products, for which we pay royalties. We also buy services from the university, like work in the heart laboratories, under contract." The heart research is mostly developmental: the company has a contract and pays for animal implantations to test devices and teach surgeons about procedures and care. It also sponsors clinical trials of the Ineraid implant, for which it paid part of the costs and provided the devices for the first five patients.

Academic people worry about situations like this. They fear that faculty who work for private companies will be motivated by the potential profits rather than by a desire for discovery per se. Some researchers feel that this issue was behind Utah's delay in approving further heart implantations after the surgery on Barney Clark. "Isn't it strange that Utah suddenly turned off the artificial heart implantations?" says Dr. Bernhard. "They did it because their peers at other universities began asking them what in hell they were doing out there. It was obviously wrong, and it's up to the academic and medical communities, not the FDA or NIH, to police themselves. Groups with a personal interest in their productions should be with commercial outfits, not at universities."

Most researchers feel that industry does have a vital cooperative role to play in advancing research. "Certain centers have spent years struggling along with just NIH grant money," says Dr. Pennington. "The instant they got involved with industry, however, they blossomed and started making giant strides. I don't know whether it's just the increased funding or better management and efficiency, but industry is absolutely essential to circulatory support research."

The University of Utah has always sought to transfer technology to the private sector at the appropriate time. This is one reason that the Utah arm is available today. "I think it's wonderful that the university can couple with industry to get a better transfer of concepts," says Jacobsen. "Both sides have something to offer, and often we don't talk very well. It's certainly positive not only to do good science and engineering but also to have somebody want your products. The industrial side helps level out funding, and it brings in people who don't have academic attitudes about performance—they remind us to optimize a product in terms of use. Our cooperation makes for better products and larger cash flow in the state."

The controversy about the Symbion-Utah ties reflects feelings that for-profit research and development, especially in medicine, ought not to be carried out in nonprofit academic settings and that physicians ought not to have financial stakes in the devices they use to treat patients. The latter objection pursued Dr. DeVries after he moved to Humana.

SYMBION AND HUMANA

Humana invests in companies like Symbion to obtain current information on developing technologies. "We got early access to the artificial heart, the Ineraid, and an insulin pump as a result of our connection to the company," says Jones. "By investing in companies doing good, useful research, we obtain information that enables us to offer appropriate technology and therapies in our hospitals."

Humana's contact with Dr. DeVries began with such a venture-capital investment. "We have an assessment group that seeks out promising medical technology," says Jones. "The Symbion investment caused us to learn about and participate in what was happening at Utah." Drs. Allan Lansing and Ronald Barbie went to Utah to train under Dr. DeVries to prepare for heart implantation surgery at Humana. After implanting artificial hearts in several calves, Dr. Lansing returned to Humana and informed Jones that the technology was extremely promising and that their group should know more about it. "That's how I got into all this," says Jones. "We have a magnificent group of doctors at the Heart Institute, and when I started hearing from them that this technology ought to be nurtured, I began looking into it."

Having meanwhile come to Louisville on a social visit, Dr. DeVries stopped by the Humana Heart Institute and was extremely impressed. He and Jones met and discussed his major problem at Utah, namely, the amount of time he had to spend raising money. The upshot was that Humana offered to pay for up to 100 artificial heart implantations in Louisville as long as the researchers made scientific progress.

Like other academicians seeking higher incomes, physicians are increasingly leaving universities for private practices. The main reason Dr. DeVries left Utah, he says, was simply to have a better job—to leave a university at which he could not do much heart surgery for a very lucrative private practice. "This had nothing to do with Humana itself," he says.

Another reason that Dr. DeVries moved was to have more time to participate in a project that means a great deal to him. Not only could he stop worrying about funding, he could also operate without having to turn down patients who could not afford the surgery, as he had at Utah.

"I felt that the move was the best thing to do for me and for the project," Dr. DeVries says. "I had studied the situation very carefully in terms of what Humana and I would get out of the artificial heart program and what their constraints on me would be. I felt very strongly that I wanted no restraints whatsoever. Humana is not in the business of doing research, while I'm experienced at it from having had an academic career. They don't tell me how to do research, and I don't tell them how to raise money. It's a very nice symbiotic relationship."

As noted, Dr. DeVries was originally on the board of directors of Symbion and also had stock in the company. Responding to criticism of this state of affairs, Utah officials had informed him that they could tolerate neither situation; they did not see a conflict of interest but wanted to avoid the appearance of one. "I didn't see a conflict of interest either, but not wanting anything as petty as that to interfere with the artificial heart experiments, I resigned the directorship immediately," he says.

At the time, the Utah IRB was still deciding whether to authorize additional implantations of the Jarvik-7 artificial heart. Dr. DeVries told them that he would not perform another implantation until he had sold his stock. "Meanwhile, I wanted to get the best price for it," he says. "I saw no reason to sell stock that was decreasing in value—I was using that investment to send my children to college." Once Dr. DeVries received approval from Utah to perform a second implantation, he began looking for buyers for the stock. "I ended up selling it for the very lowest price I could have," he says, "but I had to do it to avoid having ethical threats looming over the project." When Dr. DeVries grumbled to Dr. Jarvik about the drop in stock price, his friend responded that he was stupid to sell his shares.

By the time the FDA granted approval on November 4, 1984 for additional implantations of the Jarvik-7 artificial heart, Dr. DeVries had already moved to Louisville. Now Humana was entitled to ask Dr. DeVries to sell his stock, for a reason slightly different from that which prompted Utah's request. Humana wanted Dr. DeVries to be absolutely unfettered in his actions and his ability to choose whatever device offered patients the best promise of recovery and a good quality of life. "We have no interest in one artificial heart over another," says Jones, "and we hope that our institution can be involved in experimentation with any device that shows promise."

Like Dr. DeVries, Humana had already decided to sell its shares in Symbion but was meanwhile holding out for a good sale price. "If there had been no apparent conflict of interest, we would have kept and sold the stock just like any other investment," says Jones. "Utah owned Symbion stock throughout the first implantation, and I think we bent way over backwards in selling ours. But I saw no reason to be involved in this issue because it's not important to us."

Both Humana and Dr. DeVries specify that they sought to avoid an *apparent* conflict of interest. "I saw no conflict of interest in my stock ownership," says Dr. DeVries. "Physicians do this sort of thing time and time again—they often make money off medical devices. People make it into an issue because they don't want researchers involved in petty little things like the horrors of making money. My viewpoint is that if a principal investigator is not sure enough about his device to want to make a profit from it, he has no business using it on patients. My patients had no problem with the issue—in fact, they were really happy to think that I could make money from the artifiicial heart. But I just didn't want any accusations dirtying up the water."

It is useful at this point to clarify Dr. DeVries' present situation at Humana and in the experimental program. Dr. DeVries derives his income from work in a private practice group with Dr. Lansing. Although Dr. DeVries directs the artificial heart program at the Humana Heart Institute, he is not an employee of Humana and earns no salary from the company. He conducts his clinical tests at a foundation within the hospital that is involved with education and research. The surgeons donate their professional services, and Humana pays the costs not covered by third parties.

Both Jones and Dr. DeVries discuss these matters with firm conviction that their involvements with Symbion and the experimental series have been strictly above board. From speaking with them, one might conclude that the issue of conflict of interest has been raised by biased observers looking for nits to pick. Other physicians and institutions are engaged in similar practices without being singled out for criticism. Some researchers, however, have serious reservations about any use of products and facilities to profit themselves. "There's nothing wrong with a company deciding that a particular laboratory's research will benefit them if they fund it—hands off," says Dr. Bernhard. "It's freedom for the investigator. That's different from the researcher's taking some scheme, whether it's very good or completely harebrained, having a personal investment in that enterprise, and trying to sell it out in public because the media and the public have no idea if it's good or bad. That's what this artificial heart business has been all about—people using experimental technology for their own purposes on dying patients who are desperate enough to do anything."

"Physicians have been investing in drug and paramedical companies for years," says Dr. Pennington. "I don't necessarily think that this practice is wrong, as long as it doesn't affect the way they care for patients. But while I have done consultations for young companies that would prefer to pay me in stock, I won't accept that form of payment. I prefer to avoid the possibility of a conflict of interest."

Other researchers likewise consider the situation to be clear-cut. "I couldn't get funding to develop the TASS, so I formed my corporation Tacticon to build it," says Saunders. "I was able to get a government loan for small businesses. I'm still professionally affiliated with Smith-Kettlewell but moved my laboratory out of there because my work is a profit venture that has no business in a nonprofit medical research center."

In an article on what has been called "the new medical-industrial complex," Arnold S. Relman, M.D., editor of The *New England Journal of Medicine,* defines it as "a large and growing network of private corporations engaged in the business of supplying health-care services to patients for a profit—services heretofore provided by nonprofit institutions or individual practitioners." (The definition excludes companies that manufacture pharmaceuticals or medical equipment and supplies.) Dr. Relman feels that involvement in the medical-industrial complex undermines the strength and moral authority of physicians; he suggests that the AMA add to its ethical code "the principle that practicing physicians should derive no financial benefit from the health-care market except from their own professional services."

Apparently the problem is widespread enough to warrant such an official position. As another example, we might point out that many of the physicians who have criticized implantations of the Jarvik-7 artificial heart in humans are cardiologists and transplant surgeons, whose opinions themselves constitute conflict of interest.

Profit-Making Institutions

The second issue that has been raised by the Jarvik-7 artificial heart is that of the increasing role of profit-making institutions in health care. For-profit companies have long supplied the public with drugs and other medical devices. The pacemaker industry has by no means taken a unique direction, for nonprofit institutions are subject to the same kinds of pressures and have been involved in scandals of their own. Advocates of private enterprise in medicine feel that a company's profits from medical advances are irrelevant to the goal of helping the public.

The artificial heart is a good example of arguments on either side. Private companies are racing to develop total artificial hearts, ventricular assist devices, or both. A partial list of those working on these devices includes Thoratec, Thermedics, Novacor, 3M, W.R. Grace, Lilly, Applied Biomedical, Nimbus, and, of course, Symbion. Research and development in private companies add more funds to the artificial heart program but also increase competition. When the NIH supported the artificial heart program, it required researchers to share their results. Now, however, patent protection is decreasing the exchange of scientific information. The public is beginning to wonder if its tax dollars ought to have been spent to launch private enterprise on huge profit-making ventures like the artificial heart.

Similarly, for-profit companies now own or manage over 20% of American hospitals and are moving into affiliated areas. Although physicians have run investor-owned hospitals for many years, companies like the Hospital Corp. of America and Humana Inc. have introduced a new approach, namely, establishing hospital chains. Chains can use particularly effective methods to cut costs and increase profits; they can also raise money by selling stock. Using business techniques like market research, advertising, and promotion, they can capitalize on the familiarity of their names as symbols of nationwide reliability and excellence at reasonable prices.

The concept of hospital chains has been criticized because of its implications for medical care. Our focus is on one chain, Humana, and its connection with the artificial heart. Humana Inc. is the country's second largest investor-owned hospital chain and, thanks to the Jarvik-7 artificial heart, the best known. In fiscal 1984 the company earned $193.3 million profit on revenues of $2.6 billion; it has 87 hospitals and 143 MedFirst clinics. Like other hospitals, however, Humana has been losing patients; it seeks to reverse this trend and increase its share of the health care market. Conscious that for-profit medicine is often criticized, Humana is establishing itself as a supporter of medical innovations. It has succeeded in enhancing its business and reputation through public visibility: the first heart implantation at Humana won the Silver Anvil Public Relations award in the category of "Institutional Programs—Business" of the Public Relations Society of America. Even controversy over its role in testing the artificial heart helps increase that visibility.

Humana owns 15 selected hospitals called "Centers of Excellence," each devoted to treating patients with state of the art medicine. By attracting business from broad geographical areas, the centers help compensate for society's ongoing effort to keep hospital costs (and profits) down. The Heart Institute in particular is building a reputation that might enhance those of other Humana hospitals at comparatively small financial risk. Humana aspires to make the Heart Institute a world center for treatment that will be first in line to be reimbursed for routine heart implantations.

To accomplish these goals Humana attracted Dr. DeVries with an offer few researchers could refuse. Having arranged for medical history to be made at the Heart Institute, the company spent thousands of dollars to set up a news office and information center for reporters at the Louisville Commonwealth Convention Center. In so doing, they gained both visibility and notoriety.

One of several criticisms recently leveled at Humana is that the Heart Institute is not an appropriate site for the artificial heart implantations because it is not a nonprofit medical research center. Humana responds by citing the obvious fact that Dr. DeVries, the only surgeon in the country with permission to perform the implantations, freely chose the Heart Institute as the location in which to continue his work. Other researchers at Humana hospitals are similarly involved with investigations, clinical tests, and teaching.

"We've done primary work in skin cloning and implantation of the Ineraid device," says Jones. "We have one of the finest university hospitals in the world, plus our Centers of Excellence in which research and training are carried on in spinal cord injury, burns, diabetes, neuroscience, cardiology, and other areas." As evidence of the quality of this work, Jones points out that Humana has a 24-year history with no scandals and not the slightest hint that the work done at their institutions is not first rate. "We put the Humana name on our hospitals years ago because our corporate mission is the offering of quality that is both unexcelled and measurable," he says. "We believe that you can't find a hospital anywhere in the world that is better than a Humana hospital."

Critics of the company also worry that Humana might use the clinical tests of the Jarvik-7 artificial heart as a means of gaining a monopoly on new and beneficial procedures involving artificial heart technology. In April 1985, however, the Heart Institute sponsored a symposium attended by scientists from 19 countries, including almost every major figure in heart transplantation and implantation. "Everything we have learned thus far has been freely shared with them, and that will continue," says Jones. "We own no interest in any device that might be used in the artificial heart experiments; nor will we have any patents or monopolies. We've also been criticized for sharing too much information, so people talk out of both sides of their mouths about that."

Humana has also been accused of pushing artificial heart technology too fast by funding implantations of a device that offers only a poor quality of life—an odd charge since it is the FDA that determines whether such tests will take place. The Humana Hospital–Audubon IRB is appointed by the president of the medical staff at that hospital, but a majority of its members are independents. "Although patients are somewhat inherently coerced by the fact that they're dying, they make a free, fully informed choice to participate in the artificial heart experiments," says Jones. "It will take many tests with a variety of devices before we have a breakthrough, if any occurs at all."

Another charge is that Humana sought to sponsor the artificial heart implantations rather than some other form of medical research only to enhance the company's reputation. Jones responds by pointing out that in our pluralistic society, institutions are utterly free to allocate their resources in whatever ways seem wisest to them. Humana tries to nurture excellence wherever it finds it.

A related criticism centers on the amount of publicity that has surrounded the tests of the Jarvik-7 artificial heart. Despite its skill in handling the well-publicized implantations, Humana has a public relations staff of only eight people. "We didn't invite a single person to come here to cover the artificial heart implantations," says Jones. "Reporters flocked to Louisville because of the story's inherent interest, so we made it as easy as possible for them to work." The decisions about what information to release were made solely by the physicians and the recipients; Humana had nothing to do with it. "Simply because the implantations occurred at Humana, our name has become well known," says Jones, "but that would have been true at Utah or any other place."

In discussing the national scientific press, Jones adds that none of the criticisms directed toward Humana came from people who had gone to Louisville to observe the situation closely. "The criticisms have all come from people distanced from the situation who haven't made the effort to find out the facts because they have vested interests in their opinions," he says. "Those who were here know that as far as publicity was concerned, we have respected the wishes of the patients' families."

Other supporters of the Humana program attribute some of the criticism to professional bias and envy, pointing out that there was little objection to the publicity about the implantation performed at the state-supported University of Utah. Since physicians do not always have the time or the initiative to educate patients, good public relations can perform that function. By illustrating the uneven progress of medical experimentation, publicity can increase public understanding of the process and perhaps decrease the number of malpractice suits. Until this education is complete, however, the physicians involved must bear the scrutiny and disappointment of observers who are not accustomed to accepting stasis and setbacks.

As a way of squelching some of these various protests, Humana has established an independent, interdisciplinary panel of experts to advise the artificial heart team on how best to conduct the experiments. "We have asked them to study the program from every viewpoint and make suggestions as to how we can carry out our work with the highest ethical, legal, and scientific standards," says Jones. The panel consists of Dwight E. Harken, M.D., Clinical Professor Emeritus of Surgery at the Harvard Medical School; Albert R. Jonsen, Ph.D., Professor of Ethics in Medicine and Chief of the Division of Medical Ethics in the Department of Medicine, UCSF; and Walter Wadlington, L.L.B., James Madison Professor of Law at the University of Virginia Law School and Professor of Legal Medicine at its School of Medicine.

Meanwhile Jones is unperturbed by objections that he attributes to the major changes occurring in the way health care is delivered in America. Health care costs have climbed unbelievably rapidly, so that investigators like Dr. DeVries find their funds very much at risk. Into an increasing void steps an organization like Humana with substantial low-cost, high-quality hospital resources. Whether or not its critics approve of Humana's attitude and practices, the fact is that without the company, clinical tests of the Jarvik-7 artificial heart might not be occurring at all.

"About 99% of what we hear about our program is good," says Jones. "I think that criticism is always helpful—we need people at the edges to spot the things that need improvement. I'm a great believer in the competition of ideas, and we live in a wonderful society in which people can speak freely, so often we learn very useful things. We feel good about ourselves for carrying out our work in ways that are both honorable and effective, and we have never done anything that requires the least apology. The research is experimental, and we don't know if it will succeed, but it's being done well, and we hope that it's efficacious."

It is too early to tell how Humana's role in the artificial heart experiments will work out. The distinction between profit-making and nonprofit health care institutions is becoming increasingly blurred as the latter adopt more and more commercial tactics, including earning profits from their research. "I guess that every major university in the country owns drug stocks, for instance, while participating in drug experiments," says Jones (emphasizing that Humana owned no stock in Symbion while clinically testing its artificial heart). Physicians likewise have long tried to profit from their findings.

Jones and Dr. DeVries are justified, then, in implying that they should not have been singled out in anticipation of their engaging in practices that are quite common. (The appeal to commonality, however, reminds one of the familiar parental question, "If 'all the kids' decided to go jump off the roof, would you want to do it too?") What "everybody" does is not necessarily right. In *acknowledging* an apparent conflict of interest, to say nothing of losing money to avoid it, Dr. DeVries and Humana have ironically made the issue, however petty, *seem* quite real.

SECTION

X

FINANCES, LEGALITIES, AND ETHICS

29

Funding and the Food and Drug Administration

One reason for the publicity surrounding bionics is the dramatic shift in financing of American medical care and research that has occurred in recent years. Now that taxpayers provide most of the funding for investigational work, medical researchers in general are responsible to a larger audience. They must answer to funding agencies, the FDA, IRBs, health economists, malpractice attorneys, medical ethicists, religious leaders, journalists, patient and animal rights groups, and the rest of the public. This holds true especially for an innovative, controversial field like bionics.

The organizations in this group that cause researchers the biggest headaches are the funding agencies and the FDA. Along with "simplicity," the word one hears most often from bioengineers is "funding." The Six Million Dollar Man and the Bionic Woman did more than encourage public fantasies about becoming superheroes; they also implied that the government was interested in bionics. Most people are aware that medicine has low priority compared to the funding of military or space programs; they may not know that bionics gets short shrift when it comes to funds allotted to medical research. The creativity of bionics researchers must be applied to fund-raising almost as often as it is to prostheses and implants.

Funding Agencies

In a recent article on the political history of the artificial heart, Michael J. Strauss, Jr., M.D., of the University of Washington, Seattle, points out that public support for biomedical research increased rapidly after World War II. During the war years the government had supported successful medical research, and people expected this pattern to continue. The economy was strong enough to keep funding medical projects without jeopardizing other social programs. The public became more concerned about health, especially chronic degenerative diseases and cancers. Their confidence in the medical profession was high, for by the 1950s physicians had attained status and power in society.

As Dr. Strauss notes, these expectations and concerns were picked up by bureaucrats who translated them into public policy. Congressional committees, federal agencies like the NIH, and special interest groups of science lobbyists all looked to medical research as a way to benefit. Up through the 1960s the government doled out funds without many restrictions for applied sciences like bionics, which as we have seen can teach as much about the body as can basic research.

In the 1980s, however, funding is the single biggest obstacle to development in bionics. "Our government is more interested in the destructive than it is in the constructive," remarks one researcher. Much could be accomplished, and many billions of dollars ultimately saved, if the goverment made health its major concern. Both politicians and medical researchers are in the business of promoting public welfare, but their priorities for doing so are often mutually exclusive.

"We would never have put men on the moon if the government regulated the space program as it does the medical program," says Malchesky. "We're not targeted to look at specific diseases unless someone famous comes down with them. Industry won't fund a project unless they can control it and the work is far enough along for them to calculate its potential profits. If the government directed 2 billion dollars toward a disease, there's no question that researchers would be able to lick it. But we can't keep costs down and make medical progress at the same time."

Recent funding cutbacks make it seem as if the gap between idea and patient has seldom been wider. Researchers complain that funding agencies are too conservative to appreciate new ideas or deceptively simple devices. They fund only basic research, scientists claim, and once a researcher has been supported for a while, they redirect the money on the assumption that he or she has learned where to get alternate funding. The process of preparing grant applications takes enormous time that could be spent more profitably in the laboratory.

A good example of poor funding and enterprising ways to get around the problem can be found in the science of prosthetics. As noted, some researchers in prosthetics think that because of money problems, the field is not even worth working in—so they work in robotics instead. "Suddenly everyone has realized that *robotics* is here," says Jacobsen dryly. "Aren't you *glad* that *robotics* is here?" Jacobsen certainly is, because the country is running an enormous cash flow through robotics, which has a remarkably symbiotic relationship with prosthetics. Just about anything learned in either field can be applied to the other, so Jacobsen and other researchers apply for grants in robotics, hoping eventually to use their findings to move human beings as well as machines. "I built up a whole research program in robotics specifically for that reason," says Donath.

"The catch is the frightening way in which all these popular fields follow the same pattern," says Jacobsen. "Robotics is now spiraling upward, but without a technical breakthrough, it will all come tumbling down. A major failure will leave us with a poisoned water hole that no one will touch."

Donath was ingenious years ago when he tried to get funding for research on FNS before Petrofsky began publicizing the technique. Having aroused no interest, Donath asked himself what he could do to finance similar research. The answer came in the form of a pharmaceutical company that wanted objective rather than clinical evidence of the efficacy of its anti-inflammatory nonsteroidal drugs for arthritis. Then the National Institute of Handicapped Research got money to set up rehabilitation engineering centers and decided to fund quantitative evaluations of human function. "We submitted a proposal, and to tie it in with the drug company, I used arthritis as the model," says Donath. "Now we're trying to develop the technique to evaluate a variety of different therapies applied to the arthritic patient. In the back of my mind is the intention to apply what we gain here to the problem of multichannel stimulation."

One would like to think that there are good reasons for such worthwhile projects being overlooked. At least there are *logical* reasons that become clear if one examines the procedures by which grant applications are approved by a funding agency like the NIH.

The NIH funds grants (including training grants) and contracts by similar processes. Researchers apply for grants to work on specific projects that interest them. By law the NIH must send these applications through its peer review system composed of nongovernmental scientists who have been carefully chosen to guarantee informed evaluation and to avoid conflict of interest. The Division of Research Grants has over 60 study sections, one of which evaluates the proposed project's scientific merit, the applicant's credentials, and the chances that the project will be completed within the duration of the grant. The study section assigns a priority to the application, which then goes to the national advisory council composed of nongovernmental scientists and laypersons.

The advisory council looks over the applications and decides which to recommend for funding. "The institutes themselves also reshuffle the applications a little bit depending on program relevance, but usually we deal with the peers," says M.S. Fish, Ph.D., Special Assistant to the Deputy Director for Extramural Research and Training at the NIH. Four of the institutes presently support a small grants program that encourages innovative ideas.

Contract-supported programs like that funding artificial heart research are different from grants in that instead of applying for funding, researchers apply for a specific project in response to a governmental statement of work. The NIH publishes announcements that it wants research carried out on a certain topic, and different investigators and centers compete with each other to be assigned the contracts. The institutes (the Heart, Lung and Blood Institute, in this example) assemble panels of experts to do the primary review of contract proposals, so the process is slightly different from that for grant applications. Programs funded by contracts often extend much longer than grants; the heart program has lasted for 2 decades because no one knew for sure what the problems would be or how long it would take to solve them.

"In January 1984 the NIH completed the temporary VAD program," says Dr. Bernhard.* "They showed that several different systems could perform ventricular assist and that there is a reversible form of heart disease to which to apply it. That's precisely what the government is supposed to do—help investigators work out the problems until manufacturers become interested in making the devices, then get out."

"That's exactly right," says Fish. "We don't compete with the private sector—we're more than happy to have them pick up promising research."

"The peer review system is an excellent one that works very well," says Dr. Bernhard. "But it's a hard system, because many applications are turned down." Although the NIH budget has grown, competition for these funds has grown even faster. The NIH is now funding about 20% to 25% of applications submitted. Researchers apply for grants averaging 3 to 4 years in duration; funding is appropriated for only a year at a time, but the NIH tries to assure the applicant of support for the full period. Each year the funded researchers submit progress reports. Scientists usually do not lose funding within the grant period, but if their progress has been slow, they may not receive additional funding when their grants expire and they reapply. "Budgeting is a real problem right now—people who might otherwise be funded won't be," says Fish. "Any new grants we do fund lengthen our commitment base for the next few years, so we must limit new projects that we fund each year. We can't do everything for everybody."

It is difficult to predict what types of projects will be funded. Congress leans toward certain priorities—research in cancer was tops for many years; that disease has since been bumped by AIDS. The National Academy of Sciences is pressuring the NIH to fund more physical chemistry and basic research. "Projects we fund must be relevant to the NIH mission, which is to improve health through biomedical research," says Fish. "Congress wants us to make funding relevant to particular disease categories and to get results to patients. We have been pushed more and more toward applied and clinical activities than we used to be."

The NIH must be responsive to biomedical researchers, including those who serve the organization as reviewers. Familiar with the programs, researchers are quick to write to the NIH and Congress with any complaints. Since Congress appropriates the money allotted to the NIH, the organization must also acknowledge *its* priorities. The NIH is more or less caught in the middle.

James B. Wyngaarden, M.D., who has directed the NIH for the past few years, has been out in the field a good deal trying to identify and respond to problems in the organization. His efforts have been directed toward the ends of the applicant spectrum, that is, new investigators and well-established ones. New investigators can apply for an R-23 grant (which is specifically for them) or a traditional research grant, and the NIH is trying to encourage them to come

*Several grants in this area are, however, still active.

on board. "All other things being equal, a first application is slightly more likely to be funded," says Fish. "New investigators have been receiving a slightly higher percentage of grants per applicant than second-time applicants, and we want to improve the percentage and encourage new people in the fields."

Well-established investigators at one of the institutes can be funded with the Javits award, which gives them the option of extending their new grants for a total of 7 years so that they will not have to compete for funding at the end of 3 or 4 years. "Established investigators with these 7-year grants won't have to spend huge amounts of time writing new applications," says Fish. "We hope to expand this type of program to other institutes."

Within budget constraints, this is about the best the organization can do. "Obviously some people in the middle won't be funded," says Fish. "But we're really doing our best to solve the problems that we perceive and that are brought to our attention."

Dr. Wyngaarden is also disturbed that outside reviewers sometimes tend to be conservative and assign low priority to a project that looks like a high risk for not being completed. "This is a major concern," says Fish. "If a scientist with a good track record proposes a high risk project that may be a real breakthrough, we don't want to be conservative and stick only with less dramatic projects that promise smaller contributions. But we ourselves make these decisions only to a limited extent. For the most part, we rely on our review groups."

Fish sympathizes with disappointed grant applicants. "Before I came here, I got one of the NIH grants I requested and was turned down for the other," he says. "We try to check on complaints to see if they have substance. I think that we're reasonably responsive to comments and have excellent communication with the outside community. We have also recently adopted a formal appeals system in order to improve our responsiveness even more."

The Food and Drug Administration

Earlier we met Lillian Petrocelli, who spent several months in a hospital trying to recover from paralysis, stress bleeding, and a chronic *Pseudomonas* infection in one lung. Desperate to save her, Lillian's physicians would arrange to have her enter FDA-approved trials of various drugs. "I do recognize that the FDA has to make sure that medical products are safe, but sometimes they drag their feet a lot," Lillian says. "I'd be lying in a hospital bed suffering, and it would take the staff a few days just to get through the red tape so that they could treat me with investigational drugs. I was always signing some informed consent form or another. I think that sooner or later you really have to take a chance if you're to survive."

The average person learns of the FDA through news reports about either innovative drugs and devices that are slowly being digested by its machinery or products that it has yanked off the market. Over the past few years some of these reports have concerned implants and prostheses. In 1983 the FDA banned the sale of artificial fibers for scalp implantation. In an unrelated case the following year, a Manhattan dermatologist who uses liquid silicone injections for cosmetic surgery challenged the FDA's jurisdiction over his practice on the grounds that the agency regulates only interstate commerce, while he both obtains and purifies his material within New York state. As discussed earlier, more recently the FDA has been investigating charges by two former Shiley employees that welding defects exist in one brand of the company's heart valves.

Throughout this period the FDA has captured public attention because of its ongoing evaluation of Dr. DeVries' request to implant the Jarvik-7 artificial heart. Dr. DeVries first applied to the University of Utah's IRB in 1980 for permission to begin implantations. The following year the FDA ruled that he could implant the Jarvik-7 artificial heart in patients unable to come off the heart-lung machine after surgery. These guidelines were broadened 9 months later and again in the summer of 1982, when the FDA implemented those that led to the selection of Barney Clark: implant recipients had to have Class IV inoperable congestive heart failure, be over 18 years old, and live within a 45-minute drive of Salt Lake City.

It was not until late 1984 that Dr. DeVries received permission to perform seven more implantations on additional classes of patients. According to university spokespersons, the delay resulted from a combination of politics, personality differences, inefficiencies, and communication gaps plaguing their IRB and from Dr. DeVries' having waited several months to report to the board on the first implantation. Meanwhile, Dr. DeVries had moved the artificial heart program to Humana Heart Institute shortly after he had received FDA permission for more implantations. Initial press reports attributed his decision to his frustration with red tape. Although Humana's board granted his request to perform implantations much more speedily, commentators wondered whether he had traded the restrictions of bureaucratic regulations for others imposed by even stricter profit-oriented goals. As noted, Dr. DeVries does not feel this to be the case.

In general the FDA seems to be objective and straightforward about cardiovascular devices. "The FDA is becoming increasingly knowledgeable about heart assist devices: they recognize that mechanical heart support has a place and that it can be provided by different groups using different systems under a variety of circumstances," says Dr. Jarvik. "I don't think that there is anything unique about FDA policy on the artificial heart as opposed to other Class III medical devices, and I haven't seen any real problems with the FDA per se."

The progress of the Jarvik-7 artificial heart implantations, especially Dr. DeVries' defection to the private sector, has introduced the public to the review processes of the FDA and IRBs. This makes it appropriate to explain what the FDA regulatory process involves and why bionics researchers object to it so strongly.

The Federal Food, Drug and Cosmetic Act of 1938 authorized the FDA to regulate medical devices that were entered into interstate commerce. For violations like adulterated or misbranded devices, the FDA could seize the product; stop its production, distribution, or use; and/or recommend that anyone responsible for making or distributing it be prosecuted. The law was limited in that it required premarket approval for drugs but not for devices.

It was noted earlier that after World War II research and interest in medicine increased, and many new devices appeared. Administratively the FDA began determining that products like sutures and contact lenses were drugs so that they could regulate them. By the 1960s problems with intraocular lenses, intrauterine devices, and pacemakers had made it clear that new legislation was in order. In 1969 the "Cooper committee," headed by Dr. Cooper, studied the injury and death rates related to medical devices and recommended amendments to the Food, Drug and Cosmetics Act.

On May 28, 1976 President Ford signed into law the Medical Device Amendments (MDAs) to the 1938 act. The amendments empowered the FDA to regulate devices through most phases from their development to their use by consumers.

In December 1978 the Good Manufacturing Practice Regulations (GMPs) went into effect. These regulations created standards for manufacturing practices; by not meeting these standards, manufacturers would violate the MDAs. The GMPs gave the FDA some clout against American Technology, Inc., a now-defunct California firm that manufactured defective pacemakers. The FDA had observed since 1977 that the company's manufacturing conditions made their products suspect; before the GMPs they did not have the regulatory tools necessary to enforce the requirements of the MDAs.

The MDAs expanded the definition of devices and distinguished them from drugs but borrowed from previous legislation. "The MDAs were based in part on drug legislation because we had a history of regulating drugs to use as a precedent," says an FDA spokesman. "We've been building on our earlier legislation ever since we started regulating." Some "drugs" were classified as "transitional devices," then reassigned to their proper category of "devices." Because certain products like radiation-emitting medical devices could be regulated under more than one law, the FDA simplified matters by merging two bureaus into the Center for Devices and Radiological Health. The agency was still left to confront the sheer numbers of medical devices—over 4000, ranging from bandages to CAT scanners and beyond. Based on manufacturing procedures and use, these devices fall into five categories: over-the-counter, prescription, investigational, custom, and critical devices. The two most relevant here are investigational devices (those still being developed) and critical devices (those that sustain or support life or are intended for implantation; their failure can significantly injure the patient).

Part of the FDA's job in regulating a device is to place it in one of three classifications according to how much control its development and use require.

Class I, General Controls, includes about 30% of medical devices like surgical instruments or skin markers. These require minimum control.

Class II, Performance Standards, covers about 60% of devices, including most external prostheses. These must satisfy general controls plus applicable performance standards based on existing information.

Class III, Premarket Approval (PMA), covers the remaining 10% of devices, including most critical devices and implants. These must meet the standards of Classes I and II plus PMA. Intraocular lenses were the first products regulated as investigational PMA devices.

These classifications were established by the FDA. Individual products are examined and placed into one of the three classes by panels of experts. Assembled several years ago, the panels are still at work.

The FDA uses working rather than formal definitions of implants: devices that remain inside body cavities for at least 30 days (although some temporary devices fall into this category). The performance of an implant (like those of other medical devices) is the responsibility of both its manufacturer and the health care professionals who use it; failures are evaluated case by case.

Manufacturers who wished to save time and money could formerly apply for a Product Development Protocol (PDP), an alternative to the PMA process. Under a PDP, the manufacturer could simultaneously develop a device and collect the data necessary to satisfy the FDA that it was safe and effective. In practice, PDPs proved to be long and ponderous. A better alternative for clinical investigation has been the Investigational Device Exemption (IDE), which became effective in July 1980 and was revised in January 1981. The IDE regulates the procedures and conditions as well as the activities of sponsors, investigators, and IRBs involved in clinical investigations while exempting the device from certain of the MDAs' requirements. Several of the devices described in this book are being tested under IDEs.

An IRB is a group appointed by a medical institution to review, approve or disapprove, and monitor the progress of clinical investigations. "The IRBs don't report to the FDA, but they have specific responsibilities set out by FDA regulations," says the spokesman. "They are controlling entities at the local level that assure that medical investigations follow specific protocols and are closely supervised." They evaluate the proposed investigation's soundness, value, proof that a device works, and risks and benefits to the patients. While encouraging the testing of devices, IRBs protect patients by confirming that they are educated sufficiently to give informed consent and by maximizing their safety during clinical trials. In reviewing and evaluating investigations, IRBs apply ethical standards and consider local community attitudes.

Compared to having FDA personnel stationed across the country to police experiments, the IRB system is comparatively simple, efficient, and inexpensive. But the estimated 500 IRBs in the United States are relatively new at their jobs and do not communicate with each other very well. The result is a wide range in the amount of regulation.

Clinical investigations are always preceded by laboratory tests and extensive trials with animals. The first human users of a new device are often high risk or dying patients at the research center developing it. Testing is then broadened to include patients outside the center, often in other selected hospitals doing FDA trials. Many of the patients in this book are involved in such tests. Similar results must be shown by different investigators before the FDA will approve a device. In addition to collecting technical and medical data throughout this process, researchers also seek information on the legal, ethical, and social problems that the device implies.

On September 14, 1984 the FDA announced that it would require manufacturers and importers of medical devices to report deaths and serious injuries linked to their products. This requirement, which had been authorized by the MDAs in 1976, had been in effect for drugs for quite some time. The Public Citizens Health Research Group attributed the delay to lobbying by manufacturers; the Health Industry Manufacturers Association countered with a claim that they had supported a reporting rule.

The FDA explains that it simply takes a lot of time to implement certain regulations. "When a statute is passed and Congress says that we can establish regulations in various areas, we then must determine priorities," says the spokesman. Some statutes require additional information for their implementation. "We had to set up the classifications and put other regulations into effect, and we can't do everything at once," says the spokesman. "We must also decide whether we have the manpower actually to enforce a regulation we've written." The GMPs, on the other hand, were more urgent and already had provisions for reporting and complaints that made the agency's task easier.

The FDA is often criticized for not taking action fast enough. Delays occur partly because the MDAs were being drafted in the late 1960s, when society was biased in favor of open government. "A statute that takes into consideration every possibility creates many opportunities for hearings, for example, at different levels for various problems," says the spokesman. "The whole process of writing statutes is very lengthy, as is that of following the statutes and regulations that have been written since. If a product is doing serious damage, there's no question that we will remove it from the market immediately. The difficulty is with questionable products and conflicting opinions." The bottom line is the individual product and its risk-to-benefit ratio, which must favor the consumer.

All this work has resulted in what many researchers consider asinine legislation modeled after old, simplistic laws and drafted by attorneys who knew little about their subject matter. "The legislation and regulations having to do with medical devices and their human application are becoming difficult to deal with," says Trudell. "They strive for consumer protection—protection from what?* As a society we must assume that a scientist or physician is well-intentioned rather than a charlatan: no one I know goes out intending to hurt people. Legislatively they look for absolutes—something very rare in this universe."

American standards for testing medical devices fill several volumes; paperwork is time consuming; tests are expensive enough to raise medical costs considerably; and regulations are much stricter than those of foreign countries. Researchers believe that such regulation does more harm than good when it delays the use of effective therapies on people who need them immediately. They think also that in its desire to protect both the public and itself, the FDA carries the effort to avoid liability or harm to extremes. "Progress is never linear or smooth," says Dr. Galletti. "A regulatory approach is based on a pessimistic view of the world, a discovery approach on an optimistic one that recognizes that a few people must inevitably get hurt. I hope that like other movements of opinion, hyperregulation will die too."

Meanwhile many American bioengineers arrange to work abroad as often as possible. Others stick it out in the United States while complaining that every personnel shift at the FDA completely changes the interpretations of the rules. The FDA is also charged with bowing to special interest groups and concealing bureaucratic complications while trying to implement prior restraint on human thought, finding it too complex to be adequately legislated.

The FDA acknowledges that fewer or no restrictions on medical research in other countries make studies there go faster. Whether such studies are appropriate to a product's final analysis and evaluation in the United States is another issue. "We've recently changed our regulations to accept the results of studies performed outside the United States," says the spokesman. "At one time we were very reluctant to do so, feeling that without our regulatory structure, such studies would not be appropriate evidence of a product's safety and efficacy."

The agency comments on other complaints with qualifications. Rules interpretations, for instance, are more likely to change on the administrative rather than on the regulatory side. "Changes in circumstances rather than in personnel are more likely to have that effect," says the spokesman. "If our budget is cut, for example, we must decide administratively how to spend that money and will obviously change our method of doing business accordingly. Regulatory statutes remain the same for everyone."

The FDA dismisses complaints that it caters to special interest groups by explaining that its decisions are based on as many sources of information as it can find. "We don't operate in a vacuum," says the spokesman. "The manufac-

*The Public Citizens Health Research Group declines to comment on FDA regulation of implants and prostheses.

turers of medical devices and the professionals who use them are obviously people we must deal with."

One relevant example of these conflicting opinions is the present controversy over FDA regulation of breast implants. Since 1982 silicone breast implants have been categorized as Class III medical devices, which require premarket approval. Before then breast implants were not closely regulated even though they have been used for nearly 25 years. As late as 1978 manufacturers did not have to inform the FDA that they were marketing the implants. Now, however, the FDA wants to study breast implants and investigate their safety more thoroughly. Manufacturers and reconstructive surgeons object, feeling that the FDA is clamping down on these products for political reasons. Reconstructive surgeons feel also that the FDA is singling out breast implants from their category for closer scrutiny, that 25 years is a long time to wait to do so, and that the agency seems to be ignoring the ongoing research and the many women who have had no complications from their implants. (There is no clear evidence that silicone seepage is or is not harmful; the FDA's concern seems to be whether it is safe for women with implants to nurse their babies.) "We're concerned that the advantage of reconstruction in possibly reducing mortality rates in breast cancer might be lost if the public becomes unnecessarily worried about implants," says Dr. Dowden. The investigations would also eventually raise implant costs.

The agency's view of the matter is straightforward. "The degree of regulation depends on the degree of risk," says the spokesman. "There had been a general consensus that certain materials used in breast implants were biocompatible. But there are several different kinds of implants. We received information to the effect that some of the materials being used were causing problems, so we had to go back and regulate them. The same thing happened with materials in certain pacemaker leads. We need to make sure that patients understand their risk."

"It's better to accept a complication rate and try to do better than it is to regulate and take a liability approach," says Dr. Galletti. "It's self-destructive to judge a prosthesis implanted in 1970 on the basis of what is known in 1980 or will be known in 1990. My hope is that the public will know better eventually, and certainly with respect to breast implants that seems to be the case."

It will come as a surprise that despite the mounds of paperwork they must shuffle and the other delays they encounter, bionics researchers in general are not *quite* as hard on the FDA as they are on funding agencies. They grudgingly acknowledge that the problem is a social, legal, and political one that transcends a single agency. "The FDA elicits a lot of criticism, but they really try to do their jobs. It's difficult to walk the line between all the groups who are against implantations and those who want to push for innovations," says Geyer. "How well the FDA regulates is a very complicated issue because they are a government bureaucracy," says another researcher. "They are the target of politicians, because a good way for politicians to get publicity is to criticize a government agency, especially one involved in health care."

It is worth noting, too, that some medical researchers criticize the FDA for not regulating strictly enough, especially with respect to heart implantations. Some cardiologists believe that using the percutaneous Jarvik-7 artificial heart is reckless and premature. Others fear the implications of the Phoenix heart implantation, which did not even have FDA approval.* Medical, legal, and ethical experts have worried that this incident will set a precedent for physicians to bypass federal laws in emergencies and use dying patients to test unproven devices.

In a world without absolutes, underregulation would be no better than hyperregulation of bionic parts or any other devices. With luck the FDA might someday establish a medium that most researchers and patients will find appropriate. Meanwhile, other issues raised by bionic parts must still be addressed.

---

*Responding to the legal violation and the pressure resulting from news accounts, the FDA in March 1985 sent a letter of mild rebuke to officials of the University of Arizona Medical Center regarding the unsanctioned implantation. After more investigation, the agency sent another letter in May 1985 indicating that they would not take further action against the physicians who performed the surgery.

30

The Meaning of Bionics

The scientist we met at the matinee of *The Empire Strikes Back* perhaps felt interested and relieved when Dr. DeVries began implanting Jarvik-7 artificial hearts in human patients. Like many of his peers, our scientist may have had reservations about the ethicality of the Utah/Symbion/Humana arrangements but acknowledged that the time had come for clinical trials. Medicine progresses because of daring experiments. The extensive publicity generated by these experiments in particular may have raised hopes a little too high, but it also informed the public that even the simplest type of artificial organ is extremely difficult to design successfully. People also became aware that little if any attention had been paid to the implications of heart implantations.

Some of these implications have not been good. According to Utah and Humana, the implantation program has overwhelming public support. Shortly after Barney Clark's death, however, Dr. DeVries discovered that nearly two dozen nails had been driven into the left front tire of his car and the brake cable had been cut. (Fortunately the tire blew as his son Jon drove several friends through a parking lot; the car went into a snowbank, and no one was hurt.) Other people sent hate mail. Meanwhile the Clarks' house had been vandalized while the media reported that they were in Utah.

A more intellectual expression of this negative atmosphere comes from British journalist Malcolm Muggeridge. Describing transplantations, he writes, "I have in honesty to admit that I feel for this spare-part surgery—as I know others do—a deep, instinctive repugnance that is not capable of a wholly rational explanation. . . . It has to do with a sense that all creation preeminently deserves respect . . . [and that] . . . our present way of life is carrying us in the opposite direction. . . ." "Repugnance" evidenced by rejection is the body's immediate reaction to any implant; one may be tempted to attach symbolic as well as physiological significance to its repulsion.

Members of the medical profession have condemned the artificial heart implantations as pointless, vulgar displays of glamorous heroics.* Physicians are citing Lewis Thomas' warnings about "halfway technologies"—outlandish, expensive devices that relieve or eliminate disease symptoms without treating

*Some researchers have suggested that the first implantation of the Jarvik-7 artificial heart was intended as a successful attempt to gain venture capital for its manufacturer, Symbion.

the cause. Use of such devices would seem more like experimentation for the sake of research than treatment for the benefit of patients. "There are certainly lunatics with no conscience who will try anything on anybody who will hold still for it," says Dr. Bernhard. "People with heart failure will do anything in their desperation, and others will use them to their own advantage to try experiments or to get some publicity. I take a very dim view of that."

Although she herself thought that tampering with creation was "frightful," Mary Shelley also makes clear in her novel *Frankenstein* that from the creature's point of view, misfortune has resulted not from its very *existence* but rather from the miserable *manner* in which it must live. As Dr. Kolff points out, "If you make a good device, and someone walks down Main Street with it, all your problems are solved." The Jarvik-7 artificial heart, with its bulky external power supply, is still a crude device; were it already fully implantable, the experiments might be universally hailed as a succession of miracles. People may be objecting not to an artificial heart per se, but rather to its present imperfections. "The public thinks that the artificial heart is crazy," says Dr. Nosé. "It won't look crazy 10 years from now."

In his most frequently quoted article, Dr. Jarvik writes that to achieve its objective, the artificial heart "must be more than a pump . . . [and] . . . more than functional, reliable and dependable. It must be forgettable." He speaks from the viewpoint of the heart recipient, but given the amount of controversy that surrounds the artificial heart, the statement reverberates with additional meanings.

The artificial kidney and the heart-lung machine were accomplishments of similar stature, but they were invented before the days of MDAs and IRBs. The Jarvik-7 artificial heart has been unique in the amount of thoroughgoing scrutiny its implantations have received and the number of ethical questions it has raised. The IRB of the University of Utah had to establish both a protocol for implantation and an informed consent form for patients—ultimately 11 double-spaced typed pages that had to be approved twice with a 24-hour interval between signatures. In an article from an issue of *The New England Journal of Medicine* devoted to the artificial heart, F. Ross Woolley, Ph.D., a member of the board, writes, "There were certainly inadequacies and errors in what we did, but they represented ignorance of the unknown rather than acts of arrogance or neglect." While this thinking made it possible for Barney Clark's implantation to occur, it merely pointed to the amount of work required to answer the remaining questions raised by the artificial heart and other bionic parts. Each set of questions in turn raises a new set having to do in part with poorly defined words like "cost-benefit ratio," "quality of life," "death," and so on. They may well prove more difficult to deal with than all of the technical complexities already discussed.

We have already noted the questions of cost and conflict of interest that have arisen from the heart implantation program. Harry Schwartz, Writer in

Residence at the College of Physicians and Surgeons at Columbia University, has reported that in Washington and Utah, powerful opponents of the implantation program have tried to sabotage it because they fear its cost implications. Some observers feel that Congress has funded the artificial heart program blindly and that much of its $200 million investment has been wasted.

Others add that U.S. funding priorities seem to be askew. Many professionals feel that money channeled into artificial heart research could be better spent educating people how to avoid heart disease—in effect, that we should regress from substitutive to preventive medicine. Why pour millions into the development of an artificial heart, they reason, when New York City emergency rooms must turn away victims of heart attacks? (In passing, though, we might recall the difficulty anticipated by some Cleveland Clinic Foundation researchers in getting physicians and patients to accept artificial organ technology as preventive medicine.)

The appeal to priorities could also be focused within the field of bionics: rather than financing artificial hearts for end-stage cardiac failure patients, it might make more sense to fund prosthetic limbs or neural prostheses so that otherwise able-bodied people, anxious to be employed, could return to work. Other forms of bionics used as research tools might also indirectly lead to better treatment for larger numbers of people.

Another important cost issue is that of who will pay for bionic parts. Although all of us financed the development of artificial hearts, only the wealthy could afford them at this point. As noted, however, the alternative form of payment, exemplified in kidney dialysis, has led to prolonged deaths and indiscriminate spending because the government decided to make the treatment universally available.

Assuming that cost issues can be resolved, we must then decide who will receive bionic parts. One possible selection criterion, age, has a few precedents: patients past a certain age cannot receive heart transplants in America or kidney dialysis in Britain. Otherwise we have conspicuously avoided addressing this problem. This is hardly surprising: unless Congress wants to fund heart implants, for instance, as it does dialysis, it must invent some justification for limiting access to devices developed with federal funds. Perhaps the decision on allocation ought to be made not by bureaucrats but by appointed public representatives.

The individual patient's right to treatment must be weighed against a fair distribution of resources. These are usually allocated through some combination of medical criteria, social worth criteria, and random selection. Some researchers have complained about heart implants being tested on extremely ill patients, but otherwise the criteria for selecting recipients have seemed to be generally acceptable.

What criteria should be used, however, as government restrictions relax and implantation occurs more often? Prior health history must be examined—but should the patient be required to show evidence of having practiced preventive medicine and good health habits? Chances for recovery are obviously a factor—but should the patient be expected not only to survive but to live productively? Is the presence of a strong family support group, now required by the Humana team, a fair criterion for eliminating a solitary but self-sufficient individual determined to fight for his or her life? Much thought must go into a humane system for rationing resources.

Issues like these have already been raised by heart disease victims. The possibility of turning off the Jarvik-7 artificial heart was mentioned when Bill Schroeder had his second stroke. Thomas Creighton received a second donor heart soon after his first transplantation failed—meanwhile what happened to other patients who had been waiting in line for their *first* transplantations? Bridges to transplantation may become increasingly successful, but if a donor heart is not found, a patient's sustenance on a percutaneous blood pump could become a living nightmare.

If bionic parts can be equitably distributed, other questions will arise. One is that of how much one can reasonably expect of a given device even after it has taken years to develop. There should be some way to determine how many replacement devices can be supplied in case of a series of failures. Perhaps some devices are best restricted to use only for temporary assistance. Some might be considered "luxuries" rather than necessities.

The psychology of bionic devices, especially those used for life support, is complex enough to be considered a field in its own right. We have already noted possible psychological side effects of kidney dialysis; some early dialysis patients also suffered dementia from toxins remaining in their bloodstreams. We should therefore be prepared for similar reactions to being tethered to an artificial heart power supply or having other bionic parts that limit choices of life-style. In each case someone must decide whether the device under consideration will really enhance the quality of life for the patient (and his or her family) to an extent that overcomes its liabilities.

If bionic parts become completely implantable and thoroughly successful, on the other hand, we must anticipate the positive effects of having our bodies rebuilt. Although bioengineers are not in the business of prolonging lives, fictional characters are notorious for selling their souls to live eternally—who knows what desperate measures might be taken as bionic parts continue to improve? None of the patients interviewed for this book expressed discomfort at being a bionic person, yet others may not share that view. Many blind and deaf people do not want "normality" imposed on them. Other persons may refuse to receive even trustworthy artificial parts for religious reasons.

As bionic experiments continue, researchers must decide how much information about them should be released to the public at what stage of the proceedings. The professional privileges of research teams regarding the release of information may conflict with the informed consent of patients and the right of society to hold investigators responsible for their actions. Artificial heart candidates seem to have little choice about hiring agents to represent them in dealings with the media. Publicity can be extremely detrimental to a patient's recovery. At first Bill Schroeder enjoyed being a media hero; now, however, he may feel that he has failed the millions of people to whom he was a superpatient.

As new bionic devices join with other medical technologies to combat the hazards of modern life, disabled people like Bill will live longer and make up a larger percentage of the population. Society must accommodate these people, and medical researchers must concentrate on finding better ways to treat their disabilities.

Improving the quality of people's lives will inevitably prolong them. The number of elderly persons in our society is increasing, while there are comparatively fewer young people available to bear the heavy burden of longer-term health costs. Before deciding how to care for elderly individuals, society must learn to accept them. Ours is a narcissistic, youth- and beauty-oriented culture with little tolerance for old age or even the deformity and inconvenience that might result from disabilities, implants, or prostheses. Even today, increasing numbers of elderly persons are reported to be committing suicide or murdering their spouses to end the illness and depression that threaten them. We might hope that as their numbers increase to a majority, elderly individuals will find social forces turning in their favor.

Society might also do well to change some of the legalities of bionics. Like other medical products, bionic parts have important legal implications, particularly when they have been implanted in the body. The past 20 years have shown an increasing number of medical device product liability suits, with verdicts equally divided between plaintiffs and defendants. Product liability in general is the subject of at least one legislative effort to reform the system at the national level; medical device litigation is so widespread and complicated that it may well change the nature of the industry. Although physicians and hospitals can be named in suits involving bionic parts, the defendants are most commonly manufacturers.

Whether the device is elective or lifesaving (a distinction that is not always clear), the performance of a bionic part has two aspects: the quality of the product itself and the efficacy of the procedures surrounding its implantation. Separating these two aspects creates all kinds of legal difficulties. The patient and attorney must decide who to blame for a failed implantation—the manufacturer of the product? the physician who recommended it? the surgeon who implanted it? the hospital? all of the above? They must also decide what cause of action to use in the case.

Theories of recovery presently being applied by courts in medical product liability suits include strict liability, which is applied to products but not procedures or services. Strict liability is variously defined and has been used rather inconsistently. Negligence, another cause of action, requires considerable proof.

Breach of warranty, which courts often refuse to apply to physicians or hospitals, can be a complicated cause of action. The patient seldom sees the product's warranty (which is probably carefully limited) but would have access to it in the event of litigation. Unless the patient waives his or her rights, there may be an implied warranty that the product works as described. Although the patient does not actually purchase the bionic part, he or she can certainly sue the direct provider of it, who would then sue the manufacturer.

A medical device may not fail for years, long past the expiration of a warranty or a statute of limitations. Reflecting other "latent malpractice" cases, in December 1983 New York's highest court liberalized the state's statute of limitations on suits involving the malfunctioning of medical implants. The court declared that the 3-year period for filing damage suits should begin with the appearance of the health problem, not the date of the implantation.

Manufacturers have become the main targets of medical product lawsuits because the doctrine of strict liability, especially in states that enforce it most rigidly, makes winning a product case much easier than winning a malpractice suit. Strict liability retroactively holds manufacturers responsible for product standards. Litigation involving strict liability has both increased plaintiffs' chances of recovery and shifted the alliances of physicians. Anxious to avoid a malpractice suit, a physician may well be willing to testify against the manufacturer of a product that he or she recommended. Manufacturers, on the other hand, are afraid to blame physicians for fear that other physicians will then avoid using their products.

Manufacturers may be somewhat protected by the difficulty plaintiffs could have in obtaining jurisdiction to bring cause of action against them from other states (although "long-arm" statutes generally expedite this process). The manufacturers can also argue that they complied with FDA regulations (although there is widespread disapproval of the MDAs and of the time it takes even to implement certain regulations). Finally, they can try to protect themselves by warning patients directly, as well as through physicians, of potential risks in using certain devices.

If they continue to be singled out as targets for lawsuits, manufacturers may become reluctant to produce medical devices, including bionic parts. (Manufacturers of certain vaccines have recently chosen to leave the business rather than face the liabilities, resulting in a call for federal guarantees and loss limitations.) Switching targets would not help the situation much. If litigation were instead aimed at physicians and hospitals, malpractice premiums would increase, and professionals would have to compensate by raising patients' fees. If cases were found in favor of defendants, shifting the risk of injury onto users of bionic parts, development costs would fall, but patients would feel inadequately protected.

As many researchers have suggested, the situation may improve if we change from a conservative, litigious society to a more liberal, optimistic one willing to accept the risks that accompany progress.

Society must also accept the fact that everyone dies and that life and death should be spiritual, not technological, experiences. People are definitely moving away from life-sustaining devices toward the right to terminate their own lives, or those of their loved ones, with dignity. The controversy surrounding definitions of death can only increase if a patient receives a bionic part that will run indefinitely after the rest of the body stops. The problem becomes even worse if an implant recipient develops cancer or another terminal disease that might make natural death, postponed by the bionic part, a relief. When such decisions must be made, peaceful, dignified death is a right, not a privilege.

U.S. leadership has done a splendid job of avoiding the formulation of long-term health policies. The NIH has steadily supported artificial heart research for 20 years without establishing criteria for expected results. The controversy raised by Barney Clark's implantation subsided during the several months that passed before the FDA authorized additional heart implantations, almost as if we expected Barney's death to end these troublesome experiments and the questions they raised. Through its apparent nonchalance, the government is leaving industry the opportunity to take over directions regarding the costs and allocations of bionic parts.

The problems caused by the artificial kidney appeared slowly, while those raised by the Jarvik-7 artificial heart seemed to burst upon us—but neither device is about to disappear. Even if they did, we would still have patients like Jaimie Fiske and Baby Fae to remind us that we have also dismissed most of the issues surrounding transplantations.

"Too often people bring in ethics late in the development of a mode of therapy, either to negatively criticize or to justify what has occurred," writes Dr. Buchwald in a recent article about the Infusaid pump. "It is far more appropriate to place ethics up-front, in the beginnings of a field of therapeutics. . . . the accountability of everyone involved in this field must be stressed at the very outset." The difficulty of considering these matters well in advance is that bionics has created a new era of medicine, and we cannot foresee its evolution. We are faced with the necessity of preparing now for an unknown future. It is just as well that like other medical "breakthroughs," bionic parts really develop slowly and fail repeatedly before they succeed; we need that time to follow their progress and formulate our thoughts as their implications become clearer.

"I have a great deal of faith in the human mind and conscience," says Dr. DeVries. "If we take scientific steps and find new technologies, they will present a whole new set of problems, which we can then solve. Trying to solve the potential problems first is going backwards. I believe very strongly that technology moves at its proper speed; humans adapt to it by solving the problems as they arise. We'll never solve all the problems, so we shouldn't allow our fear of them to keep us from progressing." Instead we should prepare to be flexible and allow our thinking to evolve with the technology.

We must first decide what kind of forum to create to deal with these issues.

"All those national commissions are a perfect waste of time," says Dr. Galletti. "The artificial heart program has been examined by at least three or four different commissions within the past 10 years, and they all say the same thing. Their reports become emasculated because strong views are removed. We need strong views, even if they are contradictory. We need debate, openly and in appropriate forms in universities—not in magazines, and not in the lay press. We ought to attract more young people to take the lead in these matters and use their common sense."

Our efforts need not simply waste time. The Utah IRB has proved that confronting these issues sincerely and with good intentions can get effective results. They saw to it that a preliminary protocol was established for an innovative experiment and that a patient was fully educated and informed of his rights.

Ultimately one must remember that discussions do not always lead to unanimously accepted plans for action. "The conclusions of regulatory agencies are true intellectually but may be irrelevant in terms of an individual patient's mind-set," says Dr. Galletti. "A device may change a patient's life in a way that can't be measured by consensus, and society must cater to mind-sets that are incompatible. The human mind is more complex and diverse than the committees want to admit."

Patients are becoming better educated and more aggressive about suggesting innovative therapies. Dr. DeVries has been besieged by requests for heart implantations; at least one heart disease victim threatened to die on the steps of the University of Utah Medical Center if he did not receive an artificial heart. All the heart implant recipients volunteered and twice signed their lengthy informed consent forms. Perhaps rather than making decisions for these people, we should establish means of regulating the decision-making process, leaving that up to individual patients and their physicians.

Artificial heart implantations are in roughly the same position occupied by heart transplantations 15 years ago. It seems repeatedly necessary to belabor the obvious and point out that all commonplace medical procedures were once "extraordinary measures."

Decisions about artificial hearts and other devices will be made by healthy people who are not in an appropriate position to evaluate them. All the artificial heart recipients were incapacitated by weakness and pain before their implantations; after the surgery their lives were not only saved but also better. Disabilities have their financial and human costs too, but unlike the costs of bionic parts, they are hard to measure. The fact is that an artificial heart recipient has yet to switch off the power drive.

The patients who have received artificial hearts have been celebrated for their courage, and justly so. The artificial heart is a different extraordinary measure because it is a bionic part; substitutive medicine calls for a particular attitude. As Dr. Galletti has pointed out, drugs simply need to be administered to establish their effectiveness, while artificial organs must prove their clinical feasibility before they can demonstrate clinical usefulness. In that sense devices are much closer to operations than they are to drugs; if they are to be regulated, the evaluation process should be modeled after that of new surgical procedures. Thus far the Jarvik-7 artificial heart has demonstrated only its feasiblity for a relatively short period of time. Until it proves useful, implanting hearts may seem like walking on the moon—it can be done but might not be worth doing at the present time.

As emphasized, bionic parts are valuable in part for what they teach about the natural body. They may prove also to be unique research tools for exploring the mind and spirit. Those who approve can try them; those whom they offend might demonstrate their disapproval by becoming organ donors. But everyone should welcome their challenge. Instead of hardening into conservatism, our skepticism and questioning could lead to greater maturity. Rather than blindly accepting what medical technology offers or refusing to try bold new procedures, we can instead make intelligent choices about our health and our lives.

INDEX